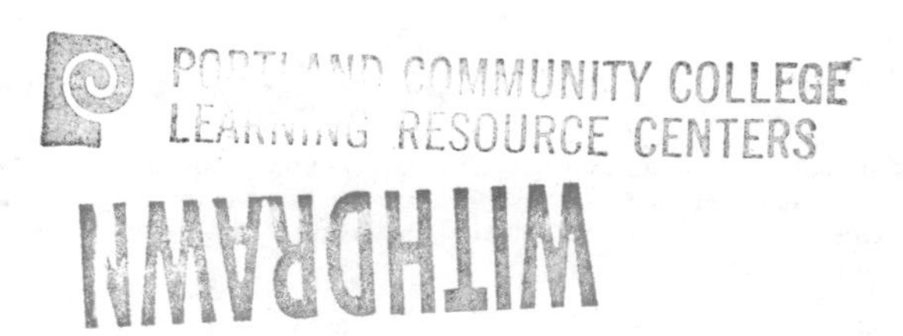

Delmar's Automotive Dictionary

South • Dwiggins

Delmar Publi

I(T)P An International Thomson

Albany • Bonn • Boston • Cincinnati • Detroit • London • Madrid • Melbourne
Mexico City • New York • Pacific Grove • Paris • San Francisco • Singapore • Tokyo
Toronto • Washington

NOTICE TO THE READER

Publisher does not warrant or guarantee any of the products described herein or perform any independent analysis in connection with any of the product information contained herein. Publisher does not assume, and expressly disclaims, any obligation to obtain and include information other than that provided to it by the manufacturer.

The reader is expressly warned to consider and adopt all safety precautions that might be indicated by the activities herein and to avoid all potential hazards. By following the instructions contained herein, the reader willingly assumes all risks in connection with such instructions.

The publisher makes no representation or warranties of any kind, including but not limited to, the warranties of fitness for particular purpose or merchantability, nor are any such representations implied with respect to the material set forth herein, and the publisher takes no responsibility with respect to such material. The publisher shall not be liable for any special, consequential, or exemplary damages resulting, in whole or part, from the readers' use of, or reliance upon, this material.

Delmar Staff
Publisher: Robert Lynch
Acquisitions Editor: Vernon Anthony
Production Manager: Mary Ellen Black
Art and Design Coordinator: Cheri Plasse

COPYRIGHT © 1997
By Delmar Publishers
an International Thomson Publishing Company

The ITP logo is a trademark under license.

Printed in the United States of America

Online Services

Delmar Online
To access a wide variety of Delmar products and services on the World Wide Web, point your browser to:
http://www.delmar.com/delmar.html
or email: info@delmar.com

thomson.com
To access International Thomson Publishing's home site for information on more than 34 publishers and 20,000 products, point your browser to:
http://www.thomson.com
or email: findit@kiosk.thomson.com

Internet

A service of I(T)P®

For more information, contact:

Delmar Publishers
3 Columbia Circle, Box 15015
Albany, New York 12212-5015

International Thomson Publishing Europe
Berkshire House 168-173
High Holborn
London, WC1V 7AA
England

Thomas Nelson Australia
102 Dodds Street
South Melbourne, 3205
Victoria, Australia

Nelson Canada
1120 Birchmont Road
Scarborough, Ontario
Canada, M1K 5G4

International Thomson Editores
Campos Eliseos 385, Piso 7
Col Polanco
11560 Mexico D F Mexico

International Thomson Publishing GmbH
Konigswinterer Strasse 418
53227 Bonn
Germany

International Thomson Publishing Asia
221 Henderson Road
#05-10 Henderson Building
Singapore 0315

International Thomson Publishing—Japan
Hirakawacho Kyowa Building, 3F
2-2-1 Hirakawacho
Chiyoda-ku, Tokyo 102
Japan

All rights reserved. No part of this work covered by the copyright hereon may be reproduced or used in any form or by any means graphic, electronic, or mechanical, including photocopying, recording, taping, or information storage and retrieval systems without the written permission of the publisher.

1 2 3 4 5 6 7 8 9 10 XXX 02 01 00 99 98 97

Library of Congress Cataloging-in-Publication Data

South, David W.
Delmar's automotive dictionary / David W. South, Boyce H. Dwiggins.
p. cm.
ISBN 0-8273-7405-4 (pbk.)
1. Automobiles—Dictionaries. I. Dwiggins, Boyce H. II. Title.
TL9.S64 1997
629.2'03—dc21 96-38136
CIP

Preface

Why an automotive dictionary? This dictionary is intended to provide a practical guide to the technical jargon of the automotive world. This dictionary provides:

- a thorough guide to parts and repair terminology
- terminology used in automotive collision repair
- terminology used in trucking and medium to heavy duty truck repair
- racing and hot-rodding terminology, including common slang terms

Every attempt has been made to make this dictionary as understandable as possible to both the automotive professional or enthusiast and to the general public. Whether you are trying to understand service information to make your own repairs or are just trying to understand your own technician, you will find this dictionary indispensable. Professional technicians may find uncommon or historical terms they are not familiar with; they may enjoy disagreeing with the definitions in this dictionary, and we hope they tell us all about it—automotive technology has become far too complex for any one person to completely master.

When we built this dictionary, we consulted with experts in linguistics to determine what to include. Much of what you see in general dictionaries of the language, such as parts of speech and pronunciation guides—certainly have their use. We decided that the overwhelming majority of terms found in automotive jargon are taken from common terms found in the language, so it didn't seem practical to include parts of speech and pronunciation guides. If you, as the user, feel differently, let us know, and we can consider revisions for future editions.

We hope that you find this dictionary to have practical value, and that you enjoy using it as much as we have enjoyed assembling these terms.

A: An abbreviation for *ampere.*

AAA: An abbreviation for the *American Automobile Association.*

AACA: An abbreviation for the *Antique Automobile Club of America.*

AAI: An abbreviation for the *Alliance of American Insurers.*

AAMA: An abbreviation for the *American Automobile Manufacturers Association.*

AARA: An abbreviation for the *Antique Auto Racing Association.*

A-arm: A triangular-shaped suspension control-arm device, with two points connected to the chassis and one to the wheel spindle. *Control arms* are sometimes called A-arms because from the top view, they are shaped like the letter *A*. Also referred to as *A-frame.*

ABCM: An abbreviation for *anti-lock brake control module.* Some refer to it as anti-lock brake computer module.

ABDC: An abbreviation for *after bottom dead center.*

abnormal operating conditions *(AOC):* Adverse. Other than normal operating conditions, such as rain, snow, sleet, and ice.

abnormal wear: The excessive wear on a drive or driven member as might be caused by improper tensioning, misalignment, or abrasive materials in the drive environment.

abnormal wear pattern: A visual indication that two or more members, such as gears, pulleys, and/or belts are improperly tensioned or aligned or that abrasive materials are present.

abort: The procedure for intentionally terminating a program when a mistake, malfunction, or error has occurred.

abrasion resistance: A measure of a wire covering's ability to resist surface wear due to mechanical damage.

abrasive: A very hard substance used for the removal of material by cutting, grinding, lapping, or polishing metals.

abrasive finishing: Any of several processes for removing scale or surface contaminants using abrasives, such as bonded grinding wheels or disks, coated abrasives, honing stones, or bonded abrasive sticks.

absolute pressure: A pressure measured from *absolute zero.*

absolute pressure sensor: A device for sensing pressures from *absolute zero.*

absolute rating: A single eternal explanation for all reality. That point at which all motion in matter ceases, such as *absolute zero.* Also used in filter ratings to indicate the diameter of the largest particle, normally expressed in micrometers (μm), that will pass through the filter. A filter media with an exact and consistent pore size theoretically has an *absolute rating.*

absolute zero: The lowest temperature on the Kelvin temperature scale (0°K), equivalent to –459.7°F (–273.2°C). Temperature measured from 0°K is an absolute temperature. All molecular motion ceases at 0°K.

absorb: To take in by capillary action, as in a sponge.

absorbent medium: A material akin to a sponge in that it can draw in fluid and retain it within its structure. In this sense, it can act as a filter to remove *(absorb)* and retain fluid.

absorptive lens: A filter lens designed to reduce the effects of glare, reflection, and stray light.

ac: An abbreviation for *alternating current.*

AC: An abbreviation for *alternating current.*

A/C: An abbreviation for *air conditioning.*

ac adapter: A transformer-type power supply that plugs into an *ac (alternating current)* power outlet and provides low voltage *ac* or *dc* to provide power for accessory equipment.

A-cam: Pattern used for grinding pistons in an oval- or cam-shape with 0.005 inch (0.127 mm) difference between the thrust face and the pinhole side.

acc: Abbreviation for *accessory.*

ACC: **1.** An abbreviation for *Automotive Communications Council.* **2.** An abbreviation for *air conditioning compressor signal switch.*

acceleration: An increase in velocity or speed.

acceleration squat: The tendency of the rear part of a vehicle to press down on the rear springs during hard acceleration.

accelerator: A control, usually a foot-operated pedal, linked to the throttle valve of the *carburetor* and used to control the flow of fuel into the engine.

accelerator pedal: A foot-operated device for controlling the flow of fuel into the engine.

accelerator pedal position sensor: A device designed to send an electrical signal to the *central processing unit* relative to the position of the *accelerator pedal* at any given time.

accelerator pump: A pump in the *carburetor* connected by linkage to the *accelerator pedal* that momentarily enriches the *air-fuel mixture* when the *accelerator* is depressed at low speed.

accelerometer: An instrument that measures a vehicle's linear or lateral rate of acceleration in *g force* or feet per second.

acceptable weld: A weld that meets all the requirements and the acceptance criteria prescribed by the welding specifications.

access fitting: A *service port* or *service valve.* Also may refer to an *access valve.*

access slot: An opening that permits access to a device, such as the openings in the *backing plates* of a brake system that allow access to the *star wheel adjuster.*

access time: The time that is required to retrieve information from a system's memory.

access valve: A *service port* or *service valve.* Also may refer to an *access fitting.*

accidental ground: A condition that exists when a wire, connected to the positive battery terminal, contacts a grounded metal part of the car.

ACCS: An abbreviation for *air conditioning cycling-clutch switch.*

accumulator: **1.** A tank located in the outlet of the *evaporator* to receive the *refrigerant* that leaves the evaporator. A component used to store or hold liquid refrigerant in an air-conditioning system. This device is constructed to ensure that no liquid refrigerant enters the *compressor.* **2.** A device that cushions the motion of a clutch and servo action in an *automatic transmission.* **3.** A register or storage location that forms the result of an arithmetic or logic operation. Commonly used when a series of calculations are to be totaled.

accumulator-dehydrator: An air conditioner accumulator that includes a *desiccant.*

accumulator-drier: A term often used for *accumulator-dehydrator.*

accuracy: The conformity of an indicated value to a value accepted as a standard.

ACCUS: An abbreviation for the *Automobile Competition Committee for the United States.*

acetic acid: An activator that is used in *RTV/silicone* sealants to make them more rubber-like in composition.

acetone: **1.** A highly flammable liquid sometimes used as a racing-fuel additive. Acetone CH_3COCH_3 helps to prevent other chemicals in a fuel mixture from separating. **2.** A powerful ketone-type lacquer solvent. **3.** Used as a solvent to clean parts.

acetylene: A highly flammable gas (C_2H_2) used for metal cutting, *welding,* and *brazing.*

acid: Hydrogen (H) compounds that yield hydrogen ions when dissolved in water. There are many types of organic and inorganic acids. Though acids are the most important and useful of all the chemicals, acids are not wanted in some environments.

ACID: An acronym for a four-mode, driving-test cycle used to test exhaust emissions or vehicle driveability; the modes are Accelerate, Cruise, Idle, and Decelerate.

acid dip: **1.** A method of paint stripping. Metal parts can be immersed in an acid dip to remove all traces of old paint and chemical impurities. **2.** To immerse stock-car-body panels in acid to reduce sheet metal thickness by *etching,* thus reducing weight.

acidity: The presence of acid-type chemicals that are identified by the acid number. Acidity within some environments, such as the *crankcase* of an engine, causes corrosion, sludge, and varnish to increase.

acid rain: Corrosive rain formed when sulfur (S) emissions from motor vehicles and industrial plants combine with hydrogen (H) and oxygen (O) in the atmosphere to form sulfuric acid (H_2SO_4). The mixture of these chemicals with water (H_2O) produces an acid solution that is found in rain. Not only is it corrosive to anything it may come into contact with, it also raises the acidity of lakes and ponds, often to the point that fish and other aquatic life cannot survive.

Ackerman principle: The geometric principle used to provide toe-out on turns. The ends of the steering arms are angled so that the inside wheel turns more than the outside wheel when a vehicle is making a turn, without scrubbing the tire treads on the road surface.

Ackerman steering: A term often used for *Ackerman principle.*

ACL: An abbreviation for *automatic chassis lubrication.*

ACM: An abbreviation for *air-control module.*

ac power supply: A source of alternating current *(ac),* such as an ac outlet, transformer, inverter, or an *alternator* (ac generator).

ACRL: An abbreviation for *American City Racing League.*

ACRS: An abbreviation for *air-cushion restraint system.*

ACR3: An abbreviation for an *Air Conditioning and Refrigeration Refrigerant Recovery, Reclaim, and Recycle* system.

acrylic: A polymer-based coating acrylic ($C_nH_{2n-2}O_2$) widely used for automotive topcoats. Its physical properties can be controlled in part by the choice of the alcohol used to make the ester.

ACSI: Abbreviation for the *Automotive Cooling System Institute.*

ACT: An acronym for *air charge temperature.*

active coil: Those coils that are free to deflect under a load.

active plate material: The sponge lead in an automotive battery that is spread over the negative-plate grid or the lead peroxide that is spread over the positive-plate grid.

active restraint: A vehicle occupant restraint, such as a lap belt and/or shoulder harness, that must be attached or connected by the person using it.

active solvent: A liquid that can dissolve a paint binder when used alone.

active spring coil: Active coils in the center of the spring operate during the complete range of spring loading. Also, see *inactive spring coil* and *transitional spring coil.*

active suspension system: Also known as *computer-controlled suspension system,* a computerized system able to control *body roll, body pitch, brake dive, acceleration squat,* and *ride height.* Suspension systems that are controlled by double-acting hydraulic cylinders or *solenoids (actuators)* mounted at each wheel. The actuators support the vehicle's weight, instead of conventional springs or air springs.

Actual Cash Value *(ACV):* The true value of a product, such as a used vehicle.

actual throat: A welding term indicating the shortest distance between the weld root and the face of a fillet weld.

ACV: **1.** An abbreviation for *air-control valve.* **2.** An abbreviation for *actual cash value.*

adapter: **1.** Any of various pieces of hardware that permits non-matching parts to connect, mesh, or function together. **2.** A device used to connect an engine and a transmission not originally designed to be used together. **3.** Welds under a spring seat to increase mounting height of the fit seal to the axle.

adaptive memory: The feature of a computer memory that allows the microprocessor to automatically compensate for changes in the dynamics of the process being controlled. Anything stored in adaptive memory is lost when power to the computer is interrupted, such as when the battery is disconnected.

adaptor: Another way of spelling *adapter.*

additive: **1.** Any material added to a lubricating grease or oil to improve its suitability for service. It may improve a property that the lubricant already possesses or give it properties that it does not naturally possess. **2.** Any material added to the cooling system to inhibit rust, increase the boiling point, and/or decrease the freezing point. **3.** Any one of a number of special chemicals added to a paint to bring about special effects.

add-on: **1.** Any device or system added to a vehicle by the dealer, independent garage, or owner. **2.** A component or device added to a computer system to increase its storage capacity, to modify its architecture, or to upgrade its performance; circuitry or system that can be attached to a computer.

add-on coolers: **1.** A method of increasing the transmission system's cooling capacity by adding an external fluid cooling unit. **2.** A device to provide cooling for the power-steering fluid.

address: A number identifying the location of a word in computer memory.

adhesion: **1.** The ability of a tire to remain in contact with the road surface. **2.** The property of oil that causes it to cling to metal surfaces, such as bearings.

adhesive bonding: **1.** A technique for bonding metals and/or plastics together during assembly of panels and bodies. **2.** A process used to attach aftermarket body kits, such as rocker panels and spoilers.

adiabatic engine: An engine having *combustion chambers* insulated with a high-temperature material. Heat loss is kept at a minimum and is retained rather than being allowed to dissipate through the cooling and exhaust systems. This results in a higher proportion of thermal energy being converted to useful power.

adjust: To bring the parts of a component, system, or device to a specific relationship, dimension, temperature, or pressure.

adjustable shock: Another term used for *adjustable shock absorber.*

adjustable shock absorber: A *shock absorber* having an external means of adjustment to calibrate it precisely for a specific operating condition.

adjustable strut: A strut with a manually operated adjustment for strut firmness. The strut adjusting knob, usually accessible without raising the vehicle, varies the strut orifice opening. Also see *travel-sensitive strut.*

adjustable torque arm: A member used to retain axle alignment and, in some cases, control axle torque. Normally one adjustable and one rigid torque arm are used per axle so the axle can be aligned. This rod can be extended or retracted for adjustment purposes.

adjusting cam: Eccentric bolts that are used to automatically or manually adjust the brake shoe-to-drum clearance. Positioned in the backing plate of drum brakes, the cam positions the shoe(s) closer to the drum.

adjusting shim: A metal *shim,* available in various thicknesses, used to change the valve clearance in some *overhead cam engines.*

adjusting sleeve: An internally threaded sleeve located between the *tie rod ends.* The sleeve is rotated to set *toe in/toe out.*

adjustment: To make a necessary or desired change in clearance, fit, or setting.

adsorb: To collect a very thin layer on the surface of a material.

adsorbent media: Generally used in filters for the removal of odors, smoke, fumes, and some impurities. The chief adsorptive granular media used for filters are activated charcoal and similar forms of carbon, Fuller's earth, and other active clays. Also see *canister filter* and *filter.*

adsorption: The attraction and/or retention of particles by molecular attraction or electrostatic forces present between the particles and a filter medium.

advance curve: A term generally relating to *spark advance curve.*

AEA: An abbreviation for the *Automotive Electrical Association.*

AERA: An abbreviation for *Automotive Engine Rebuilders Association.*

aerobic sealer: A silicone rubber sealing compound that requires oxygen for curing, such as *RTV,* used to hold parts together.

aerodynamic: The ease with which air can flow over the vehicle during higher speed operation. An aerodynamically sound vehicle has very little *wind resistance.*

aerodynamic drag: The resistance of air against an object, such as an automobile, trying to pass through it. Also referred to as *air drag* and *air resistance.*

aerodynamic force: A force created aerodynamically, such as by an *air foil.*

aerodynamic resistance: Resistance of the air against an object, such as a vehicle, trying to pass through it. The result of four factors; coefficient of drag, frontal area, vehicle speed, and air density. Also referred to as *air drag, air resistance,* and *aerodynamic drag.*

AESMC: An abbreviation for the *Automotive Exhaust Systems Manufacturing Council.*

A/F: An abbreviation for *air/fuel ratio.*

AFR: An abbreviation for *air/fuel ratio.*

A-frame: A term often used for *A-arm.*

afterboil: The boiling of fuel in the *carburetor* or coolant in the engine immediately after the engine has been stopped.

after bottom dead center *(ABDC)*: The position of a piston as it begins its compression or exhaust stroke.

afterburner: A type of engine exhaust manifold that burns any HC and CO remaining in the exhaust gas.

aftermarket: Equipment sold to consumers after the vehicle has been manufactured. Aftermarket equipment and parts are sold by catalog, dealers, independent garages, and parts houses.

afterrunning: A term often used when an engine continues to run after the ignition has been turned off. More often referred to as dieseling.

after top dead center *(ATDC)*: The position of a piston as it begins its intake or power stroke.

A/Gas: One of four classes for non-supercharged coupes and sedans. Under *National Hot Rod Association (NHRA)* rules, overall weight range is from 6.5 to 8.49 pounds per cubic inch of engine displacement.

aging: A time-temperature dependent change in the properties of certain materials occurring at room or slightly elevated temperatures following hot or cold working, or following quenching after thermal treatment.

AHRA: An abbreviation for the now-defunct *American Hot Rod Association.*

AIA: **1.** An abbreviation for the *Automobile Importers of America.* **2.** An abbreviation for *Automotive Industries Association.* **3.** An abbreviation for the *Asbestos Information Association.*

AIADA: An abbreviation for the *American International Automobile Dealers Association.*

AIAM: An abbreviation for the *Association of International Automobile Manufacturers.*

aiming screws: Self-locking screws for adjusting the headlamp in horizontal and vertical positions and for retaining the proper position.

air: The combination of gases that make up the earth's atmosphere: nitrogen (76–78%), oxygen (18–21%), and small amounts of carbon dioxide, argon, and other gases. When air is drawn into an engine, the oxygen combines with the fuel during combustion, producing carbon dioxide and water vapor.

AIR: An acronym for *air-injection reactor.*

air bag: **1.** Passive restraint with an inflatable air bag located in the steering wheel in front of the driver and in the dash in front of the right front seat passenger. **2.** An inflatable bladder used in the place of a spring in an *air suspension* system. Also see *air lift.*

air bag igniter: A combustible device that converts electric energy into thermal energy to ignite the inflator propellant. The igniter is an integral component of the inflator assembly.

air bag inflator: A term often used for *air bag igniter.*

air bag module: The air bag and inflator assembly together in a single package. This module is mounted in the center of the *steering wheel.*

air bag system: The air bag system is designed as a supplemental restraint. In the case of an accident it will deploy a bag from the steering wheel or passenger side dash panel to provide additional protection against head and face injuries.

air bleed: Holes or tubes in the carburetor to allow air to premix with gas flow.

airborne: A term used to describe contaminants floating in air through the engine. The contaminants are light enough to be suspended in the air stream.

air box: An enclosed chamber to direct air into a *carburetor* or *intake manifold.*

air brake: **1.** A moveable dynamic *spoiler* that can be raised against the wind to slow a high speed vehicle. **2.** The braking system on some heavy duty trucks that uses compressed air to expand the brake shoes by cam or wedge against the brake drums.

air carbon arc cutting: A carbon arc-cutting process that removes molten metal with a jet of air.

air charge temperature *(ACT):* The temperature of incoming air in a fuel-injection system.

air cleaner: A device connected to the *carburetor* in a manner that all incoming air must pass through it. Its purpose is to filter dirt and dust from the air before it passes into the engine.

air compressor: **1.** Engine-driven mechanism for supplying high pressure air to truck brake systems. There are basically two types of compressors: those designed to work on in-line engines and those that work on V-type engines. The in-line type is mounted forward and is gear driven, while the V-type is mounted toward the fire wall and is *camshaft* driven. **2.** A gasoline engine or electric-motor-driven mechanism for supplying shop air for the lift, air tools, and paint spraying equipment.

air conditioner: A device used for the automatic control of the temperature, humidity, cleanness, and movement of air in a given space.

air-conditioner muffler: A tank-like device usually located in the compressor discharge line to reduce compressor pumping sounds.

air conditioning *(A/C):* The process of adjusting and regulating, by heating or refrigerating, the quality, quantity, temperature, humidity, and circulation of air in a space or enclosure; to condition the air.

Air Conditioning and Refrigeration Refrigerant Recovery, Reclaim, and Recycle system *(ACR3):* Equipment that is used to recover, reclaim, and recycle refrigerant from an air, conditioning or refrigeration system.

air-control module *(ACM):* A component of the fuel control system that monitors intake air volume and meters fuel accordingly.

air control valve *(ACV):* A component used to route air from the pump to either the exhaust manifold or to the *catalytic converter.*

air cooled: Removing heat from the engine by circulating air across the cylinder block and heads.

air-cooled engine: An engine that is cooled by the passage of air, generally forced, around and over the cylinders.

air cooling: Simple method of engine cooling that relies on forced airflow over extended metal fins on the cylinder head and the block to maintain proper operating temperature.

air core: A method of constructing coils or transformers in which the wire is wound on a hollow, air-filled form instead of one using magnetic material in the core.

air cushion restraint system: A term used for *air bag system.*

air dam: **1.** A panel across a race car's front end, designed to reduce the air pressure beneath the vehicle for a better *ground effect.* **2.** Panels around the radiator to ensure that all air passes through, not around, the *radiator.*

air delivery system: The component that contains the air ducts, doors, blower, evaporator core, heater core, and controls that deliver air to the interior via the various outlets.

air door: A door in the duct system that controls the flow of air in the air conditioner and/or heater.

air drag: The resistance of air against an object, such as an automobile, trying to pass through it. Also referred to as *aerodynamic resistance* and *aerodynamic drag, air resistance.*

air duct(s): Tubes, channels, or other tubular structures used to carry air to a specific location.

air filter: A filter that removes dust, dirt, and particles from the air passing through it.

airflow sensor *(AS)*: An instrument for measuring airflow in an electronic fuel-injection system to be processed by the *electronic control module* with other sensory data to calibrate the *air/fuel mixture.*

air foil: The inverted wing of a race car designed to increase downward *aerodynamic force* and, with it, *vehicle traction.*

air/fuel mixture: The proportion of air to fuel provided by a *carburetor* or *fuel-injection system.*

air/fuel ratio *(A/F, AFR)*: The relative proportions of air and fuel entering an engine's cylinders as produced by the *carburetor* or *fuel-injection system;* the measure of the amount of air and fuel needed for proper combustion. The ideal or *stoichiometric ratio* for gasoline is 14.7:1 air to fuel by weight. A higher ratio would contain more air and less fuel, and would be considered a *lean mixture.* A lower ratio with more fuel and less air would be a *rich mixture.* The air/fuel ratio is determined by the orifice size of the main jets inside a *carburetor,* the dwell duration of the mixture control solenoid inside a feedback *carburetor,* or the orifice opening and fuel pulse duration of a fuel injector.

air gap: A small space between parts that are mated magnetically or electrically.

air hoist: A hoisting device using compressed air in a cylinder, acting against a piston, with suitable outside connections, such as a hook.

air horn: **1.** A horn that is actuated by compressed air. **2.** A tubular passage containing the choke valves in the atmospheric side of a carburetor venturi through which the incoming air must pass.

air hose: Air lines between the tractor and trailer supplying air to the trailer brakes.

air-injection reactor *(AIR)*: An *air-injection system* comprised of a vane pump, diverter valve, and check valve; a type of *emission control system* that pumps fresh air into the exhaust.

air-injection system: One that supplies fresh air to the exhaust stream, which helps oxidize HC and CO, and, gives the catalyst in the *catalytic converter* the extra air it needs to oxidize those pollutants.

air-injector system: Engine emission-control system that injects fresh air at each exhaust port. The injected air mixes with the hot exhaust gases prolonging combustion, which reduces hydrocarbon and carbon monoxide exhaust emissions.

air inlet valve: A movable door in the plenum blower assembly that permits the selection of outside air or inside air for both heating and cooling systems.

air intake and exhaust system: The parts on the automobile engine used to get the air into the engine and the exhaust out of the engine, including air cleaner, muffler, tail pipe, and associated ducting.

air intake system: A system that allows fresh clean air to enter a component such as an air conditioner, passenger compartment, or engine.

air jack: A device that uses compressed air to lift a vehicle. On some *Indy* and sports *GT* cars, the jacks are built into the chassis, permitting the whole car to be raised instantly. This enables the pit crew to change all four tires simultaneously.

airless blasting: A method of surface cleaning parts using propelled *shot.*

air lift: A tradename for a pneumatic helper spring with a *Schrader valve* that simplifies increasing or decreasing air pressure to compensate for changes in load.

air line: A hose, pipe, or tube through which air passes.

air lock: A pocket of air that blocks the normal flow of liquid in a system.

air nozzle: **1.** The tube through which air is delivered to the exhaust gas in an air-injection system. **2.** A device used to direct a stream of air into the desired area.

air outlet valve: A movable door in the plenum blower assembly that directs airflow into the heater core or into the duct-work that leads to the evaporator.

air-over-hydraulic brakes: Brakes utilizing a hydraulic system assisted by an air pressure system.

air pollution: The introduction of impurities and contaminants, many of which are caused by humans, into the atmosphere.

air pressure: **1.** *Atmospheric pressure.* **2.** The pressure produced by an air pump or compressor in a *cylinder.*

air pump: **1.** A belt- or direct-driven vane-type pump that supplies the air needed for most air-injection systems. **2.** A term often used for *air compressor.*

air resistance: The resistance of air against an object, such as an automobile, trying to pass through it. Also referred to as *air drag, aerodynamic drag,* and *aerodynamic resistance.*

air scoop: Openings at the front and/or along the side of a vehicle's bodywork to channel cool, ambient air to the radiator, engine, induction system, oil cooler, or brakes.

air shifting: Process that uses air pressure to engage different range combinations in the transmission's auxiliary section without a mechanical linkage to the driver.

air shock absorber: A term often used when referring to an *air shock* or *load-leveling shock absorber.*

air shock: Also known as *load-leveling shock absorber.* A shock operating on principles of air pressure; may also have a hydraulic section.

air slide release: An air-operated release mechanism for positioning a *sliding fifth wheel,* operated from the cab of a tractor by actuating an air-control valve.

air spring: Also known as *air bag.* An air-filled device that functions as the spring on axles that utilize air pressure in the suspension system.

air spring suspension: A single or multiaxle suspension relying on *air bags* for springs and weight distribution of axles.

air suspension system: A suspension system that uses contained compressed air for vehicle springing.

air suspension speaker: A speaker mounted in a closed box so that the enclosed air acts as a spring against the back side of the speaker cone.

air temperature sensor: A unit consisting of an *aspirator, bimetallic sensing element,* and a *vacuum modulator* used to sense in-car temperature.

airtight container: **1.** A container used to hold waste and oily rags, so that *spontaneous* *combustion* is prevented. **2.** A container in a part of a sealed system.

air to water intercooler: A *heat exchanger* on a turbocharged engine that uses *ambient air* to cool air coming from the *turbo* to the *intake manifold.*

AK steel: A term often used for *aluminum-killed steel.*

Al: Symbol for *Aluminum.*

ALCL: An abbreviation for *assembly line communications link.*

alcohol: A colorless volatile liquid; some forms can be used as a fuel for racing engines.

ALDL: An abbreviation for *assembly line data link.*

algorithm: Step-by-step specification of the solution to a problem, usually represented by a flow chart that is finally translated into a readable and understandable program.

align bore: A stationary or portable boring machine used to correct an engine's out-of-round and/or warped main bearing housings.

align hone: A stationary machine operation that corrects an engine's out-of-round and/or warped main bearing housings using a special honing mandrel.

alignment: The act of lining up or the state of being in a true line.

alignment gap: The distance between two adjacent auto-body panels.

alignment pin: A pin or pins used to align one part with another, such as the pins used to align a cylinder head on an engine block.

alignment stud: A stud or studs used to align one part with another, such as the studs used to align a cylinder head on an engine block.

align ream: A machine or hand process to enlarge the inside diameter of *bushings* to the proper size.

alky: A performance term used for alcohol, usually methanol, used as a fuel for racing car engines.

alkyd: A coating based on a polyester binder. The polyester binders are chemical combinations of molecules that contain more than one acid or alcohol group. Alkyds are widely used in water-based house paints and automotive primers.

Alliance of American Insurers *(AAI):* An association of insurance companies that write auto, liability, property, and worker's compensation coverage insurance.

allotropic: A term that applies to elements that appear in more than one form, though their atomic composition is the same. For example, the properties of graphite and diamond are the same as the element carbon (C). Since their physical properties are entirely different, both are called an allotropic of carbon.

alloy: A metal containing additions of other metallic or nonmetallic elements to enhance specific properties such as strength and corrosion resistance.

alloy powder: Powder prepared from a homogeneous molten alloy or from the solidification product of such an alloy.

alloys: Light-weight aluminum or magnesium alloy wheels. Also see *mags.*

alloy steel: Steel containing specified quantities of alloying elements added to effect changes in mechanical or physical properties.

all-season tire: Tires with special tread designed to improve traction on snow or ice (generally provides 37% higher average snow traction compared with non-all-season tires), while providing acceptable noise levels on smooth road surfaces.

all-terrain cycle: A small, three-wheeled, off-road vehicle by Honda.

all-terrain vehicle: A small, lightweight, four-wheeled vehicle with high flotation tires designed primarily for off-highway use.

all-wheel drive *(AWD):* A vehicle drivetrain with a center differential having all four wheels under power at all times.

alphanumeric: Set of all machine-processable alphabetic letters (a to z), numeric digits

(0 to 9), and special characters (such as those that appear on a typewriter).

alpha particle: A particle that is a by-product of radioactive decay. It has a positive electrical charge that is twice the negative charge of an electron.

alt: An abbreviation for *alternator.* Also, may be used as an abbreviation for alternate or altitude.

altered: A drag-racing vehicle with a stock-looking coupe, roadster, or sedan body, but without normal street equipment. The engine and/or cockpit may be moved to the rear for better weight distribution.

alternating current *(ac or AC):* The type of electrical current actually produced in an *alternator.*

alternator: An electricity-generating device that converts mechanical energy to electrical energy in the form of alternating current. Diodes rectify the alternating current into direct current.

altitude: The distance of a point above sea level. Important to automotive emissions control because the higher the altitude, the fewer oxygen molecules per given volume of air, which alters the effective compression and *air/fuel ratios.*

altitude compensation system: An altitude barometric switch and solenoid used to provide better driveability at more than 4,000 feet (1,219 meters) above sea level.

alum: **1.** A crystallized double sulfate of aluminum (Al) and potassium (K). **2.** More commonly, an abbreviation for aluminum (Al).

aluminized coating: A metal spray process used to coat engine components subject to high temperatures for long periods of time to increase heat dissipation to the ambient atmosphere.

aluminized valve: A valve with a thin layer of aluminum sprayed on the valve face and, sometimes, on the top of the valve head to provide a thin, hard, corrosion-resistant coating.

aluminum *(Al):* A versatile engineering and construction material that is light in weight, yet some of its alloys have strengths greater than steel. It has high resistance to corrosion and is an excellent conductor of heat and electricity.

aluminum cylinder block: An engine block cast of aluminum or aluminum alloy, usually with cast iron sleeves installed as cylinder bores.

aluminum-killed steel: Steel alloy in which aluminum has been added to "kill" it in the molten stage and refine its grain structure. A process of stopping molten steel from bubbling and combining with oxygen after being poured into ingots. Also see *silicone-killed steel.*

amber lens: The color lens used for turn signals and flashers on modern motor vehicles.

ambient air: The air that surrounds an object.

ambient air temperature: The temperature of the air that surrounds an object.

ambient compressor switch: An electrical switch that energizes the air conditioner compressor clutch when the outside ambient air temperature is 47°F (8.3°C) or above. Similarly, the switch turns off the compressor when the air temperature drops below 32°F (0°C).

ambient sensor: A sensor used on computerized automatic temperature-control systems that senses the outside air temperature and uses this information as an input to the system; a thermistor used in automatic temperature-control units to sense ambient temperature. Also see *thermistor.*

ambient switch: A switch used to control compressor operation by turning it ON or OFF. The switch is regulated by ambient temperature.

ambient temperature: The temperature of the air surrounding a vehicle.

ambient temperature sensor: A sensor that measures the outside air temperature as it enters the evaporator.

American Automobile Association *(AAA):* A motor club providing travel information, emergency road service, and other services to its members.

American Automobile Manufacturers Association *(AAMA)*: A trade association of Chrysler, Ford, and General Motors that sponsors research, disseminates information, and lobbies on behalf of the American automotive industry in the United States.

American City Racing League: A racing series sanctioned by the Sports Car Club of America *(SCCA)* for three-car teams representing specific cities and running 2000 *spec cars* with 2.0 liter Ford engines.

American Hot Rod Association *(AHRA)*: A drag-racing sanctioning body that is no longer in existence.

American International Automobile Dealers Association *(AIADA)*: An association of auto dealers and their employees who sell and service automobiles manufactured in the United States and abroad.

American National Standards Institute *(ANSI)*: A privately funded organization that promotes uniform standards in such areas as measurements.

American Petroleum Institute *(API)*: A petroleum-industry lobbying and public information group.

American Society of Mechanical Engineers *(ASME)*: An association to which mechanical engineers can belong. Through *ASME,* members can keep current on new technologies and procedures in the engineering field.

American Society of Test Engineers *(ASTE)*: An association to foster improved communications among those involved in the testing industry.

American Society of Testing and Materials *(ASTM)*: A professional organization to develop and promote standards for products, materials, systems, and services.

American Trucking Association *(ATA)*: A national federation of commercial trucking associations.

American wire gauge *(AWG)*: A standard for the measurement of the size of wire. The higher the gauge number, the smaller the wire.

ammeter: An instrument used to determine the amount of amperage (current draw) in a circuit by the strength of the magnetic field that is created by the current flowing through the wire.

ammeter shunt: A low-resistance conductor used to increase the range of an ammeter. It is shunted (placed in parallel) across the ammeter movement and carries the majority of the current.

amp: An abbreviation for *ampere.*

ampacity: The current-carrying capacity of conductors or equipment, expressed in *amperes.*

amperage: The amount of current, expressed in *amperes.*

amperage capacity: An indication of the length of time a battery can produce an *amperage,* or the amount of amperage that a battery can produce before being discharged.

ampere *(A)*: A unit of measure for current.

amp/hour: Amperes per hour; a standard measure for a rate of current flow.

anaerobic sealants: A chemical sealant placed on a gasket in an engine to aid in sealing and to position the gasket during installation.

anaerobic sealer: Liquid or gel that bonds two parts together in the absence of air.

analog computer: A computer that measures continuously changing conditions, such as temperature and pressure, and converts them into quantities.

analog instrument: An instrument having a needle on a dial used for taking measurements, such as temperature and engine *RPM.*

analog-to-digital converter: Mechanical or electrical device used to convert continuous analog signals to discrete digital numbers.

anaroid tube: A thermo-mechanical device in a *fuel-injection system* that regulates the amount

PORTLAND COMMUNITY COLLEGE
LEARNING RESOURCE CENTERS

of fuel being injected according to differences in temperature and pressure in the *intake manifold.*

anchor: 1. A slang expression for *brakes.* 2. A mounting point on a vehicle structure for a stressed, non-structural component, such as a seat belt or a seat.

anchor end: That end of a *brake shoe* that is attached to a fixed point on the *backing plate.*

anchor pin: A steel pin rigidly attached to the *backing plate* of *drum brakes. Return springs* are attached to the anchor pin and to the *brake shoes* to hold the shoes against the anchor pin in a non-applied position. In an applied position, it prevents the shoes from rotating with the drums.

anchors: Performance term for *brakes.*

aneroid bellows: An accordion-shaped temperature sensor charged with a small amount of volatile liquid. Temperature change causes the bellows to contract or expand, which, in turn, opens or closes a switch, such as a thermostat.

angle block: A cylinder block that does not have a *deck* at 90 degrees to the cylinders.

angle mill: A machining operation to mill the *deck* surface at a shallow angle on the exhaust side of the engine block in order to increase the *compression ratio* by decreasing the combustion chamber volume.

angle plug head: A cylinder head having spark plugs that are angled toward the exhaust valves.

anneal: 1. A heating and cooling of steel in the solid state, usually requiring gradual cooling. 2. To heat aluminum to 640°F (338°C) and then cool it to 450°F (232°C) to soften it to make it ductile.

annealing: A heat-treatment process to reduce hardness or brittleness, relieve stresses, improve machinability, facilitate cold working, or produce a desired microstructure or property.

annulus: Any type of *ring gear,* including the ring part of the *ring and pinion* in the *rear end,* and the gears in the *planetary gear set* of an *automatic transmission.*

anode: The positively charged *electrode* in an *electrolytic cell* toward which *current* flows.

anodize: A protective oxide coating to a metal surface using the metal as the anode in an electrical cell and allowing an electrolyte to act upon it.

ANSI: An abbreviation for the *American National Standards Institute.*

ant: An abbreviation for *antenna.*

antenna: A wire or other conductive metallic structure used for radiating or receiving electromagnetic signals, such as those for radio, television, or radar.

anti-backfire valve: A valve that is used in the *air-injection system* to prevent *backfiring* in the *exhaust system* during *deceleration.*

anti-drive: A suspension geometry that resists a vehicle's tendency to *drop* or *dive* on the front springs when braking.

anti-foam agent: An additive that reduces foaming caused by the churning action of the *crankshaft* in the engine oil.

anti-fouler: A device installed on a *spark plug* in an oil-burning engine to reduce the fouling of the plug.

antifreeze: A chemical solution added to the coolant (water) to prevent freezing; usually ethylene glycol and anticorrosion chemicals. Ethylene glycol resists evaporation, but the anticorrosion elements in the antifreeze may be used up in one year, depending on the amount and type of driving.

antifriction bearing: The term applies to almost any ball-roller or taper-roller bearing.

anti-icers: Chemicals added to gasoline to prevent it from freezing.

anti-knock additive: A compound that may be added to gasoline to increase its *octane,* thereby decreasing its knock.

anti-knock index: The average of the *motor octane number (MON)* and the *research octane number (RON);* a measure of a fuel's anti-knock characteristics.

anti-lock brake: A type of braking system that senses the speed of each wheel and, in conjunction with a computer, controls the hydraulic braking pressure, thereby eliminating wheel lockup.

anti-lock brake control module: The computer used to control an anti-lock braking system.

anti-lock brake system: A term often used for *anti-lock brake.*

anti-oxidant: A substance that prevents or slows down oxidation of plastic material that is exposed to air and the elements.

anti-percolation valve: A device used to prevent fuel from evaporating from the fuel bowl of a carburetor while the engine is running. It is connected to the throttle linkage, so it is closed when the throttle is open, and open when the throttle is closed. With the engine off, hot fuel vapors boil out through the vent line and into the *charcoal canister.*

antique: Any automobile built before 1930.

Antique Auto Mobile Club of America *(AACA)*: An association of collectors, hobbyists, and others interested in the preservation, maintenance, and restoration of antique automobiles.

Antique Auto Racing Association *(AARA)*: An association of persons interested in the history of auto-racing; preserving, restoring, and driving antique race cars.

anti-rattle hardware: Clips, springs, and washers to prevent disc pads or *brake shoes* from rattling and vibrating.

anti-rust additive: A solution that is added to coolant to retard rust formation inside a cooling system.

anti-seize compound: A thread compound designed to protect threaded connections from damage due to rust or corrosion.

anti-siphon system: A small passage in a *carburetor* to prevent fuel from siphoning from the *float bowl* into the engine.

anti-skid control: Accessory for the brake system that operates on rear wheels, or all four wheels, to prevent wheel lockup during braking. Braking pressure is reduced to wheel(s) that are about to lock up and skid, by electronic controls.

anti-smog device: A part or parts designed to reduce or eliminate vehicle exhaust emissions.

anti-squat: Suspension geometry that will resist a vehicle's tendency to *drop* or *squat* on the rear springs when accelerating.

anti-static agents: Treatment used during, and/or after the molding process to minimize static electricity in plastic materials.

anti-sway bar: A suspension component, often called a *sway bar,* intended to prevent side-to-side body movement in relation to the axles and wheels.

anti-theft system: Deterrent systems designed to scare off would-be car thieves by sounding alarms and/or disabling the *ignition system.* Common components used in an automobile anti-theft system include an electronic control module, door switches at all doors, trunk key cylinder switch, hood switch, starter inhibitor relay, horn relay, and alarm.

AOC: An abbreviation for *abnormal operating conditions.*

APAA: An abbreviation for the *Automotive Parts and Accessories Association, Inc.*

apex: The innermost point of a turn or corner on a roadway or race course.

apexes: The peaks on the rotor, formed by the meeting of two adjoining rotor faces in a rotary engine.

apex seal: A seal used to retain the combustion pressure at all three tips of the rotor in a rotary engine.

API: An abbreviation for *American Petroleum Institute.*

API degrees: Another term used for *API gravity.*

API gravity: The scale for the density (gravity) of a liquid petroleum product, expressed in *API degrees.* The lighter the product the higher the number.

A B C D E F G H I J K L M N O P Q R S T U V W X Y Z

A-pillar: The structural support on either side of the windshield, just ahead of the front doors.

API ring: The information printed, in a ring, on the top or side of an oil or lubricant container providing the *API* specifications and ratings of the contents.

A-post: A term often used for *A-pillar.*

appearance money: The payment of money to a popular driver to encourage competition in a race, so his or her participation can be advertised in advance, thereby trying to attract a larger paying crowd.

application cable: The line or cable that engages the vehicle's *emergency brake system.*

apply devices: Devices that hold or drive members of a *planetary gear set.* They may be hydraulically or mechanically applied.

apply side: The side of a *piston* on which force or pressure is exerted to move the piston to do work.

appraiser: An insurance company's representative who estimates a vehicle's damage and authorizes payment to the collision repair/refinishing shop. Also see *estimator.*

approach angle: The maximum angle, in degrees, of a line running upward and forward from the front tire contact point to the lowest obstruction under the front of the vehicle. Also see *departure angle.*

APRA: Abbreviation for *Automotive Parts Rebuilders Association.*

apron: The inner edge of a race track.

aquaplaning: A tire unable to remain in contact with the ground or pavement in wet weather that rides on the water itself. Also known as *hydroplaning* or, more simply, *planing.*

ARA: An abbreviation for: **1.** *Automotive Recyclers Association.* **2.** *Automotive Retailers Association.*

arbor: A tapered metal shaft used to secure a cutting tool or a part being turned on a lathe.

arbor press: A manual- or power-operated press used to force arbors or mandrels into or out of holes and for similar assembly or disassembly operations.

arc: **1.** To run *flat out* on an oval track. **2.** The discharge of electrical current across a gap between two electrodes.

arc blow: In welding, the deflection of an arc from its normal path because of magnetic forces.

arc braze welding: A braze welding process that uses an arc to provide the heat.

arc cutter: A term often used for *thermal cutter.*

arc cutting: The thermal cutting process that severs or removes metal by melting it with the heat of an arc between an electrode and the work piece.

arc force: The axial force developed by an arc plasma during a welding procedure.

arc gouging: A thermal gouging that uses an arc-cutting process to form a bevel or groove.

architecture: The physical structure of a computer's internal operations, including its registers, memory, instruction set, input/output structure, and so on.

arcing: A term that applies to the spark that occurs in an electrical circuit in an air gap, such as a *spark plug.*

arcing time: The time elapsing from the severance of the circuit to the final interruption of current flow.

arc length: The distance from the tip of the welding electrode to the adjacent surface of the weld pool.

arc spraying: A thermal spraying process using an arc between two consumable electrodes of surfacing materials as a heat source, and a compressed gas to atomize and propel the surfacing material to the substrate.

arc welding: A welding process that produces fusion of work pieces by heating them with an arc with or without the application of pressure and with or without filler metal.

arithmetic-logic unit: A basic element of the *central processing unit (CPU)* in a computer. That portion of the *CPU* where arithmetic and logical operations are performed.

arm: **1.** To turn on a theft-deterrent system. **2.** Crankshaft throw. Also see *long arm* and *short arm.*

armature: **1.** A part moved by magnetism. **2.** A part moved through a magnetic field to produce current.

Armco: The trade name of a particular type of guard rail or barrier used on public roadways and race courses.

arming: In an automobile anti-theft system, arming means placing the alarm system in readiness, enabling it to detect an illegal entry. Arming is accomplished when the *ignition switch* is turned off and the doors are locked.

armored ring groove: A metal ring groove cast into a piston during manufacturing to increase resistance to wear.

arnoid gauge: An instrument used to align the center of a crankshaft journal with the centers of a crankshaft grinder.

aromatic: A type of solvent based on benzene ring molecules. Aromatics are often used as diluents in acrylic lacquers. Typical examples are benzene, xylol, and toluol.

arnoid hydrocarbons: Compounds having carbons linked in a closed ring by alternating single and double bonds.

articulated vehicle: Large trucks or buses with two or more wheeled units, so designed for ease of cornering.

articulating upper coupler: A bolster plate kingpin arrangement that is not rigidly attached to the trailer but provides articulation and/or oscillation about an axis parallel to the rear axle of the trailer.

articulation: **1.** The action of a chain joint in flexing from the straight to an angle and back to the straight as the joint passes around a sprocket or other path. **2.** Vertical movement of the front driving or rear axle, relative to the frame of the vehicle to which they are attached.

AS: An abbreviation for *airflow sensor.*

ASA: An abbreviation for the *Automotive Service Association.*

asbestos: A fiber mineral that is heat resistant and nonburning. Once used in brake linings, gaskets, and clutch facings, it is no longer used due to health hazards.

Asbestos Information Association *(AIA):* An association to provide industry-wide information on asbestos and health, and on industry efforts to eliminate existing hazards.

ASE: An abbreviation and registered trademark of the *National Institute for Automotive Service Excellence.*

ASIA: An abbreviation for the *Automotive Service Industry Association.*

A-shim: A valve spring-adjuster insert with a thickness of 0.060 inch (1.524 mm) used to balance spring pressure and to correct installed height.

ASME: An abbreviation for the *American Society of Mechanical Engineers.*

aspect ratio: Also known as *tire profile.* A measurement of a tire; the percentage of the tire's height to the width.

asphalt eater: A top-performing drag car.

aspirator: A device that uses suction to move air, accomplished by a differential in air pressure; a one-way valve attached to the exhaust system of an engine that admits air during periods of vacuum between exhaust pressure pulses. Used to help oxidize hydrocarbon (HC) and carbon monoxide (CO), and to supply additional air that the catalytic converter may require. Can be used instead of a belt-driven, air-injection pump in some applications.

aspirator valve: A device used to draw out fluids by suction. In this case, a pollution device is used to draw fresh air by suction into the exhaust flow to reduce emissions.

A B C D E F G H I J K L M N O P Q R S T U V W X Y Z

assembly: Any unit made up of two or more parts.

assembly line communications link: An electrical connector used to check a vehicle's operating system while it is still on the assembly line.

assembly line data link: The information processed for use in assembly-line diagnostics.

assembly lube: A special lubricant used to coat parts that rub or rotate against each other during initial assembly.

Association of International Automobile Manufactures *(AIAM)*: A trade association of United States subsidiaries of international automobile companies.

assy: An abbreviation for *assembly*.

ASTE: Abbreviation for *American Society of Test Engineers*.

ASTM: Abbreviation for *American Society of Testing and Materials*.

asymmetrical cam: A *camshaft* having different profiles for the intake and exhaust lobes.

asymmetrical rear-leaf spring: A spring on which the rear axle is not located in its center.

AT: An abbreviation for *automatic transmission*.

A/T: An abbreviation for *automatic transmission*.

ATA: An abbreviation for the *American Trucking Association*.

Atari dashboard: A digital instrument panel; so called due to its resemblance to an Atari video game.

ATC: **1.** A trade name by Honda for a three-wheeled *all-terrain cycle*. **2.** An abbreviation for *automatic temperature control*.

ATC servo programmer: A mechanically operated switch to control blower speed whenever the blower switch is in the AUTO position on some car lines.

ATDC: An abbreviation for *after top dead center*.

ATF: An abbreviation for *automatic transmission fluid*.

atmo: **1.** A racing engine running on atmospheric pressure. **2.** An abbreviation for *atmosphere*.

atmosphere: The mass of air that surrounds Earth.

atmospheric ozone: As ultraviolet (UV) rays from the sun reach Earth, they are combined with smog and other pollutants to produce atmospheric ozone. Atmospheric ozone, unlike stratospheric ozone, is considered harmful. Whenever possible it is to be avoided.

atmospheric pollution: Impurities and contaminants in the atmospheric environment, many of which are caused by humans.

atmospheric pressure: The pressure of the atmosphere at any given location. At sea level, it is 14.696 *psia* (101.33 *kPa absolute*).

atmospheric pressure sensor: A device designed to send an electrical signal to the *central processing unit* relative to the atmospheric pressure at any given time.

atom: A basic unit of matter consisting of *protons, neutrons,* and *electrons*.

atomization: The breaking down of a liquid into small particles, like a mist, by the use of pressure.

atomized: A liquid is atomized when it is broken into tiny droplets of the liquid, much like a mist or spray form.

ATRA: An abbreviation for *Automatic Transmission Rebuilders Association*.

A-train: A combination of two or more trailers in which the dolly (converter or turntable), is connected by a single pintle hook or coupler, and the drawbar connection is at the center, between each vehicle.

attenuation: The decrease in the strength of a signal as it passes through a control system.

attrition: The wearing down by rubbing or friction; abrasion.

ATV: An abbreviation for *all-terrain vehicle.*

auto: Shortened form for *automatic* or automobile.

autocross: A form of automotive competition that is held on a tight, closed course. Vehicle handling and agility is stressed, rather than flat out speed.

auto control: Another term used for *automatic control.*

auto ignition: Short for *automatic ignition.*

automatic: Having the power of self-motion: self-moving: or self-acting.

automatic adjuster: A drum-brake mechanism that adjusts the lining clearance as wear occurs. It is commonly actuated on reverse stops or when the parking brake is set.

automatic chassis lubrication *(ACL):* A system where the chassis is automatically lubricated at predetermined intervals.

automatic choke: A mechanism that positions the choke valve automatically in accordance with engine temperature.

automatic control: **1.** A dial on the instrument panel that is set at a desired temperature level to control the condition of the air automatically. **2.** Any system that reacts to a predetermined condition rather than responding to external commands. **3.** Also known as *auto control.*

automatic door locks: A passive system used to lock all doors when the required conditions are met. Many systems lock the doors when the gear selector is placed in drive, the ignition switch is in RUN, and all doors are properly shut.

automatic headlight dimming: Automatically switches the headlights from high beams to low beams under two different conditions: when light from oncoming vehicles strikes the photocell-amplifier, or light from the taillights of a vehicle being passed strikes the photocell-amplifier.

automatic ignition: A condition where the engine continues to run after the ignition has been shut off. Often called *dieseling* or *running on.* Also known as *auto ignition.*

automatic level control: A shock-absorber system operated by air pressure and provided as an accessory on some vehicles to automatically maintain the correct *riding height* under various load conditions.

automatic on/off with time delay: **1.** To turn on the headlights automatically when ambient light decreases to a predetermined level. **2.** To allow the headlights to remain on for a certain amount of time after the vehicle has been turned off.

automatic steering effect: The tendency of a vehicle to travel straight out of a turn when the *steering wheel* is released.

automatic temperature control *(ATC):* An air-conditioner control system designed to maintain a pre-selected, in-car temperature and humidity level automatically.

automatic tensioning: The *constant tension* of a device, maintained at a proper value by some automatic means; to minimize the attention required.

automatic transmission *(AT, A/T):* A transmission in which *gear ratios* are changed automatically.

automatic transmission cooler: A device, often found in the *radiator,* through which automatic transmission fluid circulates to be cooled by surrounding air or engine coolant.

automatic transmission fluid *(ATF):* A red, petroleum-based fluid used to transfer power and control, lubricate, cool, and clean the *automatic transmission.*

Automatic Transmission Rebuilders Association *(ATRA):* A trade association for transmission repair shops, technicians, and suppliers of transmission repair equipment, parts, and tools.

automation: **1.** Semi-automatic or automatic material handlers, loaders, unloaders, and other labor-saving devices. **2.** Automatic cycle control of machines or equipment by tracer, cam, plugboard, numerical control, or computer. **3.** The application of machinery and equipment to perform and control semi-automatically or automatically and continuously all operations in a manufacturing plant.

Automobile Competition Committee for the United States *(ACCUS):* With representatives of *NASCAR, SCCA, IMSA,* and *USAC,* the American affiliate of *FISA,* coordinating major United States racing events with the international calendar.

automobile emissions: Certain impurities that may enter the atmosphere during vehicle operation, such as *hydrocarbons* (*HC*), *carbon monoxide* (*CO*), and *oxides of nitrogen* (NO_x).

Automobile Importers of America *(AIA):* A professional organization of importers of cars and trucks into the United States.

automotive air conditioning: The process of transferring heat from inside to outside the passenger compartment. The cooled air is also dehumidified, purified, and circulated.

automotive air pollution: Evaporated and unburned fuel, and other undesirable by-products of combustion that escape from a motor vehicle into the atmosphere.

automotive battery: An electro-chemical device that stores and provides electrical energy for the operation of a vehicle.

automotive body shop: A term often used for body shop.

Automotive Communications Council *(ACC):* A professional association of advertising, marketing, and communications executives.

automotive cooling system: The many components that operate to absorb and dissipate heat developed in the combustion process, thus maintaining the desired engine-operating temperature.

Automotive Cooling System Institute *(ACSI):* A *Motor and Equipment Manufacturers Association* subgroup made up of cooling-system product manufacturers.

Automotive Electrical Association *(AEA):* A trade association absorbed in 1991 by the *Automotive Service Industry Association (ASIA).*

automotive emissions: Another term for *automotive air pollution.*

Automotive Engine Rebuilders Association *(AERA):* An association of machine shops and others dedicated to engine rebuilding.

Automotive Exhaust Systems Manufacturers Council *(AESMC):* An association that provides technical information and lobbying efforts on behalf of the exhaust-system-replacement market.

Automotive Industries Association *(AIA):* A Canadian aftermarket trade group of distributors, suppliers, wholesalers, and retailers.

Automotive Parts and Accessories Association, Inc. *(APAA):* A trade association for aftermarket retailers, wholesalers, manufacturers, and distributors.

Automotive Parts Rebuilders Association *(APRA):* A trade association of automotive parts rebuilders and suppliers of remanufactured parts.

automotive power brakes: A brake system having a vacuum and atmospheric air-operated power booster or hydraulic power boost to multiply braking force.

Automotive Recyclers Association *(ARA):* International association of automotive recyclers, owners, and dealers in used car and truck parts.

Automotive Retailers Association *(ARA):* An association of the automotive retailers, sales and service, including collision repairs, mechanical repairs, used car sales, auto wrecking, and towing.

Automotive Service Association *(ASA):* A trade organization for body, mechanical, and transmission shop owners.

Automotive Service Industry Association *(ASIA):* A trade association for those who manufacture, distribute, and sell parts, tools, and equipment to the *do-it-yourself (DIY)* market and professional aftermarket.

Automotive Warehouse Distributors Association *(AWDA):* An association of major manufacturers and warehouse distributors of aftermarket parts.

Autronic eye: A trade term for *automatic headlight dimmer.*

aux: **1.** An abbreviated form of *auxiliary.* **2.** An auxiliary item of equipment, such as aux fuel tank.

auxiliary: A spare; an extra; a back-up system.

auxiliary air valve: A device that allows air to bypass a closed throttle during engine start-up and warm-up.

auxiliary drum parking brake: The incorporation of an auxiliary parking-brake drum inside a rear rotor on some four-wheel drive disc-brake systems.

auxiliary seal: A secondary seal mounted outside the seal housing: **1.** To prevent refrigeration oil from escaping and entering the clutch assembly. **2.** To aid in the prevention of the loss of fluid from a system.

auxiliary section: The section of a transmission housing the auxiliary drive gear, main shaft assembly, countershaft, and synchronizer assembly, where range shifting occurs.

auxiliary shaft: A separate shaft, in an overhead cam engine, that drives devices such as the *fuel pump, oil pump,* and *distributor.*

auxiliary springs: **1.** A second or third valve spring with a different resonant frequency to cancel out harmonic vibrations that limit engine speed. **2.** The spring(s) added to a vehicle, generally in the rear, to support a heavy load.

auxiliary venturi: A small *secondary venturi* mounted inside the main venturi of a *carburetor* to provide increased air velocity. May also be called a *booster venturi.*

available voltage: The maximum voltage that can be produced in the secondary circuit of a conventional ignition system.

avalanche diode: A semiconductor designed to operate in the breakdown region to produce a constant voltage across the diode for regulating through the current though it may vary.

A V–8: A Ford model A retrofit with a late model V–8 engine.

avgas: Slang for aviation gasoline; generally higher octane than automobile gasoline.

AWD: An abbreviation for *all-wheel drive.*

AWDA: An abbreviation for *Automotive Warehouse Distributors Association, Inc.*

AWG: An abbreviation for *American wire gauge.*

axial compressor: A compressor designed so that the cylinders are arranged parallel to the output shaft.

axial leads: Wires that extend from the ends of an electronic component, such as a capacitor, along the axis of the unit.

axial load: A type of load placed on a bearing that is parallel to the axis of the rotating shaft.

axial motion: Motion that occurs along the axis of a revolving shaft or parallel to the axis of a revolving shaft.

axis: The center line of a rotating part.

axle: **1.** A cross member supporting a vehicle on which one or more wheels are mounted. **2.** A pair of wheels at either end of a vehicle. Some brake repair shops and turnpike tolls charge a "per axle" rate.

axle bearing: A *bearing* that supports an *axle* or *half shaft* in an *axle housing.*

axle boot: The flexible cover that retains grease and/or oil in a *transmission* or a *constant velocity joint.*

axle carrier assembly: A cast-iron framework that can be removed from the rear-axle housing for service and adjustment.

axle gears: Bevel gears that transfer power from the differential pinion gears to the splined axle shafts.

axle hop: The tendency of a live axle housing to rotate with the wheels slightly and then snap back during hard acceleration. This action may be repeated several times, creating a loss of traction until the driver releases the accelerator.

axle housing: Designed in the removable carrier or integral carrier types to house the drive pinion, ring gear, *differential,* and *axle shaft* assemblies.

axle ratio: The ratio between the rotational speed *(rpm)* of the drive shaft and that of the driven wheel; gear reduction through the *differential,* determined by dividing the number of teeth on the ring gear by the number of teeth on the drive pinion.

axle seat: Suspension component used to support and locate spring on the axle. Also known as a *spring chair.*

axle shaft: Alloy steel shaft that transfers *torque* from the differential side gears to the drive wheels. This shaft also supports vehicle weight on most passenger cars.

axle-shaft end thrust: A force exerted on the end of an *axle shaft* that is most pronounced when the vehicle turns corners and curves.

axle-shaft tubes: Tubes that are attached to the axle housing center section to surround the axle shaft and bearings.

axle tramp: The tendency of a live *axle housing* to rotate with the wheels slightly and then snap back during hard acceleration. This action may be repeated several times, creating a loss of traction until the driver releases the accelerator.

axle windup: A term often used for *axle tramp.*

azeotrope: A mixture of two or more liquids, such as refrigerants, that when mixed in precise proportions, behave like a compound.

B

b: Black.

babbitt: An alloy used to line bearings made up of tin, antimony, copper, and other metals.

baby moons: Small chrome-plated wheel covers.

backbone frame: A chassis structure having one boxed member running down the center. It is usually divided into two parallel members at each end to support the power train and suspension system.

backfire: **1.** An explosion in the *exhaust system* of a motor vehicle caused when an unburned *air/fuel mixture* is ignited, usually upon deceleration. **2.** An explosion of the *air/fuel mixture* in the *intake manifold,* which is evident at the carburetor or throttle body and may be caused by improper ignition timing, crossed spark plug wires, or an intake valve that is stuck open. **3.** The momentary recession of the flame into the welding tip, cutting tip, or flame-spraying gun, followed by immediate reappearance or complete extinction of the flame.

backfire suppression valve: An anti-backfire valve used in the *air-injection system* of an exhaust emission control.

backfiring: **1.** The pre-explosion of an *air/fuel mixture* so that the explosion passes back around the opened intake valve, through the *intake manifold,* and through the *carburetor.* **2.** The loud explosion of over-rich exhaust gas in the *exhaust manifold* that exits through the *muffler* and *tailpipe* with a loud popping noise.

back flush: The use of a reverse flow of water, with or without a cleaning agent, to clean out the cooling system of a vehicle.

backing plate: Stamped steel plate upon which the wheel cylinder is mounted and the brake shoes are attached; a metal plate that serves as the foundation for the brake shoes and other drum brake hardware.

backlash: The excessive clearance between the meshing teeth of two gears.

backlight: The rear window of a vehicle.

back motor: A mid- or rear-mounted engine.

back plane: The main circuit board of a system, containing edge connectors or sockets so other printed circuit boards can be plugged into it.

back pressure: **1.** Resistance of an exhaust system to the passage of exhaust gases. This can have an adverse effect on performance, fuel economy, and emissions. Excessive back pressure may be caused by a clogged catalytic converter, or a dented or crimped pipe. **2.** The excessive pressure buildup in an engine crankcase.

back pressure EGR: Some emissions-control systems use a back-pressure sensor or diaphragm to monitor back pressure so that exhaust gas recirculating flow can be increased when the engine is under maximum load, and producing maximum back pressure.

back pressure EGR valve: A back-pressure-dependent EGR valve. See the specific application; *negative-back-pressure EGR valve* or *positive-back-pressure EGR valve,* as applicable.

back seat: The position of a valve stem when turned to the left (ccw) as far as possible back seating a two-seat service valve.

back staging: Placing a competition vehicle at the start of a drag race behind the usual staging position. Also referred to as *shallow staging.*

backup light: Lamps that illuminate the area behind the vehicle and warn others of the driver's intention to back up. All vehicles sold in the United States after 1971 are required to have such lights.

A B C D E F G H I J K L M N O P Q R S T U V W X Y Z

backyard mechanic: An amateur mechanic or one with little training. Often called a *shade-tree mechanic.*

bad: Good.

bad car: A performance term for an extremely fast car.

badge engineering: The act of producing the same car under more than one name.

bad sector: A sector on a computer disk that will not read or write correctly. Usually due to a minor physical flaw in the disk.

baffle: **1.** A barrier used to reduce noise in an enclosed system, such as the exhaust system. **2.** A barrier to prevent splashing of liquid in a tank.

bail: The spring-steel wire loop used to secure a cover, such as on a master cylinder reservoir.

Baja: A Spanish term for "lower." The term more generally refers to Mexico's Baja race, which is held in a rugged desert region creating one of the world's toughest off-road courses.

Baja Bug: A Volkswagen Beetle modified for off-road use, developed in the late 1960s by Drino Miller.

balance: **1.** To have all rotating parts in a state of equilibrium. **2.** To balance all rotating parts statically and dynamically in an engine to ensure maximum performance.

balance control: A control in a stereo amplifier that adjusts the relative output volume from each of the stereo channels.

balance pipe: **1.** A pipe that connects the exhaust pipes in a dual exhaust system to equalize the pressures. **2.** A pipe that connects the venturis of dual carburetors.

balanced carburetor: A carburetor in which the float bowl is vented to the air horn to compensate for the possible effects of a clogged air filter.

balanced valve: A type of hydraulic valve that produces a pressure change that is proportional to the variations of spring pressure or the movement of a mechanical linkage.

balancer: A heavy crankshaft pulley that aids in overall crankshaft balance as it rotates.

balance shaft: Found primarily in I–4 and V–6 engines, a rotating shaft incorporating a *harmonic balancer* or *vibration damper* designed to counteract the natural vibrations of other rotating parts, such as the *crankshaft,* in an engine.

balance tube: **1.** A tube to connect the exhaust pipes in a dual exhaust system to equalize the pressures. **2.** A tube to connect the venturis of dual carburetors.

balancing: The process of proportioning weight or force equally on all sides of an object. Most *crankshafts,* for example, are balanced both *statically* and *dynamically.*

balancing coil gauge: An indicating device, such as a fuel gauge, that contains a pair of coils in the instrument-panel unit.

balk ring: A rotating device found in a manual transmission that prevents premature engagement of gears while shifting.

ball-and-nut steering gear: Another term for *recirculating ball-and-nut steering gear.*

ball and socket: A term often used for *tie rod end.*

ball-and-trunion joint: A type of *universal joint* that is combined with the *slip joint* in one assembly.

ballast: Material that is added to a racing car chassis to change the weight distribution and/or increase the overall vehicle weight to the minimum class requirement.

ballast resistor: A term often used for *ignition resistor.*

ball bearing: An anti-friction bearing with an inner and outer race having one or more rows of balls between them.

ball check valve: A one-way valve having a ball and seat.

ball joint: **1.** A joint or connection where a ball moves within a socket, allowing a rotary motion while the angle of the rotation axis changes. **2.** A suspension component that attaches the steering knuckle to either control arm featuring a ball-and-socket joint to allow pivoting in various directions. Also known as a *spherical joint.*

ball joint angle: The inward tilt of the steering axis from the vertical.

ball joint centerline: An imaginary line drawn through the centers of the upper and lower ball joints.

ball joint free play: The allowable *radial* and *axial* motion between the ball-joint housing, checked with the load removed.

ball joint inclination: The inward tilt of the top of the steering axis centerline through the ball joints as viewed from the front of the vehicle.

ball joint internal lubrication: A ball joint assembly may be pre-lubricated and factory sealed or it may have provision for periodic scheduled lubrication.

ball joint preload: A term relating to certain ball joints, often spring-loaded, having constant friction between the ball-and-joint housing socket.

ball joint seal: A Neoprene rubber seal that fits over a ball-joint stud against the housing to retain lubricating grease and keep unwanted foreign debris, such as sand or dirt, out.

ball joint slack: A term often used for *ball joint free play.*

ball joint suspension: A type of front suspension in which the *wheel spindle* is attached directly to the upper and lower *suspension arms* through the *ball joints.*

ball nut: In a *recirculating ball nut steering gear,* the ball nut has internal threads that are meshed to the threads of a worm with continuous rows of ball bearings between the two. The ball bearings are recirculated through two outside loops called ball guides. The sliding ball nut has tapered teeth cut on one face that mate with teeth on the cross-shaft sector.

balloon foot: A term used to describe a slow driver; one who tends to back off the throttle early.

ball stud: A stud with a ball-shaped end.

band: **1.** A hydraulically controlled device installed in an automatic transmission around a clutch drum, used to stop or permit drum turning. **2.** A manual or hydraulic device installed around a drum to provide a braking action.

banjo housing: A banjo-shaped case that houses a final drive live axle.

BAR: An acronym for the *Bureau of Automotive Repair.*

barb fitting: A fitting that slips inside a hose and is held in place with a gear-type clamp or pressed-on ferrule.

bare electrode: A filler metal electrode that has been produced as a wire, strip, or bar with no coating or covering other than that which is incidental to its manufacture or preservation.

bare out: To strip a car body to its basic shell, generally in preparation for its restoration.

BARO: An abbreviation for **1.** Barometric pressure. **2.** Barometric-pressure sensor. **3.** Barometer.

barometer: An instrument used for measuring barometric pressure.

barometric pressure *(BARO):* The pressure exerted by the weight of the earth's atmosphere, equal to one bar, 99.97 *kPa,* or 14.5 *psi* at sea level. Barometric pressure changes with the weather and with altitude. Since it affects the density of the air entering the engine and ultimately the *air/fuel ratio,* some computerized emissions-control systems use a *barometric pressure sensor* so that the *spark advance* and *exhaust gas recirculate (EGR)* valve flow can be regulated to control emissions more precisely.

barometric pressure sensor: A device that senses barometric pressure and sends an electrical signal to the *CPU* for optimum engine control.

barrel *(bbl):* **1.** A term sometimes applied to the cylinders in an engine. **2.** A term used to refer to the number of throttle bores in a *carburetor.*

barrel faced ring: A compression piston ring with a rounded contact face.

barrel finish: The rounded surface on a piston skirt.

barrel roll: A vehicle rollover, sideways.

barrier hose: A hose constructed with a special liner to prevent refrigerant leakage through its walls. Most air-conditioning systems in vehicles manufactured after 1988 have barrier-type hoses.

base circle: The low portion of each cam on the *camshaft,* concentric with the *journal,* which is not part of the lobe.

base coat: The initial layer of paint.

baseline: An initial reference point.

base metal: **1.** The largest proportion of metal present in an alloy. **2.** A metal that readily oxidizes, or that dissolves to form ions. **3.** The metal or alloy that is welded, brazed, soldered, or cut. **4.** After welding, that part of the metal that was not melted.

base station: The bottom section of a station buck that serves as a reference point in metal working.

base 10: A base unit in the metric system. All metric units are increased or decreased in units of 10. One meter, for example, has 10 decimeters, 100 centimeters, or 1,000 millimeters.

basic fuel metering: The amount of fuel delivered to the injectors in a continuous-flow, fuel-injection system, based on airflow sensor readings of engine load and *rpm.*

basic fuel quality: The amount of fuel delivered to the injectors, in a pulsed, fuel-injection system, based on airflow sensor readings of engine load and *rpm.*

basket of snakes: A tuned exhaust system with individual intertwined headers.

bass compensation: A circuit in an audio amplifier that increases the output of low-end audio frequencies at low listening levels to compensate for the human ear's loss of sensitivity under these conditions.

bastard: A term often used when referring to odd sizes or shapes.

BAT: An abbreviation for *battery.*

batch: A group of records or programs that is considered a single unit for processing on a computer.

bathtub: **1.** An auto body design that resembles an inverted bath tub. **2.** A *combustion chamber* in an engine with an area that resembles an inverted bath tub with its valves seated at its base.

BAT–: The negative battery terminal.

BAT+: The positive battery terminal.

battery *(BAT):* A device for storing energy in chemical form so it can be released as electricity.

battery acid: An electrolyte used in a battery; a mixture of water (H_2O) and sulfuric acid (H_2SO_4).

battery backup: Auxiliary power that is provided to a computer so volatile memory information is not lost during a power failure, or when otherwise disconnected from its normal power source.

battery capacity: The energy output of a battery measured in *amp/hours.*

battery cell: That part of a battery made from two dissimilar metals and an acid solution. A cell stores chemical energy for use later as electrical energy.

battery charge: The restoration of chemical energy to a battery by supplying a measured flow of electrical current to it for a specified time.

battery charger: An electrical device that is used for restoring a battery to its original state of charge by passing a current through the battery in a direction opposite of the discharge current flow.

battery charging: The act of charging a battery.

Battery Council International *(BCI):* A professional association of manufacturers, suppliers, and distributors of lead-acid batteries.

battery efficiency: A battery's ability to vary the current it delivers within a wide range, depending on the temperature and the rate of discharge.

battery element: 1. A cell. 2. A group of unlike positive and negative plates assembled with separators.

battery maintenance: Generally, *preventative maintenance,* such as visual inspection, adding water, cleaning top and terminals, tightening hold down, and testing.

battery rating methods: There are several rating methods: *cold cranking power rating, reserve capacity rating,* and *twenty-hour rating.*

battery terminal: A means of connecting the battery to the vehicle's electrical system. The three types of battery terminals are: post or top, side, and "L."

bayonet socket: A lamp socket having two lengthwise slots in its sides, making a right-angled turn at the bottom. A lamp with two pins may be installed by pushing it into the socket and giving it a slight clockwise turn.

BBDC: An abbreviation for *before bottom dead center.*

bbl: An abbreviation for *barrel.*

BBM: An abbreviation for *break before make,* as in a switch.

BCI: An abbreviation for *Battery Council International.*

BCM: An abbreviation for *body control module.*

BDC: An abbreviation for *bottom dead center.*

B-cam: A pattern used to grind pistons in an oval or cam shape with a 0.006 *inch* (0.152 mm) difference between the face and pinhole side.

B-dolly train: A combination of two or more trailers in which the dolly is connected by two or more pintle hooks, couplers, and drawbar connections located between each vehicle, making a rigid connection. The resulting connection has one pivot point.

bead: 1. That part of a tire that contacts the rim of a wheel. 2. A narrow half-round pattern where metal has been welded together.

bead filler: A rubber piece positioned above the *bead* to reinforce the sidewall and act as a rim extender.

bead lock: A plate that clamps a tire *bead* to a wheel rim used on circle-dirt-track race cars.

bead wire: A group of circular wire strands molded into the inner circumference of a tire to anchor the tire on the rim.

beam axle: A shaft that does not transmit power but provides a means of fastening wheels to either, or both, ends.

beam solid-mount suspension: A tandem suspension relying on a pivotal mounted beam, with axles attached at the ends for load equalization.

beans: Performance term for *horsepower.*

bearing: A term used for *ball bearing* or *bushing.*

bearing block: The outside surface of a bearing that seats against the *housing bore.*

bearing bore: A term often used for *housing bore.*

bearing cap: A device that retains the needle bearings that ride on the trunnion of a U-joint and is pressed into the yoke.

bearing cap register: The cut-out portion of the *engine block* that keeps the *bearing cap* aligned to the *housing bore.*

bearing crush: The additional height, manufactured into each bearing half, to ensure complete contact of the bearing back with the housing bore when the engine is assembled.

bearing groove: A channel cut into the surface of a bearing to ensure oil distribution.

bearing id: The inside diameter of a bearing.

bearing od: The outside diameter of a bearing.

bearing oil clearance: A space that is provided between a shaft and a bearing so oil can flow through it.

bearing pre-lubricator: A pressurized oil tank attached to the engine-lubrication provisions to maintain oil pressure when the engine is not running.

bearing shell: One half of a single rod or main bearing set.

bearing spacer: A device that is used to hold a *bearing* in the *housing bore.*

bearing spin: A bearing that has rotated in its housing or block, generally due to failure as a result of lack of lubrication.

bearing spread: The small extra distance across the parting faces of a bearing half in excess of the actual diameter of the bearing bore.

bearing upper: The *bearing shell* that is positioned opposite the *bearing cap.*

beater: A car used for everyday transportation.

beef: A term used for *beef up.*

beef up: To strengthen and/or reinforce.

beehive spring: A spring that is wound in the shape of a beehive.

before bottom dead center: The position of a piston approaching the bottom of its intake or combustion stroke.

before top center *(BTC)*: A piston as it is approaching the top of its stroke.

before top dead center *(BTPC)*: Any position of the *piston* between *bottom dead center* and *top dead center* on its upward stroke.

Belleville washer: A circular disk formed into a conical shape. When a load is applied, the disk tends to flatten, constituting a spring action. May be referred to as a *coned-disk spring.*

bell housing: A term often used for *clutch housing.*

bell-mouthed drum: A well-worn brake drum that is deformed so that its open end has a larger diameter than its closed end.

bellows: A flexible chamber that can be expanded to draw a fluid or vapor in and compressed to pressurize the fluid or vapor.

belly pan: Body panel(s) covering the bottom of a competition vehicle to reduce the *coefficient of drag.*

belly tank: A tear-shaped, World War II aircraft auxiliary fuel tank used as the body for a *lakester.*

belt: **1.** In a tire, the belt(s), generally steel, restrict ply movement and provide tread stability and resistance to deformation, providing longer tread wear and reducing heat buildup in the tire. **2.** A device used to drive the water pump and/or other auxiliary devices, such as the *alternator,* off the engine.

belt-clamping action: As related to a continuous variable transmission, the action taking place when the V-pulley sheaves clamp the drive belt.

belt cover: **1.** A nylon cover positioned over the belts in a tire that helps to hold the tire together at high speed, and provides longer tire life. **2.** A rayon, cotton, or nylon cover to protect the interior of a drive belt from the environment and absorb the wear that occurs at the belt-sheave interface.

belt dressing: A prepared solution, generally in spray form, formulated for use on automotive V-belts to reduce or eliminate belt noise. It is not generally recommended for serpentine belts.

belt-driven cooling fan: A rigid or flexible cooling fan is driven by a belt from the *crankshaft* in vehicles that have longitudinally mounted engines. Usually the belt-driven cooling fan is mounted to the front of the water pump pulley/hub.

belted bias tire: A tire in which the belts are laid on the bias, cris-crossing each other.

belted radial tire: A tire in which the belts are laid parallel to each other and run perpendicular to the bead.

beltline: A line down the side of a car defined by the top edge of the lower body and the bottom edge of the roof and window assembly.

belt pitch: The distance between two adjacent tooth centers as measured on the pitch line of a *synchronous belt.*

belt ride dimension: A measurement of the distance from the top of the belt to the top of the sheave groove to determine proper belt fit in the groove.

belt tension: The tightness of a drive belt, generally measured in ft-lb (N•m).

belt width: The distance across a belt measured at the widest point.

bench bleeding: A procedure used to bleed the air from a new or rebuilt master cylinder before it is installed in the vehicle.

bench race: To talk about racing, generally just after an event.

bench seat: A full-width seat that can accommodate two or three persons in a vehicle.

bench test: The testing of an engine or component, out of the vehicle, for ease of observation and study.

bend: **1.** A curve or angle that has been *bent.* **2.** To form a curve or angle by *bending.*

bending: The forming of a material, usually metal, into a particular shape.

bending sequence: The order in which several bends are made so as not to be blocked by a previous bend.

bending stress: A stress, while bending, that involves both tensile and compressive forces that are not equally distributed.

Bendix drive: A type of starter motor drive that engages, and disengages, the starter and flywheel.

Bendix Folo-Thru drive: A starter motor drive engaged by initial rotation of the armature, causing the drive pinion to be twisted outward on a threaded sleeve until it is meshed with the flywheel gear. The gears are disengaged automatically at a predetermined speed of about 400 *rpm.*

Bendix screw: The helix-grooved shaft of a *Bendix drive.*

bent: Not flat or straight, intentionally or unintentionally.

bent eight: A V–8 engine.

bent six: A V–6 engine.

bent stovebolt: A Chevrolet V–8.

benzine: A highly flammable liquid, C_6H_6, sometimes found in refined gasoline. Its use, however, is restricted to 3.0% in some areas due to its toxicity.

benzole: A mixture of *aromatic hydrocarbons* with a high percentage of benzine, used as a solvent and as a fuel additive.

berm: **1.** A curb-like buildup on the outside of turns on a circular dirt track. **2.** The curb-like buildup of dirt along the edges of an unpaved road.

bevel gear: A form of spur gear in which the teeth are cut at an angle to form a cone shape, allowing a gear set to transmit power at an angle.

bezel: A trim ring, usually around a gauge, to secure the glass cover.

bhp: An abbreviation for *brake horsepower.*

bias: A diagonal line of direction.

bias-belted tire: A tire that has the ply cords placed diagonally across the tire from bead-to-bead, with alternate ply layers cris-crossing diagonally in opposite directions.

bias-ply tire: A term used for *bias belted tire.*

big arm: **1.** The throw of a *crankshaft* that has been *stroked.* **2.** Sometimes used to identify an engine with a stroked *crankshaft.*

big block: A V–8 engine that displaces more than 400 cubic inches; a *muscle car powerplant.*

big bore: A term often used for *big block.*

big end: The *crankpin* end of a *connecting rod.*

bigfoot lifter: A term used for *mushroom lifter.*

big red wrench: Slang for an oxyacetylene-cutting rig.

bigs: Large rear tires.

bigs and littles: A combination of large rear tires and small front tires.

A B C D E F G H I J K L M N O P Q R S T U V W X Y Z

big three: The three major United States automobile manufacturers, Chrysler, Ford, and General Motors.

billet: A solid bar of metal.

billet camshaft: A *camshaft* machined from a *billet* of steel.

billet crankshaft: A *crankshaft* machined from a *billet* of steel, usually used for racing applications.

bimetal: A temperature-sensitive strip made up of two metals having different heat expansion rates.

bimetal engine: A *powerplant* with *block* and *head* made of different metals, such as a *cast iron block* and an *aluminum head.*

bimetallic: Two dissimilar metals fused together; these metals expand and contract at different temperatures to cause a bending effect.

bimetallic sensing element: Another term used for *bimetallic sensor.*

bimetallic sensor: A sensing element having a bimetallic strip or coil.

bimetallic temperature sensor: A *bimetallic thermostat.*

bimetallic thermostat: A thermostat that uses a *bimetallic strip* instead of a *bellows* or *diaphragm* for making or breaking electrical contact points.

bimetal spring: A spring made of *bimetal,* such as a *choke spring.*

bimetal strip: A temperature regulating or indicating device that works on the principle that two dissimilar metals with unequal expansion rates, welded together, will bend as temperature changes.

binary: A system in which numbers are represented as sequences of zeros and ones, which is the basis of digital computing.

binary digit: A character that represents one of the two digits in the number system that has a *radix* of two.

binary numeral: The binary representation of a number.

binder: The paint material that forms the film, so called because it binds the pigment and additives into a solid, durable film.

binders: A term used for *brakes.* See *hard on the binders.*

binnacle: A console, separate from the dashboard, for switches, instruments, and controls, on or near the steering column.

birdcage: **1.** The plate on an *axle housing* for attaching suspension components. **2.** The chassis space frame with many small pieces of structural tubing.

bit: **1.** A character that represents one of the two digits in the number system that has a *radix* of two. **2.** That part of the soldering iron, usually made of copper, that directly transfers heat to the joint.

bite: Performance term for tire traction.

bk: Black.

bl: Blue.

black book: A guide used to determine the value of a used vehicle.

black box: A term often used for a *central processing unit (CPU).*

black flag: The signal for a driver to return to the pits for consultation with race officials.

black light: An ultraviolet-light system used to detect flaws in metal parts.

black smoke: The exhaust that is produced when the *air/fuel* mixture is too *rich.*

block water: Liquid refuse that must be stored on an *RV* until it can be disposed at a *dump station.*

blacky carbon: A term used in disdain for gasoline by drivers using it as a *fuel.*

blank: A term often used for *billet.*

bleed: **1.** To drain fluid. **2.** To remove an air bubble or air lock. **3.** To draw air into a system.

bleed air tanks: The process of draining condensation from air tanks to increase air capacity.

bleeder current: A continuous load placed on a power supply by a resistance load that helps improve regulation and safety.

bleeder jar: A glass or transparent plastic container used to detect the escape of air while bleeding brakes.

bleeder screw: A small, hollow screw or valve found at drum-brake wheel cylinders, in disc brake calipers, and adjacent to the outlet ports of some master cylinders. It is opened to release pressure and to bleed air and fluid from the hydraulic system.

bleeding: **1.** The slow releasing of pressure in the air-conditioning system by recovering some of its liquid or gas. **2.** The act of removing air from a hydraulic brake system. **3.** A small leak. **4.** When one paint color shows through another.

bleeding sequence: The order of bleeding brake systems or other components.

bleed orifice: A calibrated orifice in a vacuum system that allows *ambient air* to enter the system to equalize the vacuum.

blend air: The control of air quality by blending heated and cooled air to the desired temperature.

blend air door: A door in the duct system that controls temperature by mixing heated and cooled air in correct proportions to achieve the desired effect.

blinky: The timing light at the end of a quarter-mile strip.

blip: A quick punch to the throttle to *rev* the engine momentarily.

blistering: **1.** Tire tread separating from the carcass due to high heat. **2.** Bubbles or pinholes that appear in paint. **3.** To go exceptionally fast, as in a *blistering pace.*

blistering pace: To go exceptionally fast.

Bloc-Chek: A device to detect the leakage of exhaust gas into the cooling system.

block: Main casting of the engine that contains the cylinders; often made of cast iron or aluminum.

block check: A device used to detect oil or gasoline in the cooling system.

blow: A performance term for the failure of a transmission or engine, such as "blow the engine."

blowby: Byproducts of combustion, mostly hydrocarbons, that leak out of the combustion chamber, past the piston and piston rings, into the crankcase during the compression and power strokes. In modern engines, blowby vapors are drawn into the intake through the *positive crankcase ventilation (PCV)* system and are burned in the engine.

blower: **1.** A performance term for a supercharger or turbocharger. **2.** A two-stroke, diesel-engine, intake air compressor. **3.** The fan motor in a heater-air conditioning system.

blower belt: A reinforced cover with retaining straps on supercharged engines, required by *NHRA* to prevent parts from scattering should the supercharger explode.

blower circuit: All of the electrical components that are required for blower and blower speed control.

blower motor relay: A relay found in the *blower circuit.*

blower relay: An electrical device used to control the function or speed of a blower motor.

blower speed controller: A solid-state control device that operates the blower motor and, sometimes, the compressor clutch based on signals from the microprocessor.

blower switch: A dash-mounted device that allows the operator to turn the blower motor ON/OFF and/or to control its speed.

blown engine: **1.** An engine that has a supercharger or turbocharger. **2.** A seriously damaged engine.

blown head gasket: A broken head gasket that leaks water, oil, or air and reduces engine performance.

blow off: To defeat a competitor.

blow off the doors: To defeat a competitor by a wide margin.

blow torch: A jet-powered car for drag or lake competition.

blu: Blue.

Blue Book: A term given the Kelly Blue Book, a bimonthly publication that lists used car wholesale and retail prices.

blueprint: To rebuild an engine to *original equipment manufacturer (OEM)* design specifications.

blue smoke: An exhaust produced when oil has contaminated the *air/fuel mixture.*

blue streak: The trade name of a Goodyear high-performance tire.

bluing: A fluid that produces a blue surface on metal to assist in laying out work on it.

BMC: An abbreviation for *Brake Manufacturers Council.*

bmep: An abbreviation for *brake mean efficient pressure.*

bn: Brown.

bobbed: Trimmed and shortened.

bobtail: **1.** A vehicle body with a short-rear overhang. **2.** The tractor of a tractor-trailer rig. **3.** The results of a modification of a fad car.

bobweight: The weight attached to a rod journal to simulate the reciprocating mass when balancing a crankshaft.

body: The assembly of sheet metal sections and other parts that provide the enclosure of a vehicle.

body and frame: The parts of a vehicle that support all components. A frame that supports the engine and drive train and a body that houses the entire vehicle.

body control module *(BCM)*: A component of self-diagnostic systems used to control vehicle function based on monitored inputs.

body in white: **1.** A new, unpainted, and untrimmed body. **2.** A body that has been completely stripped.

body knocker: A person who does body work.

body mounting: **1.** Rubber cushions, at strategic locations, to dampen noise and vibration. **2.** To place a car body on the *chassis.*

body panels: Sheets of material joined together to form a car body.

body pitch: The tendency of a vehicle to *dive* or *squat.*

body roll: A term often used for *roll.*

body shapes: There are six basic body shapes; *sedan, hardtop, convertible top, liftback* or *hatchback, station wagon,* and *sports* or *multipurpose.*

body shop: A collision-and-damage repair and painting facility.

body strength: Body strength depends on the type of vehicle and body structure; factors such as door size and the presence or absence of a center pillar. Also, front body pillar, quarter panels, and roof panels affect how much of an impact is absorbed.

bog: **1.** To lose power and momentarily faulter when coming off the line. **2.** To stall or slow due to soft dirt or sand in off-road racing. **3.** A mud hole.

bogey: A common way of spelling *bogie.*

boggin': Competitive racing of individual off-road vehicles through a long pre-measured mud hole.

bogie: The axle-spring suspension arrangement on the rear of a tandem axle tractor.

boiling: A rapid change in the state of a liquid to a vapor by adding heat, decreasing pressure, or both.

boiling point: The temperature at which a liquid begins to boil.

boiling the hides: Smoking the tires during a drag race.

boil tank: A very large tank of boiling solution used for cleaning large parts, such as engine blocks.

bolster plate: The flat, load-bearing surface under the front of a semitrailer, including the kingpin, which rests firmly on the fifth wheel when coupled.

bolster plate height: The height from the ground to the bolster plate when the trailer is level and unladen.

bolt: A metal rod, usually with a head at one end and a thread at the other, used to secure parts and assemblies.

bolt circle diameter: The diameter of an imaginary line running through the bolt hole centers.

bolt grade: The strength of a bolt.

bolt hardness: The hardness of a bolt is identified by the number of lines on the head of the bolt. The more lines, the stronger the bolt.

bolt on: An aftermarket accessory that can be installed without modification.

bolt stretch: **1.** When a bolt is elongated and the shank and/or thread diameter is smaller than specifications. **2.** When a bolt is torqued as specified and, as a result, is stretched a predetermined amount.

bolt torque: The turning effort required to offset resistance as the bolt is being tightened.

bomb: Performance term for a vehicle that is capable of extraordinary performance.

bonded lining: Brake lining cemented to shoes or bands, eliminating a need for rivets.

bonding: The process of connecting two or more materials using chemicals or heat, or electrical or mechanical forces.

bonding lining: The lining attached to a brake shoe with an oven-cured adhesive.

Bondo: A tradename for a plastic two-part body filler.

bone yard: A facility that sells used parts for vehicles.

bonnet: **1.** A safety helmet. **2.** British term for a vehicle hood.

Bonneville: The salt flats in Northwestern Utah; one of the world's most famous courses.

bookkeeper: One who keeps record books, prepares invoices, writes checks, makes bank deposits, checks bank statements, and is responsible for tax payments.

boondockin': Traveling in a remote area, usually with a 4×4 or other *off-road vehicle.*

boondocks: A remote, undeveloped area, often the setting for unauthorized, and illegal, auto competition.

boonies: **1.** A slang term for *boondocks.* **2.** A race driver is "off to the boonies" if spun off course.

boost: The increase in intake-manifold pressure, produced by a turbocharger or supercharger.

booster battery: A charged battery connected to a discharged battery in order to start the engine.

booster cables: A term used for *jumper cables.*

booster venturi: A device in a *carburetor* that mixes fuel with incoming air.

boot: **1.** Slang for a tire. **2.** A flexible rubber or plastic cover used over the ends of master cylinders, wheel cylinders, transmissions, or constant velocity joints, to keep out water and other matter. **3.** The British term for the rear deck or trunk of a car.

boots: Performance term for *tires.*

Borden tube: A thermo-mechanical device in the fuel injection system that regulates the amount of fuel being injected according to differences in temperature and pressure in the intake manifold.

borderline lubrication: **1.** Poor lubrication as a result of greasy friction. **2.** Moving parts coated with a thin film of lubricant.

bore: 1. The diameter of an *engine cylinder.* 2. To increase the diameter of a *cylinder.*

bore align: To machine an engine's *main bearing journals* to assure they are in perfect alignment.

bore centers: The center-to-center distance between two *bores.*

bored and stroked: A combination of an enlarged *cylinder bore* and a lengthened *piston stroke* to increase an engine's displacement.

bore in the water: To *bore* a cylinder out far enough to brake through the water passage.

boss: 1. The reinforced extension on a part that holds a mounting pin, bolt, or stud. 2. A slang term for outstanding quality.

bottom: 1. The underside; the lowest part. 2. When a vehicle's *chassis* hits the lowest point allowed by its *suspension system.* 3. When the *springs* are fully compressed.

bottom dead center *(BDC)*: Piston position at bottom of stroke.

bottom end: The crankshaft main bearing and connecting-rod bearing assembly in an engine.

bottom out: If a race car settles down tightly on its springs on an oval track as it travels through a baked turn, centrifugal force tends to push the car downward, toward the track's surface, causing its chassis to "bottom out."

bottom U-bolt plate: A plate that is located on the bottom side of the spring or axle and is held in place when the U-bolts are tightened.

bottom valve: A *shock absorber* component to control the flow of oil into the reservoir during compression and rebound.

bounce back: A condition that occurs when particles of paint sprayed on a body panel bounce away from the surface.

bound: The opposite of *rebound.* The inward travel of the piston rod in a shock absorber.

boundary layer: The thin layer of air along the inner walls of an intake port.

Bourdon tube: A curved tube that straightens as the pressure inside it is increased. The tube is attached to a needle on a gauge, that senses the movement of the tube and transmits it as a pressure reading.

Bowden cable: 1. A wire cable inside a metal or rubber housing used to regulate a valve or control it from a remote place. 2. A small steel cable inside a flexible tube used to transmit mechanical motion from one point to another.

boxer: A horizontally opposed engine.

box stock: Standard, conforming to *OEM* specifications.

B-pillar: The structural support just behind the front door.

B-post: A term often used for *B-pillar.*

br: Brown.

bracket racing: A handicap system that allows two cars from different classes to compete in drag racing events.

brain bucket: A helmet.

brain fade: A mental lapse often used as an explanation why a driver looses control of a race car.

brake: 1. A system used to stop or slow a vehicle and/or prevent it from moving when stopped or parked. 2. British term for *station wagon.*

brake actuator: A unit that converts hydraulic pressure, air pressure, vacuum, electrical current, or another form of energy to a force that applies a brake.

brake anchor: A pivot pin on the brake backing plate against which the brake shoe bears.

brake assembly: An assembly of the components of a brake system, including its mechanism for the application of friction forces.

brake band: A round, flat metal band with a friction surface on its inner diameter; used primarily in emergency brake systems.

A B C D E F G H I J K L M N O P Q R S T U V W X Y Z

brake bias: An excessive brake force at either end of the vehicle causing the brakes at that end to lock before the other end, often leading to loss of control.

brake bleeding: Procedure for removing air from lines of a hydraulic brake system.

brake booster: A device that uses a supplementary power source to reduce pedal force in a hydraulic brake system.

brake chamber: A unit in which a diaphragm converts pressure to mechanical force for actuation of a brake.

brake control valve: A unit that sends a signal to the computer when the brakes are being applied.

brake cylinder: That part in which a piston converts pressure to mechanical force for actuation of the brake shoes.

brake disc: The parallel-faced, circular, rotational member of a brake, acted upon by the friction material of the shoes.

brake disc minimum thickness: A term sometimes used to indicate *disc minimum thickness.*

brake dive: A term used for *dive.*

brake drag: A light, but constant, contact of the brake shoes with the drum or rotor, resulting in early failure due to excessive heat.

brake drum: The cylindrical part that rotates with the wheel and surrounds the brake shoes.

brake drum glaze: A brake drum with an excessively smooth surface that lowers friction and, therefore, efficiency.

brake equalizer: A device used in the parking-brake systems to equalize the pull of both rear-brake cables.

brake fade: A condition whereby repeated severe application of the brakes, over a short time period, cause an expansion of the brake drum and/or loss of frictional ability, which results in impaired braking efficiency.

brake fanning: Applying and releasing the brakes in rapid succession on a long downgrade.

brake feel: A term relating to the driver's ability to determine the amount of braking force exerted or required during a stop.

brake fluid: A specially formulated liquid used to transmit brake pedal pressure from the master cylinder to the wheel cylinders.

brake flushing: The removal of fluid from a brake system, washing out sediment by flushing with denatured alcohol or clean brake fluid, and refilling the system with fresh brake fluid.

brake grab: The sudden increase in the braking of a wheel, often caused by a contaminated lining due to a leaking wheel cylinder.

brake hop: A condition that occurs when the swing-arm length of the rear suspension is too short, causing the rear wheels to repeatedly leave the ground during braking.

brake horsepower *(bhp)*: A measurement of *horsepower (hp)* delivered at the *engine crankshaft.* A *prony brake* or an *engine dynamometer* is used to determine brake *horsepower.*

brake hose: A flexible tubular conduit for the transmission of fluid pressure in the brake system.

brake lights: Red lamps at the rear of the vehicle that light up when the brakes are applied.

brake light switch: An electrical switch, operated either mechanically or hydraulically when the brakes are applied, that causes the brake lights to light up.

brake line: Small-diameter, rigid-steel tubing, or flexible, rubber, reinforced hose, used to channel brake fluid from the master cylinder to the wheel cylinders or calipers when the brakes are applied.

brake lining: A special friction material that will withstand high temperature and pressure, used for lining the brake shoes or pads, either by riveting or bonding.

A B C D E F G H I J K L M N O P Q R S T U V W X Y Z

brake lining pad: The friction lining and plate assembly that is forced against the rotor to cause braking action in a disc brake.

brake machining: The practice of turning a drum or a rotor on a special lathe to remove surface imperfections such as scoring, and to eliminate runout and other dimensional problems.

brake mean effective pressure: The average effective pressure in an engine's cylinders at a specific brake horsepower and speed.

brake pad: **1.** The pad of friction material applied by the caliper to the disc, to slow or stop a vehicle. **2.** A term often used for *brake lining.*

brake power assist: A device installed in a hydraulic brake system that reduces pedal effort and affords better braking.

brake shoe: The curved metal part, faced with brake lining, that is forced against the brake drum to produce braking action.

brake shoe heel: The end of brake shoe opposite the anchor pin.

brake shoe toe: The end of brake shoe nearest the anchor pin.

brake specific fuel consumption: The measure of an engine's *fuel efficiency* during *dynamometer* testing.

brake stopping distance: The distance traveled from the application of the brakes to the point at which the vehicle comes to a complete stop.

brake system: The system in a vehicle that is used to slow or stop the vehicle.

brake torque: The engine output at the flywheel as measured on a *dynamometer.*

brake warning light: An instrument-panel light that warns of brake-system malfunction.

branch circuit: A portion of a wiring system extending beyond the final over-current protective device in a circuit.

braze: A weld produced by heating an assembly to above 840°F (450°C) but below the solidus of the base metal.

break: **1.** To damage a car in any way. **2.** To open, as a set of points.

break before make *(BBM)*: A switch, such as a headlamp dimmer, that breaks one circuit before making another.

breaker: A spring-loaded switch in a conventional distributor ignition system that opens (breaks) and closes (makes) the primary circuit.

breaker cam: The rotating part, located near the top of the conventional distributor, that has lobes that cause the breaker points to open and close.

breakerless system: An electronic ignition system that does not use conventional breaker points to time and trigger the primary voltage but retains the distributor for secondary voltage distribution.

breaker point gap: The space between the fully opened breaker points in a conventional distributor.

breaker points: In a conventional ignition system, an electrical switch that opens to interrupt current flow in the primary circuit. Often called *contact points.*

breaker trigger system: Any ignition system that utilizes conventional breaker contacts to time and trigger the system.

break-in: The operation of a new vehicle at a constant and even speed to assure even initial wear of all engine parts.

breakout: **1.** The point at which conductor(s) are taken out of a multi-conductor assembly. **2.** The point at which a branch circuit departs from the main wiring harness.

breakover angle: A term often used for *ramp angle.*

break rule: A regulation in drag racing that permits a car, defeated in an elimination, to return to competition if the car that beat it is not able to get to the line for its next round.

breathing: The ability of an engine to draw in air and exhaust gases. The better the breathing, the better the performance.

brickyard: The *Indianapolis Motor Speedway.*

bridge rectifier: A full-wave rectifier in which the diodes are connected in a bridge circuit allowing current flow to the load during both positive and negative alternation of the supply voltage.

Brinell test: A technique for testing the hardness of a metal.

brinelling: A steel-shim head gasket torqued in place on an engine block.

British Imperial System: A system of measurement used in the United Kingdom.

British thermal unit: A measure of heat energy where one Btu is the amount of heat necessary to raise 1 pound of water 1°F.

brittleness: The quality of a material that leads to fracture without appreciable deformation.

brn: Brown.

broach: A tool used for reshaping or resizing parts.

brody: A controlled, rear-wheel skid often used on dirt tracks for fast cornering.

bronze: A copper-rich, copper-tin alloy with or without small proportions of other elements such as zinc and phosphorus often used for bearings or bushings.

bronze guide: A valve guide made of bronze alloy.

bronze guide liner: A thin valve guide, easy to install, but must be broached to stay in place.

bronzewall: A type of thread-like, valve-guide repair insert that must be reamed to size after installation.

brush: A block of conducting material, such as carbon, held against an armature *commutator* or rotor slip ring to form a continuous electrical path.

brush holder: Adjustable arms for holding the *commutator* brushes of a generator against the *commutator,* feeding them forward to maintain proper contact as they wear and permitting them to be lifted from contact when necessary.

BSFC: An abbreviation for *brake-specific fuel consumption.*

B-shim: A valve spring adjuster with a 0.030 inch (0.76 mm) used to balance spring pressure and to correct installed height.

BTC: An abbreviation for *before top center.*

BTDC: An abbreviation for *before top dead center.*

B-train: A combination of two or more trailers in which the rear trailer is connected at a single pivot point, commonly a fifth wheel, which is mounted on an extension of the frame of the lead trailer.

bubble: The slowest qualifying position for a race.

bubble balancer: A wheel balancer using an air bubble to show static balance. The tire and wheel are considered balanced if the bubble is centered in its area.

bubble gum machine: Slang for a blue and/or red police-car roof light.

bubble memory: A method by which information is stored as magnetized dots (bubbles), that rest on a thin film of semiconductor material.

bubble top: A transparent roof on an automobile.

buck: Jargon for 100 miles per hour.

bucket: **1.** The passenger compartment of a roadster. **2.** Overhead camshaft valve lifters. **3.** An individual driver or passenger seat.

buckling: The bowing or lateral displacement of a compression spring.

bucks down: To have not.

bucks up: To have.

bucks up, bucks down: To have and to have not.

buff: An enthusiast.

A **B** C D E F G H I J K L M N O P Q R S T U V W X Y Z

A B C D E F G H I J K L M N O P Q R S T U V W X Y Z

buff book: A magazine for car enthusiasts.

buffing: A surface-finishing process that produces a smooth, lustrous appearance generally free of defined line patterns.

bug: **1.** A minor flaw, imperfection, or malfunction. **2.** A Volkswagen Beetle.

bug catcher: A scoop-like intake on a fuel-injection or blower system.

buggy: A small, lightweight, off-road vehicle.

build: As its name implies; for example, to modify an engine for racing it is to build it.

buildup time: The time required to form a magnetic field around the primary winding of the coil when current is allowed to flow through it in a conventional ignition system.

bulb: The glass envelope that contains an incandescent lamp or an electronic tube.

bulge: A high spot or crown in stretched metal.

bulkhead connector: A connector for wires or hoses located where they must pass through a partition or lead to another area.

bulletproof: Indestructible.

bullnose: The smooth nose of a hood when the ornament has been removed and the holes filled in.

bull ring: A dirt, oval track, generally one-half mile or less.

bump: To be forced out of a racing lineup by a faster qualifier.

bumpin': Cruising in a lowered vehicle, such as a low rider. Some have hydraulics that assist in bouncing up and down.

bump shop: An auto body repair facility.

bump spot: The slowest qualifying position for a race.

bump steer: The tendency of the steering to veer suddenly in one direction when one or both of the front wheels strikes a bump.

bump stick: The camshaft.

bump stop: A block, usually rubber, to limit suspension system deflection when a tire hits a bump.

bunch of bananas: A tuned exhaust system with intertwined headers.

Bunting bronze: A trade name for a type of bronze used for bushings.

Bureau of Automotive Repair *(BAR):* A state agency that regulates the auto service and repair industry in California.

burette: A glass container used to measure liquid in cubic centimeters (cc).

burn: A visible discoloration or sub-surface damage from an excessively high temperature, generally produced by grinding.

burnout: The moment of final oxidation or combustion of fuel in an engine.

burn rubber: To accelerate at a rate that the traction tires leave black streaks on the pavement.

burnt: Permanently damaged material caused by heating conditions producing incipient melting or intergranular oxidation.

burn time: Time required for a given amount of *air/fuel mixture* in the *combustion chamber* to burn.

burr: A feather edge left on metal after being cut or filed.

bus: **1.** A large enclosed vehicle for carrying passengers, usually for hire. **2.** Slang for a family car.

bushing: A one-piece sleeve, usually bronze, inserted into a bore to serve as a bearing.

business coupe: An inexpensive two-door body type with no rear seat, last available in the mid 1940s.

butane: A gaseous, highly flammable fuel (C_4H_{10}) that becomes liquid when it is compressed.

butterfly: A type of valve used for the *choke* and *throttle valve* in a *carburetor;* a moveable plate that controls the amount of air permitted to enter the *carburetor.*

butt gap: The distance between the ends of a *piston ring.*

butt weld: A type of weld that joins two pieces of metal by fusion.

Butyl: The trade name of a synthetic rubber.

buzz box: A small car with a noisy engine.

buzzer: An electric sound generator that makes a "buzzing" noise when activated. It operates on the same principle as a vibrating bell. Sometimes used to warn the driver of possible safety hazards, such as when the seat belt is not fastened.

buzzword: A word or phrase that happens to be a popular cliché of a group of people, such as race car drivers.

bye: A solo run during drag eliminations.

bypass: **1.** A passageway between the head and block or behind the water pump that allows a water pump to circulate coolant throughout the cylinder head and engine block before the thermostat opens. **2.** A valve that is used to regulate pressure or control the quantity of a liquid or gas.

bypass capacitor: A *capacitor* that provides a low impedance path (usually to ground) to remove unwanted signals from the main signal path.

bypass control valve: A valve used in a bypass system.

bypass line: A line or hose used in a bypass system to transfer liquid or gas.

bypass tube: A tube directly in front of the thermostat. The coolant bypasses the radiator through this tube when cold.

A B **C** D E F G H I J K L M N O P Q R S T U V W X Y Z

C: An abbreviation for: **1.** Capacitance. **2.** Celsius. **3.** Coulomb. **4.** A symbol for *Carbon.* **5.** To raise the arch of a chassis frame to provide added clearance over the rear axle and lower the overall height of the frame side rails.

CAAA: An abbreviation for *Clean Air Act Amendments.*

cab: **1.** The driver/passenger compartment of a truck or off-road vehicle. **2.** Short for taxicab.

cab behind engine *(CBE):* A type of truck with the cab behind the engine.

cable: An assembly of two or more wires that may be insulated or bare.

cable clamp: A device used to clamp around a cable to transmit mechanical strain to all elements of the cable.

cabling: The helical wrapping together of two or more insulated wires.

cab over engine *(COE):* A type of truck with the cab above the engine.

cabrioet: A European term for convertible.

CAFE: An acronym for *Corporate Average Fuel Economy.*

cage: **1.** The metal structure that separates the balls of a roller bearing assembly. **2.** A shortened term for *roll cage.*

caged roller clutch: A one-way clutch having the rollers and springs contained as a unit.

calibrate: To check, test, or adjust the initial settings of a unit or system.

calibration: The adjustment of a device or instrument so that output is within a designated tolerance for specific input values.

California Air Resources Board *(CARB):* A California agency responsible for regulations intended to reduce air pollution, especially that created by motor vehicles.

caliper: **1.** Non-rotational components of disc brakes that straddle the disc and contain hydraulic components forcing the brake pads against the rotor to slow or stop the vehicle.

calorie: A measurement of heat; the amount of heat required to raise 1 gram of water (H_2O) 1°C.

cam: **1.** The eccentric element of a one-way roller clutch that carries the *profiles* through which the rollers transmit torque. **2.** An abbreviation for *camshaft.*

cam and kit: A specially ground camshaft, complete with a set of compatible camshaft valve train components, including lifters and springs.

cam angle: A term often used for dwell angle.

camber: The outward or inward tilt of the wheels, in degrees, on a vehicle as viewed from the rear or front.

camber angle: The amount, measured in degrees from the vertical, that the top of a tire is tilted outward (positive) or inward (negative).

camber compensator: A device that is used to maintain the proper camber of the rear wheels of a vehicle equipped with *swing axles.*

cam button: A device that keeps a *camshaft* properly positioned in an engine.

cam duration: The amount of time, measured in crankshaft degrees, that a camshaft holds an *exhaust* or *intake valve* open.

camelback: A strip of new rubber used to recap a tire.

camel hump heads: High performance heads by Chevrolet, identified by two humps on the outside end of the casting.

cam follower: A term often used for *valve lifter.*

cam ground: Pistons machined to a slightly out-of-round shape to permit them to expand with engine heat without getting stuck against the cylinder walls.

cam-ground piston: A piston ground slightly oval or elliptical in shape as a means to compensate for expansion caused by heat.

cam lift: The distance, in thousandths of an inch, a cam lobe raises the *valve lifter* off the *base circle.*

cam-lobe face and nose taper: The slant, about 0.002 inch (0.051 mm), designed across the cam-face contacting surface, from the cam front to rear edge, to promote lifter rotation.

cammer: An engine with an *overhead camshaft.*

cam sensor: A camshaft-mounted sensor that signals when cylinder number one is at *TDC.*

camshaft: A shaft having lobes driven by the crankshaft via gears, chains, or belts that, in turn, opens and closes the intake and exhaust valves at proper intervals.

camshaft duration: The amount of time, measured in degrees of crankshaft rotation, an *intake* or *exhaust valve* is held open.

camshaft lift: How far a valve is pushed open, as determined by the height of the cam lobe and the geometry of the rocker arms on a push rod engine, or the cam followers on an overhead cam engine.

camshaft plug: A plug found in the rear of the engine block, at the camshaft, to retain and direct oil to the rear camshaft bearings.

cam-shaped pocket: The recess in an overrunning clutch race, large at one end, tapering to small at the other end.

Canadian-American Challenge Cup *(Can-Am):* A former annual series of *SCCA* sanctioned road races, held in the United States and Canada, for unlimited sports cars with no restriction on engine displacement.

can: A term often used for *nitro* or *nitromethane,* a highly combustible liquid used as the main ingredient in drag-racing fuel.

Can-Am: An abbreviation for *Canadian American Challenge Cup.*

candy-apple red: A popular trade name for a gold-color-base, clear-coated, metallic-red paint job.

canister: A container filled with charcoal in an evaporative control system used to filter and trap fuel vapors.

canister filter: A canister-like device containing a filter media which removes suspended particles of contaminants from air, fuel or oil.

canister purge solenoid: A solenoid valve that admits fuel vapors to the canister for processing.

can tap: A device used to pierce, dispense, and seal small cans, such as *refrigerant.*

canted valves: A cylinder-head layout where the *intake valves* are at one angle, while the *exhaust valves* are at another angle.

cantilever tire: A tire with a tread wider than the rim.

can valve: A term often used for *can tap.*

cap: **1.** A cover or lid for a container. **2.** That part of the distributor that directs electrical energy to the *spark plugs.* **3.** A small metal part, usually of hardened steel, that acts as an interference between the valve-stem end and the rocker arm. **4.** The half round, removable part of a connecting rod or main bearing.

capacitor: An electrical device for the temporary storage of electricity. Used in a conventional ignition system in the distributor to reduce arcing across the points and in the electrical charging system to reduce radio interference.

capacitor discharge ignition *(CDI):* A term used for a *capacitor discharge ignition system.*

capacitor discharge ignition system: An *ignition system* that stores its primary energy in a *capacitor.*

capacity: **1.** The ability to perform or to hold. **2.** Refrigeration produced; measured in tons or *Btu* per hour.

cape top: The fixed enclosure over the rear seat.

capillary action: The force by which liquid, in contact with a solid, is distributed between closely fitted surfaces of the joint, while being brazed or soldered.

capillary attraction: Another term for *capillary action.*

capillary tube: **1.** A tube with a calibrated inside diameter and length used to control the flow of refrigerant. **2.** A tube that connects the remote bulb to the thermostatic expansion valve. **3.** A tube that connects the remote bulb to the thermostat.

CAPP: An abbreviation for *Clean Air Performance Professionals.*

captive fasteners: A preassembled fastener used in conjunction with mating fasteners such as bolts, studs, nuts, or screws.

CAR: A term used for *carcinogenic effects.*

carb: An abbreviation for *carburetor.*

CARB: An abbreviation for *California Air Resources Board.*

carbide: **1.** A compound of solid elements, usually metal, with carbon. **2.** A mixture of very hard metals, such as tungsten carbide.

carbolic acid: A very toxic acid (C_6H_5OH); an ingredient used in cold-dip tanks.

carbon: **1.** A natural element. **2.** A by-product of *combustion.* **3.** Any form of graphite or undefined carbon.

carbon arc cutting and welding: **1.** A cutting process that uses a carbon electrode. **2.** A welding process that uses an arc between a carbon electrode and the weld pool.

carbon canister: A canister filled with *carbon,* used to absorb and store fuel vapors that are normally discharged into the air.

carbon dioxide *(CO_2):* A harmless, odorless gas composed of carbon (C) and oxygen (O); a product of complete combustion.

carbon fiber: A very strong, lightweight, synthetic fiber often used in race cars, such as Formula One and Indy cars, because of its lightweight and high strength.

carbon-fouled plug: A *spark plug* having a dry, fluffy black deposit; the result of over-rich carburetion, over choking, a sticking manifold heat valve, or clogged air cleaner.

carbonize: The formation of carbon deposits as a result of by-products of combustion in an engine.

carbon monoxide *(CO):* An odorless gas composed of carbon (C) and hydrogen (H), formed by the incomplete combustion of any fuel containing carbon. This major air pollutant is potentially lethal if inhaled, even in small amounts.

carbon tracks: A condition where there is a carbon build up, or track, running from one point to another, acting as an electrical circuit and thereby causing a short circuit.

carbonyl chlorofluoride *(COC_lF):* A toxic by-product of *Refrigerant-12,* and other *CFCs;* toxic if allowed to come into contact with an open flame or heated metal.

carbonyl fluoride *(COF_2):* A toxic by-product of *Refrigerant-12,* and other *CFCs;* toxic if allowed to come into contact with an open flame or heated metal.

carburetion: The action that takes place in a *carburetor* while converting liquid fuel to a vapor and mixing it with air to form a combustible mixture.

carburetor: A vacuum-dependent device used to mix fuel with air to form a vapor that is ideal for combustion.

carburetor emission device: An item attached or adjacent to the *carburetor* that establishes operating conditions intended to reduce exhaust emissions.

carburetor-heated air: A system in which heated air is directed from the *manifold* to the *carburetor* for improved performance with a leaner *air/fuel mixture.*

carburetor icing: A condition caused by high-velocity fuel flow; ice is formed on the outside of the *carburetor.*

carburetor insulator: A spacer between the *carburetor* and *intake manifold* to reduce engine heat to the *carburetor.*

carburetor kickdown: A moderate pressing of the accelerator pedal to change engagement of the fast-idle screw from high step to low step of the cam.

carburetor kit: A set of *gaskets* and parts necessary to rebuild a *carburetor.*

carburetor restrictor plate: A term often used for *restrictor plate.*

carburetor spacer: A steel, aluminum, or plastic plate used to raise the *carburetor* above the normal opening of the *intake manifold.*

carburetor tag: A tag affixed to the *carburetor* to identify the model for parts and specifications for service.

carburetor vacuum: A ported vacuum, obtained from a *carburetor* source above the throttle plate, present after the throttle is partially open.

carburetter: A British spelling of *carburetor.*

carburettor: A British spelling of *carburetor.*

carburize: To treat or combine with *carbon.*

carburizing flame: A reducing oxyfuel-gas flame in which there is an excess of fuel gas, resulting in a carbon-rich zone extending around and beyond the cone.

Car Care Council: A non-profit organization to educate the general public about the importance of preventive maintenance.

carcass: **1.** A tire casing to which the rubber tread and sidewall are bonded. **2.** The inner part of the tire that holds the air for supporting the vehicle. **3.** The empty remains of a stripped vehicle.

carcass plies: That which surrounds both beads and extends around the inner surface of the tire to provide load-carrying capabilities on the rim.

carcinogen: A substance or agent that produces or incites cancer.

carcinogenic: A term often used for *carcinogen.*

carcinogenic effects: Causing cancer or increasing the incidence of cancer in the population.

Cardan joint: A *universal joint* having two yokes at right angles to each other, with a cruciform-shaped joint in the middle.

Cardan universal joint: A term used for *Cardan joint.*

cargo weight rating *(CWR):* A truck's carrying capacity, in pounds.

Carolina stocker: A car built for drag racing, without regard for any recognized rules.

Carrera Panamericana: **1.** The legendary Mexican road races held in the early 1950s that ran the full length of Mexico. **2.** Vintage car races held on the public highways of northern Baja, California.

carrier: A part that holds, positions, moves, or transports another part or parts.

carrier bearing: A bearing that supports the ring-gear carrier in the *differential.*

carrier housing: Cast-iron rear axle assembly section that contains the working parts of the *differential.*

Carryall: A tradename once used for a large station wagon built on a truck chassis; predecessor of the Suburban.

carry the wheels: To do a wheel stand; a drag-racing term.

Carson top: A removable, non-folding, padded soft top, used on many customized convertibles and roadsters.

CART: An acronym for *Championship Auto Racing Teams.*

cartridge filter: A filter media that includes yarns, felts, papers, resin-bonded fibers, woven-wire cloths, and sintered metallic and ceramic structures for cleaning impurities from air or liquid. Performance obtained by a disposable cartridge filter may range from 500 µm to 1 µm or less.

A B C D E F G H I J K L M N O P Q R S T U V W X Y Z

cartridge roll: A rolled piece of sandpaper used to deburr or blend sharp edges, such as when porting and polishing a head.

Car Wash Owners and Suppliers Association, Inc. *(COSA)*: A trade association of car wash manufacturers, operators, and suppliers.

cascade: Two devices in tandem; the output of one device connected to the input of the other.

case harden: A heat-treating process that hardens the outer surface of metal, while leaving the core soft and ductile.

casing: **1.** The outer part of a tire assembly made of fabric or cord to which the rubber is vulcanized. **2.** The outer housing or shell containing an assembly.

CAS Registration Number: The Chemical Abstract Service Number used to identify a chemical.

castellate: Formed to resemble a castle battlement.

castellated nut: A nut with six raised portions or notches through which a cotter pin can be inserted to secure the nut.

caster: The angle between the steering-spindle axis and the wheel vertical as viewed from the side.

caster angle: The amount, measured in degrees from the vertical, that the upper ball joint is located behind or ahead of the lower ball joints.

casting: A metal object formed to the required shape by pouring or injecting liquid metal into a mold.

casting flash: A thin metal exuding at the parting edges of a casting mold, evident when the part is removed from the mold.

casting number: A part number that has been cast into a part during manufacture.

cast iron: A term used for a family of cast ferrous alloys containing at least 2% *carbon,* plus *silicon* and sulfur; may or may not contain other alloy elements.

cast-iron guide: A valve guide made of cast iron.

castor oil: A lubricant made from the castor bean.

cat: An abbreviation for *catalytic converter.*

catalog: An illustrated or ordered list of items and descriptions with sufficient data to identify the item.

catalysis: The action of a *catalyst.*

catalyst: A lead-sensitive substance, such as *platinum,* palladium, or rhodium, that accelerates or enhances a chemical reaction without being changed itself. When used in a *catalytic converter,* it can reduce the level of harmful pollutants in the exhaust.

catalytic converter: An automotive exhaust-system component, made of stainless steel, containing a *catalyst* to reduce oxides of *nitrogen* (NO_X), and/or *hydrocarbon* (HC), and carbon monoxide (CO), in tailpipe emissions.

catch can: A container on a race car's *radiator* or *fuel tank* to prevent liquid from spilling on the ground during a pit stop.

catch tank: A term often used for *catch can.*

catenary effect: The curve that a length of chain assumes between its suspension points.

cathode: The negatively charged cell from which current flows in an electrolytic cell.

cathode ray tube *(CRT)*: A vacuum tube used in electronic equipment and some electronic readouts to display information.

caulking compound: A thick, viscous material used as a sealer at joints, such as around the windshield.

caustic: **1.** A salt-based chemical for cleaning engine parts. **2.** A cleaner that may be used for most metals, except aluminum.

caution flag: A yellow flag displayed to race-car drivers to indicate a slow down. Also used to indicate no passing, due to a problem or mishap on the race track.

cavitation: The presence of air in a liquid during pumping, which can inhibit the flow of the liquid.

CBE: An abbreviation for *cab behind engine.*

CBU: An abbreviation for *completely built up.*

cc: **1.** An abbreviation for *cubic centimeter.* **2.** The measure of the volume of a *combustion chamber.* **3.** An abbreviation for *close cup,* a method of determining the flash point of a flammable liquid.

C-cam: The pattern used to grind pistons in an oval or cam shape, with a 0.009 inch (0.23 mm) difference between the thrust face and pinhole side.

CCC: An abbreviation for *computer-command control.*

CCCA: An abbreviation for the *Classic Car Club of America.*

CCEC: An abbreviation for *constant current electronic control.*

CCFOT: An abbreviation for *cycling clutch fixed orifice tube.*

CC-grade oil: An *American Petroleum Institute (API)* specification standard for diesel motor oil.

cc-ing: To measure or calculate the volume of a *combustion chamber* in cubic centimeters.

C-clip: **1.** A term often used to describe an *outside snap ring.* **2.** A C-shaped clip used to retain the drive axles in some *rear-axle assemblies.* **3.** A clip used to secure a pin in linkage, such as for carburetion.

CCOT: An abbreviation for *cycling clutch orifice tube.*

ccw: An abbreviation for *counterclockwise.*

C_d: A symbol for *coefficient of drag.*

CD-grade oil: An *API* performance-specification standard for *diesel motor oil.*

CDI: An abbreviation for *capacitor discharge ignition.*

CEC: An abbreviation for *combustion emission control.*

cellular core: A type of radiator core.

cellulose fiber: Transparent material made of camphor and guncotton, formed into thread-like filaments.

Celsius: The metric scale for temperature where water (H_2O) freezes at 0°C and boils at 100°C.

cementation: A process for introducing elements onto the surfaces of metals by high-temperature diffusion.

cemf: An abbreviation for *counterelectromotive force.*

center bolt: A term that generally refers to a *leaf-spring center bolt.*

center electrode: The insulated part of a *spark plug* that conducts electricity toward the electrode gap to ground.

centering cones: Tapered pieces of metal, designed to slide onto a shaft to align and hold parts perpendicular to the axis of the shaft.

centerline: **1.** To bore align. **2.** To blueprint. **3.** The axis of an object. **4.** Same as *intake centerline* when referring to a *camshaft.* **5.** A line indicating the exact center.

center link: A steering linkage that is connected to the tie-rod ends that transfers the swinging motion of the gear arm to a linear, or back-and-forth, motion.

center-mount components: The modular installation of a system, such as heating or air conditioning, whereby the evaporator is mounted in the center of the firewall, on the engine side, and the heater core is mounted directly to the rear in the passenger compartment.

center of gravity *(CG)*: The exact point around which an object, such as a vehicle, is perfectly balanced in every direction.

center of wheelbase: The exact point midway between the front and rear wheels of a vehicle.

center-point steering: A steering geometry in which the steering axis passes through the center of the tire contact points.

center to center: The distance between two centers, usually cylinder bores.

centigrade: **1.** Former name for 100 point Celsius scale, the point at which water (H_2O) boils (100°C). **2.** A term often used incorrectly to indicate a metric temperature value. The proper term for a metric temperature value is *Celsius.*

centimeter: A metric unit of linear measure equal to 0.3937 inch.

centipoise *(cP)*: A metric unit of dynamic *viscosity.* It is used by the paint industry to measure the viscosity of paint, and by the oil industry to indicate the low-temperature operating characteristics of oil.

centistroke: A metric unit of kinetic *viscosity* used to indicate the high-temperature operating characteristics of oil.

central port injection: An early *fuel-injection system* installed on the 4.3L Chevrolet Vortec V–6, using one throttle-body, injection-style injector to pulse fuel directly to individual nozzles at the intake ports.

central processing unit *(CPU)*: The component of a computer system with the circuitry to control the interpretation and execution of instructions.

centrifugal: A term often used to describe *centrifugal force.*

centrifugal advance: A mechanical means of advancing spark timing in a conventional *distributor* with flyweights and springs.

centrifugal clutch: A clutch that utilizes a *centrifugal force* to apply pressure against a friction disc in proportion to the speed of the clutch.

centrifugal filter: A rotating filter that relies on *centrifugal force* to separate impurities from the fluid, usually oil.

centrifugal filter fan: A fan found on the air-pump drive shaft used to clean the air entering the air pump.

centrifugal force: The outward force, away from the center (axis) of rotation, acting on a revolving object, increasing as the square of the speed.

centrifugal supercharger: A mechanically driven, forced-induction system using *centrifugal force* to increase air pressure.

centrifugally disengaging: A *one-way roller clutch* in which the rollers disengage with the race, in over-running conditions.

centrifugally engaging: A *one-way roller clutch,* in which the rollers make or maintain contact with the race in over-running conditions.

centrifuge brake drum: Combining the strength of steel with the friction characteristics of cast iron by spraying a lining of *cast iron* on the inside of a steel drum while hot.

century mark: 100 miles per hour.

Ceramic: A material composed of silica and earth elements used as an insulator, as in *spark plugs.*

ceramic insulator: The non-conductive material used, for example, in *spark plugs,* to insulate the center electrode from the ground.

cetane: A primary reference fuel ($C_{16}H_{34}$) for describing the ignition quality of diesel fuel.

cetane number: The number, generally from 40 to 60, that relates to the ignition quality of diesel fuel.

CFC: An abbreviation for the *chlorofluorocarbon* family of refrigerants.

CFI: **1.** An abbreviation for *continuous fuel injection.* **2.** An abbreviation for *central fuel injection.*

cfm: An abbreviation for *cubic feet per minute.*

CG: An abbreviation for *center of gravity.*

chain casing: An enclosure, housing, or guard for a *chain drive* containing the lubricating system and providing protection from contamination.

chain drive: A drive system consisting of a drive, and a driven sprocket and chain, such as a motorcycle or bicycle drive.

chain guide: A device used to support and guide a chain, such as a timing chain, to reduce or prevent whip.

chain length: The distance between the joint centers at each end of a taut chain strand.

chain pitch: The center-to-center dimension between chain links or joints.

chain tensioner: A device used to maintain chain tension, such as a guide rail or a *hydraulic pressure* piston with a rubbing shoe.

chamfer: To remove a hard edge from a part.

chamfer face: A beveled surface on a shaft or part that allows for easier assembly.

champ car: **1.** A championship car. **2.** Early term for an *Indy car.*

Championship Auto Racing Teams *(CART):* The sanctioning body of such events as the annual *Indycar World Series.*

championship trial: The traditional designation for the annual series of races for *Indycars,* sanctioned by *AAA, USAC,* and *CART,* now known as the *Indycar World Series.*

change of state: The rearrangement of molecular matter as it changes between any of the three physical states: solid, liquid, or gas.

channel: To lower the body over the *chassis.*

Chapman strut: A type of *rear suspension* having a telescoping strut that is attached to the *chassis* at the top and to two links at the bottom, restricting lateral and longitudinal movement.

charcoal canister: A container, usually located in the engine compartment, containing activated charcoal. The charcoal absorbs or traps vapors from a vehicle's sealed fuel system, generally when the engine is turned off. This is a basic component of *evaporative emissions control systems.*

charge: **1.** To fill a battery with electrical energy. **2.** To fill an air conditioner with a specific amount of *refrigerant* or oil by volume or weight. **3.** The amount of *fuel/air mixture* to be burned in a cylinder. **4.** To drive aggressively.

charger: **1.** A term often used to describe a *battery charger.* **2.** A top performance driver; one who pushes his car to the limit.

charge temperature sensor: A sensor that sends a signal to the computer causing it to vary the temperature of the intake stream.

charge the trailer: To fill the trailer air-brake system with air.

charge tolerance: The accuracy, plus/minus, permitted in the specified amount of liquid or *gas* that is charged into a system.

charging: **1.** The act of placing a *charge* of *refrigerant* or oil into an air-conditioning system. **2.** The act of refreshing a battery.

charging cylinder: A container with a visual indicator, for use when a critical or exact amount of fluid must be measured.

charging hose: A small-diameter hose, between the system and source, that is constructed to withstand high pressures.

charging station: A unit containing a *manifold* and *gauge set, charging cylinder, vacuum pump,* and *leak detector* used to service air conditioners.

chase: A process to restore damaged threads.

chasing threads: A manual process using a tap or die to restore threads in a nut or on a bolt.

chassis: The lower structure of a vehicle to which the body and running gear are attached.

chassis dyno: A term often used for *chassis dynamometer.*

chassis dynamometer: A drive-on device, used to measure net road *horsepower* and *torque,* delivered by the drive wheels.

chassis lubrication: **1.** An element of *preventive maintenance.* **2.** The procedure of applying the correct type and amount of grease to the *chassis lubrication* points at recommended intervals. **3.** Adding or changing fluids. **4.** Miscellaneous services including tire pressure and safety checks.

chassis tuning: Adjusting the running-gear geometry of a vehicle to compensate for different road conditions.

chassis waddle: A term often used when describing *lateral runout* or *tire waddle.*

cheater slick: A tire that is constructed of the same sticky rubber as a racing tire, but has a shallow tread cut into it to make it street legal.

check: To verify that a component system, or measurement, complies with specifications.

check ball: **1.** A device that maintains air or fuel pressure at a predetermined level. **2.** A device that permits the flow of fluid or vapor in one direction only.

check-engine light: A warning light, generally located in the instrument cluster, that indicates a potential engine or system problem.

checker: A term often used for *checkered flag.*

checkered flag: **1.** A flag waved at a driver, in closed-course competition, to indicate that he/she has completed the race. **2.** The first driver shown the flag is the winner of an event.

check relay: A term often used for check-valve relay or vacuum-check relay.

check-valve: **1.** A valve that permits the passage of a gas or fluid in one direction, but not in the other. **2.** A device located in the *liquid line* or *inlet* to the *drier* of some systems to prevent *refrigerant* flow in the opposite direction when the unit is turned off.

check-valve relay: An electrical switch to control a solenoid-operated *check valve.*

cheek: The plate-like part of a *crankshaft* that connects the journals, often serving as a counterweight.

chemical fire extinguisher: A type of fire extinguisher that uses dry chemicals that displace oxygen, thereby extinguishing a fire.

chemical gasket: A liquid or putty-like substance, similar to *RTV,* used as a substitute for a solid gasket.

chemical hazards: Hazard concerns primarily, but not exclusively, from solvents, fuels, asbestos, and antifreeze.

chemical instability: An undesired condition that exists when a contaminant causes a fluid, such as *refrigerant,* to break down into a harmful chemical.

chemical milling: A term often used for *acid dip.*

chemical reaction: The formation of a new substance when two or more substances are brought together.

chemiluminescence: The emission of light energy, other than by burning, during a chemical reaction.

chem mill: A term used for *chemical milling.*

cherry: **1.** In unusually fine shape. **2.** As good or better than new. **3.** Not used before.

chicane: An artificial series of turns on a straight track, in road racing events, that are marked by pylons or temporary curbs.

child-safety latch: A power-door lock system that prevents the door(s) from being opened from the inside, regardless of the position of the door-lock knob.

chilled cast iron: *Cast iron* that has been hardened using dry ice.

chimney effect: The tendency of air or gas to rise when heated.

chip: **1.** A nick in paint work. **2.** A microprocessor part. **3.** Metal removed during a milling or machining process.

chirp rubber: To quickly shift gears during hard acceleration, so the tires momentarily break loose and leave slight streaks of rubber on the pavement.

chirp the tires: A phrase sometimes used for *chirp rubber.*

chit box: A recreational vehicle, such as a mobile home or trailer.

chizer: A term often given to a Chrysler product.

chlorofluorocarbon *(CFC)*: A family of chemicals that includes the automotive air-conditioning *refrigerant, R-12.* Designated *CFC,* they have been blamed for a deterioration of the Earth's protective ozone layer and have been phased out of production by international agreement.

choke: A manually or thermostatically controlled device mounted to a *choke shaft* having vanes at the mouth of a *carburetor* that closes when the engine is cold. This increases the gasoline content in the *air/fuel mixture* that aids in starting when fuel evaporation is low.

choke heater: A device that warms the thermostatic coil of an automatic *choke,* causing it to open quickly. Later-model *carburetors* often have an electrical heating element and/or a timer circuit.

choke piston: A vacuum-controlled *piston* used to partially open the *choke* when the engine starts.

choke plate: A butterfly valve that closes at the *inlet* of the *carburetor* to enrich the *air/fuel mixture,* as when starting the *engine.*

choke pull-off: A *vacuum motor* that opens the *choke plate* during full acceleration.

choke rod: A rod connected to the *choke plate.*

choke shaft: A shaft at the mouth of a *carburetor* on which the *choke plate* is mounted.

choke valve: A term often used for *choke plate.*

chop: **1.** To lower the *greenhouse* of a vehicle. **2.** To cut in front of another vehicle in a *closed-course* race.

chop, channel, and enamel: The full restyling treatment of a car.

chopped flywheel: To machine the surface of a *flywheel* to lighten it.

chop shop: A facility where stolen vehicles are stripped, dismantled, or otherwise prepared for illegal sale.

Christmas tree: The electrical countdown system used in *drag racing.*

chrome: A simple term for *chromium.*

chrome carnival: A term that applies to a *rod* and *custom* show.

chrome molly steel: A steel alloy that contains *chromium* and *molybdenum.*

chrome rings: *Piston rings* that are plated with *chromium.*

chrome steel: A steel alloy that contains chromium.

chromies: A term used for *chrome-plated* wheels.

chromium: **1.** A basic element, Cr. **2.** A metal used in alloys to provide a durable and hard surface. **3.** An alloy used to plate metal to provide a shining surface.

chrome plate: To apply a thin layer of chrome for appearance.

Chrondek: The trade name of a popular brand of racing timing equipment.

Chryco: A term often used for Chrysler Corporation.

chute: **1.** A fast, straight stretch of track on an oval track or road course. **2.** A parachute, as used to slow drag and lake cars. **3.** The starting position for a *dragster.*

ci: An abbreviation for *cubic inch.*

cid: An abbreviation for *cubic-inch displacement.*

circle burning: An oval race track.

circlip: A split-steel snap ring that fits into a groove, to hold various parts in place.

circuit: **1.** A complete path for an electric current. **2.** A race course.

circuit board: A generic term used for *printed circuit board.*

circuit breaker: An electrical switch-like protective device that automatically opens to

A B C D E F G H I J K L M N O P Q R S T U V W X Y Z

interrupt the circuit if current exceeds its rated limit.

CIS: An abbreviation for *continuous injection system.*

CIS-E: An abbreviation for *continuous injection system-electronic.*

clad: A metal or material covered with another metal by bonding.

cladding: The application of a surfacing material, to impare corrosion and/or heat resistance.

clad metal: A metal that is covered with another metal or alloy of different composition, applied to one or both sides by casting, drawing, rolling, surfacing, chemical deposition, or electroplating.

clamp: A screw-, cam-, or lever-actuated device for temporarily holding parts together.

class: A group of competition cars with basically the same specifications and performance potentials.

class F red insulating enamel: A paint that is used to seal the interior of an engine and to aid in rapid oil return to the crankcase.

classic: **1.** A fine car. **2.** An important racing event, such as the *Indy 500.* **3.** Certain cars built between 1925 and 1948.

Classic Car Club of America *(CCCA):* An organization that is dedicated to the preservation of specific American and European luxury cars manufactured between 1925 and 1948.

classifying vehicles: Any or all of the methods used to classify vehicles, such as by weight or fuel consumption. See the appropriate heading for a specific definition.

clean: **1.** To flush. **2.** To purge. **3.** Free of dirt, grime, or grease.

Clean Air Act: A term used for Clean Air Act Amendments *(CAAA).* A Title IV amendment, signed into law in 1990 by President George Bush, that established national policy relative to the reduction and elimination of ozone-depleting substances.

Clean Air Performance Professionals *(CAPP):* An association of repair shops and technicians promoting inspection and maintenance programs to help protect our environment.

clean room: An enclosed, ventilated or air-conditioned area, free of airborne particles where delicate components, such as engines and automatic transmissions, can be assembled with minimal risk of contamination.

clearance: The space between mating parts, such as between a journal and a bearing, that allows freedom of movement or prevents interference.

clearance ramp: The area of a mechanical-lifter camshaft lobe that makes the progression from the base circle to the edge.

clearance volume: The total-volume measurement above a *piston* at top dead center, *(TDC),* including the area of the combustion chamber.

clear coat: A hard, transparent coating that is applied to a painted surface to enhance the illusion of visual depth and/or protect the surface.

clearing time: The time it takes a circuit breaker to sense an over current, until circuit interruption.

Cleveland V–8: A popular 351 cid *V–8 engine* manufactured in Cleveland by Ford.

clip: A major body repair where the front or rear of a vehicle is replaced with the front or rear of another vehicle of the same make and model.

clock: **1.** A device for telling the time of day. **2.** A term often used for *speedometer.* **3.** A device that generates a basic periodic signal used to control timing of all operations in a synchronous system or computer.

clock spring: A device, located between the steering column and steering wheel, that conducts electrical signals in an air-bag system to the module, while allowing steering-wheel rotation. This provides electrical continuity in all steering-wheel positions.

close coupled: A limited rear-seating space, such as in a club coupe or *two-plus-two.*

closed-camber head: A cylinder head having a *combustion chamber* with a very large *quench* or *squish* area.

closed course: An oval track or road race circuit.

closed-end spring: Coil springs having end loops next to the coils.

closed loop: **1.** The basic principle of electronic engine management in which input from an oxygen sensor allows the engine-control computer to determine and maintain a nearly perfect *air/fuel ratio.* **2.** A computer condition in which the *air/fuel ratio* is being controlled on the basis of various inputs to the computer. **3.** A continuous circuit from beginning to end, and beyond.

closed-loop fuel system: A computerized air/fuel metering system based on monitoring the temperature and composition of the exhaust gases.

close ratios: A transmission, usually used in drag racers, with close spacing between the speeds of the gears, allowing for minimum engine *rpm* reduction when shifting.

close the door: **1.** To immediately pull in front of an opponent after overtaking, preventing him/her from repassing. **2.** To move over while entering a curve to block an opponent just as he/she pulls alongside on the inside to pass, preventing him/her from doing so. Also known as *close the gate.*

close the gate: A term often used for *close the door.*

cloud point: **1.** The low temperature at which diesel fuel begins to produce wax crystals. **2.** The temperature at which wax begins to separate from oil.

club coupe: A two-door, four- or five-passenger vehicle with limited rear seating.

clunker: **1.** A poorly performing car. **2.** The *family buggy.*

cluster gear: A set of three or four gears on a common shaft in a manual transmission.

clutch: **1.** A device for connecting and disconnecting the power flow between the engine and standard transmissions, used during starting, shifting, and stopping. **2.** A device used to connect two collinear shafts to a driving mechanism such as a motor, engine, or line shaft, and to disconnect them at will. **3.** An electromagnetic clutch used to engage and disengage the compressor, to turn the air conditioner on and off.

clutch armature: That part of an electromagnetic clutch that is attached to the compressor crankshaft and is pulled into contact with the rotor when engaged.

clutch cable: A cable that actuates the clutch fork of a manual-transmission system.

clutch can: A term often used for *bell housing.*

clutch coil: A unit consisting of many windings of wire fastened to the front of the air conditioning compressor. When current is applied, a magnetic field is set up that pulls the armature into the rotor to engage the clutch. Also known as *clutch field.*

clutch-control unit: A computer that controls clutch operation.

clutch-cycle pressure switch: A pressure activated switch that controls the air-conditioner-compressor clutch action to prevent evaporator icing.

clutch-cycle switch: An electrical switch, pressure or temperature actuated, that cuts off the air conditioning compressor at a predetermined evaporator temperature.

clutch-cycle time (total): The time between when an air-conditioner clutch engages and when it disengages, then reengages; a time equal to one on and one off cycle.

clutch disc: Circular-shaped component, with a friction facing on each side, that transfers power from the *flywheel* and pressure plate to the splined *clutch shaft.*

clutch field: A unit consisting of many windings of wire fastened to the front of the air-conditioner compressor. When current is applied, a magnetic field is set up that pulls the armature in to engage the *clutch.* Also known as *clutch coil.*

clutch fork: A lever in the *clutch* that actuates the release bearing.

clutch gear: A gear or gears found on the *clutch shaft.*

clutch housing: Cast-iron or aluminum shell that surrounds the *clutch* assembly located between the engine and transmission.

clutch hub: A special hub used in certain limited slip-differential applications, such as single-pack types located between the splined discs and side gear.

clutch off: To get a fast start by engaging the *clutch* suddenly, such as at the start of a drag race.

clutch packs: A series of *clutch discs* and plates, installed alternately in the *clutch housing,* to act as a driving or driven unit.

clutch pedal: A pivoting component inside the vehicle that the *driver* depresses with his/her foot to operate the *clutch.*

clutch piston: An assembly in the *multiple-disc clutch drum* that is moved by oil pressure to engage the *clutch* and returned to a released position by mechanical-spring force.

clutch plate: **1.** A pressure plate that forces the *clutch disc* against the *flywheel.* **2.** A term often used for *clutch armature.*

clutch-release bearing: A component, attached to the *clutch-release fork,* that contacts and then moves the release levers when the *clutch pedal* is depressed.

clutch-release fork: A pivoting *clutch housing* component that transfers motion from the free-play adjusting rod on the clutch linkage to the attached clutch-release bearing.

clutch-relief check valve: A valve that releases to prevent the buildup of pressure in a *multiple-disc clutch* assembly.

clutch rotor: That freewheeling portion of the air-conditioning system clutch, in which the *drive belt* rides, that mates to the armature when power is applied to the coil.

clutch-safety switch: A term often used for *neutral start switch* or *neutral safety switch.*

clutch shaft: The main shaft on which the *clutch* is assembled.

clutch slippage: A term given to the condition when, although engine speed is increased, increased torque is not transferred through to the driving wheels.

cm: An abbreviation for *centimeter.*

CNG: An abbreviation for *compressed natural gas.*

CO: A symbol for *carbon monoxide.*

coach: **1.** A bus. **2.** A motor home. **3.** An enclosed, two-door sedan of yesteryear. **4.** A large, four-wheeled carriage; the forerunner of the modern vehicle.

coach builder: A manufacturer of fine custom-automotive bodies.

coachwork: The product of a coach builder.

coating: **1.** A paint that is used to provide beauty or weather protection to a vehicle or to its parts. **2.** Any material applied to a metal surface to provide protection against the elements.

coaxial cable: A cable consisting of two conductors concentric with, and insulated from, each other.

cockpit: The driver's compartment of a vehicle.

code installation: In general, an installation that conforms to the state and federal regulations and codes to insure safe and efficient conditions.

Code of Federal Regulations: Regulations that are generated, published, and enforced by the United States government.

COE: An abbreviation for *cab over engine.*

coefficient of drag (C_x): A measure of the air resistance of a moving vehicle; a measure of how much air is moved as the vehicle moves from one point to another.

coefficient of friction: The measure of the resistance of one surface moving against another.

coefficient of water/oil distribution: The ratio of the solubility of a chemical in water compared to its solubility in oil.

cog: A gear, particularly the final drive gear.

coil: A term often used to describe a *spring* or an *ignition coil.*

coil bind: A condition where springs are compressed to the point that the coils touch.

coil failure: 1. A defective ignition coil. 2. Also see *coil spring failures.*

coil over shock: A suspension component that consists of a shock absorber inside a coil spring.

coil-preload springs: Coil springs, located in the pressure-plate assembly, made of tempered-steel rods formed into a spiral that resists compression.

coil spring: A spring-steel bar or rod that is wound into the shape of a coil to provide an up-down springing effect. Found on most vehicle suspensions, these springs are used to support the car's weight, maintain height, and correctly position all other suspension parts, but are little help in supporting side-to-side or lateral movement.

coil-spring clutch: A clutch that uses coil springs to hold the pressure plate against the friction disk.

coil-spring failures: The inability of a coil spring to compress and/or rebound, due to: constant overloading, continual jounce and rebound action, metal fatigue, or a crack, break, or nick on the surface layer or coating.

coil-spring rear suspension: A rear-axle assembly that is attached to the frame through a link-type suspension system. Coil springs are mounted between the lower suspension arms and the frame, while the shock absorbers are mounted between the back of the suspension arms and the frame.

coil-spring seat: The formed mounts that determine the coil-spring position on the car frame and rear axle housing. Seats may have sound-insulating pads.

coil tower: The high-voltage, center terminal of a conventional ignition coil.

coke bottle: The shape of an auto body tucked slightly inward at the center, like a Coca-Cola bottle.

COLA: An acronym for *cost of living allowance.*

cold: 1. An object that is not hot or warm, generally below body temperature, 98.6°F (37°C). 2. The absence of all heat, –259.67°F (–162°C).

cold-cranking amps: A common term used for *cold-cranking power rating.*

cold-cranking power rating: The number of amperes that a fully charged battery will deliver for 30 seconds at 0°F (–17.8°C) without the terminal voltage dropping below 7.2 volts.

cold drawn: A process where metal is drawn or rolled into a particular shape or size.

cold-engine lockout switch: A switch that sends a signal to the body-control module, or controller, to prevent an action, such as blower-motor operation, until the coolant in the engine has risen to a predetermined temperature.

cold manifold: An *intake manifold* that does not have a preheat passage.

cold patch: A process used to repair a punctured tire or tube, without the aid of heat.

cold rate: The number of minutes a battery will deliver 300 amperes at 0°F (–17.8°C), before the cell voltage drops to below 1.0 volt.

cold-rolled steel *(CRS):* Carbon steel that is worked into shape while cold.

cold soak: To place a component in a cool area to allow it to cool to ambient temperature for twelve or more hours.

cold-solder joint: A loose or intermittent electrical connection, caused by poor soldering techniques.

cold-start injector: An electronic fuel-injection system that supplies extra fuel to the engine for cold starting.

cold-start test: A prescribed federal test procedure for measuring emissions before an engine has warmed up after a 12-hour *cold soak* at 68°F to 78°F (20°C to 25.6°C).

cold-start valve: A valve that permits additional air into the *intake manifold* during a cold start on a fuel-injected engine.

cold-weather modulator: A thermostatically controlled check valve that traps vacuum in the vacuum motor circuit when the car is accelerated hard at any temperature below 55°F (12.8°C), to prevent hesitation by allowing heated air to enter the engine.

cold weld: A method of repairing small cracks in blocks and heads by using tapered plugs to fill the cracks.

cold working: The deformation of metallic material at a temperature below the recrystallization temperature, resulting in strain hardening of the material.

collapsible steering column: An energy-absorbing steering column that is designed to collapse if the driver is thrown into it due to a heavy collision.

collector: **1.** A device that collects exhaust gases from the primary tubes and channels them into a single exhaust pipe. **2.** The tank of a radiator that receives the fluid before it passes through the radiator.

collector tank: The tank that collects coolant from the engine, containing a baffle plate to aid in even distribution of coolant through the core.

collet: A term used for *valve keeper.*

collision shop: A specialty paint and body shop that restores a wrecked or damaged vehicle to its pre-accident condition.

color code: A means of identifying *conductors* or *vacuum hoses* by the use of color.

color-code chart: A chart listing the colors of wire insulation and, sometimes, wire sizes for a particular automobile.

color sanding: Color-blending by lightly sanding to smooth surface imperfections, using 1000 grit or higher paper.

combination brake system: A dual-brake system that uses disc brakes at the front wheels and drum brakes at the rear wheels.

combination valve: An H-valve, used in some early air-conditioning systems, combining a suction throttling valve and an expansion valve.

combined emissions-control system: An early General Motors transmission-controlled spark system that uses the solenoid valve's plunger as an auxiliary throttle stop.

combustion: The burning of the *air/fuel mixture* in an *engine.*

combustion chamber: Area above a *piston* at *TDC,* primarily distinguished by a recessed cylinder head, where combustion takes place.

combustion emission control *(CEC):* An exhaust emission-control system that combines a transmission-controlled spark system and a deceleration throttle-position device.

combustion knock: A term often used for *knock.*

combustion pressure: The pressure in the cylinder from expanding gases immediately after the *air/fuel mixture* is ignited, which is about four times greater than compression pressure.

combustion recess: An indented area on the rotor face where part of the burning of the *air/fuel mixture* occurs.

comfort: A pleasing and enjoyable feeling due to the removal of excessive heat, moisture, dust, and pollen from the air.

Comfortron: A trade name for an early automatic-temperature-control system.

common point: A connection point, such as for several conductors or levers.

commutating pole: An electromagnetic bar inserted between the pole pieces of a generator to offset the cross magnetization of the armature currents.

commutator: That part of a starter or generator that transfers electrical energy to the armature or rotor.

compact high-pressure tire: A term often used for *compact spare.*

compact spare: A weight- and space-saver tire, especially designed for temporary use.

compact spare tire: A spare tire and wheel that is much smaller than the other tires on the vehicle, and is designed for short-distance driving at relatively low speeds.

compact tire: A term used for *compact spare tire.*

companion cylinders: Two cylinders in an engine that are at *TDC* at the same time.

comparator: An instrument for comparing specific critical measurements to a fixed standard.

compensating coil: A coil that serves to compensate for the mechanical friction in the moving coil of an electrical meter or gauge.

compensating port: A device used to maintain the proper level of brake fluid in the brake lines.

complete circuit: An uninterrupted electrical circuit or fluid circuit in which electrical current may travel to and from the battery, or fluid or *vapor* may circulate continuously through the system.

completely built up *(CBU)*: The complete building, rebuilding, or modification of a component or vehicle, as in a kit car.

component isolation: To isolate a component from the rest of the system or circuit for testing or replacement.

component location table: A table or chart, used with an electrical schematic, that describes or illustrates the actual location of the part being investigated.

composite headlight: A halogen headlight system that uses a replaceable bulb, allowing vehicle manufacturers to produce any style of headlight lens they desire.

composite materials: The bonding of different materials, usually for strength, such as carbon fiber and *fiberglass.*

composite spring: A term used for *fiber-composite spring.*

composite washer: A flat washer made of different elements.

composition gasket: A gasket made of a combination of materials.

compound low gear: A combination of low gear in the transmission and low range in the transfer case in a four-wheel-drive vehicle with a two-speed transfer case.

compound gauge: A gauge that registers both above and below atmospheric pressure; used on the low side of an air-conditioning system.

compressed natural gas: A gas, primarily methane, used as a motor fuel.

compression: **1.** The process of squeezing a vapor (gas) into a smaller space. **2.** The upward stroke of a piston that compresses the *air/fuel mixture* into the *combustion chamber* prior to *ignition.* **3.** Short for *compression ratio.*

compression braking: The slowing of a vehicle utilizing a diesel engine, such as that provided by a *Jake brake.* It is a misnomer that a gasoline engine will slow the vehicle by compression braking; actually, vacuum causes the braking effect.

compression height: A distance, as measured from the crown of the piston to the center of the wrist pin.

compression ignition: The operating system of a diesel engine, where heated air is used to ignite the fuel.

compression intake valve: A term used for *compression valve* or *intake valve.*

compression-loaded ball joint: A suspension ball joint, mounted above and resting on the knuckle, so the vehicle weight forces the ball into the joint.

compression pressure: The highest pressure developed during the *compression stroke* in an engine, as checked with a compression gauge.

compression ratio *(CR)*: A measurement of how much the *air/fuel mixture* is compressed inside an engine cylinder. If compressed to 1/10 of its original volume, the compression ratio is 10 to 1.

A B C D E F G H I J K L M N O P Q R S T U V W X Y Z

compression ring: A piston ring that seals pressure during the compression and power strokes. There are usually two compression rings per piston.

compression seal: A metal seal found in a direct *fuel-injection system,* to resist *compression pressures.*

compression stroke: The movement of the piston from *BDC* to *TDC,* immediately after the intake stroke.

compression valve: A calibrated valve, located at the base of the shock absorber, providing variable resistance to fluid flow during compression.

compressive strength: The maximum compressive stress that a material can withstand without significant plastic deformation or fracture.

compressor: **1.** A component of the refrigeration system that pumps refrigerant and increases the pressure of the refrigerant vapor. **2.** A device used to pump air.

compressor crankshaft seal: A term used for *compressor shaft seal.*

compressor-discharge pressure switch: A pressure-operated electrical switch that opens the compressor-clutch circuit during high-pressure conditions.

compressor-protection switch: An electrical switch installed in the rear head of some compressors to stop the compressor in the event of a loss of refrigerant.

compressor-shaft seal: An assembly consisting of springs, snap rings, O-rings, seal sets, a shaft seal, and a gasket mounted on the compressor crankshaft to permit the shaft to be turned without a loss of refrigerant or oil.

computer: A machine capable of following instructions to alter data in a desirable way and to perform most of these operations without human intervention.

computer-aided manufacturing: The use of computer technology in the management, control, and operation of manufacturing.

computer-command control *(CCC):* A term given a computer that controls the function and operation of an automotive system, or sub system.

computer-controlled brakes: A system having a sensor on each wheel, feeding electrical impulses into an on-board computer. As the vehicle is stopped, each wheel is stopped or slowed down at the same rate, reducing sideways skidding during rapid braking.

computer-controlled suspension system: A system in which a computer-controlled actuator is positioned in the top of each shock absorber or strut. The shock absorber or strut actuators rotate a shaft inside the piston rod, and this shaft is connected to the shock valve.

computerese: The jargon and other specialized vocabulary of those working with computers and information-processing systems.

computer-generated code: A term more commonly known as *trouble code.*

computerize: **1.** To equip a business or organization with computers in order to facilitate or automate procedures. **2.** To convert a manual operation into one that is performed by a computer.

computerized air suspension: A type of suspension system equipped with rubber air bags controlled by an air compressor to maintain a specific ride height determined by vehicle load and road-surface conditions.

computerized automatic temperature control: A microprocessor control system that monitors incoming data and adjusts the temperature and humidity of the air inside the passenger compartment.

computerized engine control: A microprocessor-based, engine-management system that utilizes various sensor inputs to regulate spark timing, fuel mixture, emissions, and other functions. Most systems include on-board, self-diagnostic capability and store fault codes to help diagnosis of system problems.

concave fillet weld: A fillet weld having a concave face.

concave side: An inward-curved depression.

concealed headlight: A headlight system that enhances a vehicle's style and aerodynamics by hiding the lamps behind electrically- or vacuum-controlled doors when not in use.

concentricity: The condition in which two or more features, in any combination, have a common axis.

concours d'elegance: French for "contest of elegance," a showing of luxury cars in a plush setting.

condensate: Moisture that is removed from air, such as that collected on the surface of an air-conditioning-system evaporator.

condensation: **1.** The moisture removed from *ambient air.* **2.** The process of a substance changing state from a vapor (gas) to a liquid.

condenser: **1.** A liquefier; the component of a refrigeration system in which refrigerant vapor is changed to a liquid by the removal of heat. **2.** An improper but often used term for a *capacitor.*

condenser comb: A comb-like device used to straighten the fins on the *evaporator* or *condenser.*

condenser temperature: The temperature at which compressed gas in the condenser changes to a liquid.

condensing pressure: **1.** High side or head pressure, as read from the gauge at the high-side service valve. **2.** The pressure present from the discharge side of the compressor to the metering device inlet.

conditioned air: Air that is cool, dry, and clean.

conductance: The ability of a material to transmit an electrical charge.

conductor: Any material that will conduct an electrical charge, such as a copper wire.

conductor placement: A term used for *wire placement.*

conduit: A tubular raceway, such as tubing used to protect a wiring system or branch circuit.

cone: **1.** The conical diaphragm attached to the voice coil of a speaker that produces sound. **2.** The conical part of an oxyfuel gas flame adjacent to the tip orifice.

coned-disk spring: A term used for *Belleville washer.*

configuration: The shape or form of anything, such as an engine.

conformability: The ease or difficulty of different materials to be shaped or worked.

conicity: A term that refers to *tire conicity.*

connecting link: A link in which a removable plate facilitates connecting or disconnecting the ends of a length of chain.

connecting rod: A component used to attach the piston, with pin, to the crankshaft rod journal.

connecting-rod cap: The half-round, lower, bolt-on portion of the *connecting rod.*

con rod: Short for *connecting rod.*

constant-current charging: A charging system in which a constant flow of current is fed into the battery.

constant-current electronic control *(CCEC)*: A type of engine-computer control.

constant mesh: Gears that mesh continually, such as a *planetary gear,* eliminating the clashing or grinding that may occur when other types of gears are shifted together.

constant-ratio steering gear: A steering-gear system having the same gear ratio when the wheels are near the straight-ahead position as during extreme turns.

constant tension: A system or device that is designed to be under perpetual pressure or stress.

constant-velocity joint: Two universal joints, closely coupled, so their acceleration-deceleration effects cancel each other out.

constant voltage: The common type of power in which all loads are connected in parallel, with different amounts of current flowing through each load.

A B **C** D E F G H I J K L M N O P Q R S T U V W X Y Z

constant-voltage charging: A method of charging where the voltage to the battery is constant, and the current decreases as a fully-charged condition is reached.

consumption: A term most generally used when referring to *fuel consumption.*

contact area: The area of a member that comes into contact with another member, such as a belt to a pulley, or a tire to the ground.

contact patch: The area of contact of a tire with the road surface when the tire is supporting the vehicle weight.

contact points: A term often used for *contact area* or *breaker points.*

container tube: A shock-absorber component that is used to house the internal parts.

contaminants: **1.** Chemicals or impurities in oil or fuel that reduce its effectiveness and efficiency, such as water, carbon, acids, dust, and dirt particles. **2.** Chemicals or impurities that make ambient air impure, especially those produced from the combustion process.

contaminated: A term used when referring to a system that is known to contain foreign substances, such as incompatible or hazardous materials.

contingency money: A payment made by accessory and equipment manufacturers to top rated drivers, and some crew members, for using and/or displaying their products in a race.

continuity: The ease or ability of an electrical circuit to transfer energy from one point to another.

continuous code: A series of computer diagnostic codes that relate to engine-control functions.

continuous combustion: The constant combustion of an *air/fuel mixture.*

continuous duty: A demand on an energy-consuming system that requires operation, at a constant load, for an indefinite period of time.

continuous fuel injection *(CFI):* A type of fuel-injection system that continuously sprays a stream of fuel into the engine.

continuous-injection system *(CIS):* A continuous-flow, mechanically controlled, fuel-injection system.

continuous-injection system—electronic *(CIS-E):* A continuous-flow, electronically controlled, fuel-injection system.

continuous variable transmission *(CTV):* A stepless transmission that uses a sheave clutch to transmit engine torque.

continuous weld: A weld that extends continuously from one end of a joint to the other.

control arm: The main link between the vehicle frame and the wheels that acts as a hinge to allow the wheels to go up and down independently of the *chassis.*

control cable: **1.** A rigid wire, generally sheathed, used for the remote control of a device, such as a parking brake. **2.** An insulated wire used to supply electric current to a motor, controls, or other electrically operated circuits or devices.

control head: The master controls (such as temperature and fan speed) that the vehicle driver uses to select the desired system condition.

controlled-leak governor: A governor assembly that reduces the leakage of line pressure as a vehicle's speed increases.

control link: A term used for *control arm.*

control plunger: A device that is regulated by the fuel-injection-system airflow sensor to regulate fuel delivery to the injectors.

control points: Holes, points, or flat surfaces that are used to align body parts during assembly or reconstruction.

control pressure: The fuel pressure required for a fuel-injection system to function.

control valve: The mechanism, located inside the power steering gearbox or on the steering relay rod, that controls the amount of power assist relayed to the steering linkage via a power piston.

convection: The transfer of heat by motion of the heated material, such as air.

conventional cab: A truck having a cab behind the engine.

conventional frame: A chassis frame that is separate from the body.

conventional-frame designs: One of three designs: Ladder frame; x-frame, or hourglass frame; or, perimeter frame.

conventional strut: An assembly in which the coil spring that supports the vehicle weight sits on a lower-spring seat, which is part of the strut housing, to maintain vehicle height.

conventional theory: That electrons flow from the positive to the negative point in an automotive electrical system. Actually, flow is from negative to positive.

conventional tire: A term used for *bias ply tire.*

conversion: The substitution of one element or component for another, as in substituting *R-134a* for *R-12* in an automotive air-conditioning system; changing the fuel-type requirements of an engine; or changing ignition systems.

converter dolly: An axle, *frame, drawbar,* and *fifth-wheel* arrangement that converts a semi-trailer into a full trailer.

converter pressure: The operating pressure within the torque converter.

converter-signal pressure: The pressure that signals the torque-converter, lockup clutch engagement.

convertible: An automobile having a retractable or removable roof.

convertible top: A vinyl or canvas soft-top roof system for two- or four-door vehicles that can be manually or automatically lowered and raised.

cook: **1.** To overheat, generally to the point of severe or irreparable damage. **2.** To perform well, as in "s/he's cookin'."

coolant: The fluid, consisting of water and antifreeze, that circulates throughout a liquid-cooling system, around hot engine parts, to remove the heat and prevent damage.

coolant circulation: The movement of liquid throughout the cooling system by water pump action to move heat, generated by engine combustion, to the upper radiator section.

coolant fan: An electrically- or mechanically-driven fan to increase air flow across the radiator to facilitate heat removal.

coolant-fan relay: A control device that is used to activate the electric-coolant fan at a pre-determined, high-coolant temperature.

coolant jacket: Hollow passages surrounding the cylinders in the block and the combustion chambers in the cylinder head.

coolant passages: Coolant passages, called *coolant jackets* or *water jackets,* surround each cylinder in the block to provide a means to carry away unwanted engine heat.

coolant pump: A term often used for *water pump.* A centrifugal-type pump used to circulate coolant through the *cooling system.*

coolant-recovery system: A cooling system having a semisealed pressure cap, with a radiator-overflow hose leading into a separate plastic reservoir. This saves coolant during hot operation and returns it to the *radiator* when the system cools.

coolant-recovery tank: A storage container that is used in a coolant-recovery system.

coolant reservoir: A tank used for storing excess coolant; connected to the cooling system with a small-diameter overflow hose.

coolant temperature-override switch: A temperature-controlled vacuum device that prevents overheating during idle speed, associated with late ignition, by advancing the timing to increase engine-idle speed.

coolant-temperature sensor: A thermistor that sends a signal to the electronic control unit relative to the coolant temperature in a computerized engine-control system.

coolant thermostat: A unit found in the coolant outlet of the engine to help prevent over-cooling conditions, especially during short trips.

cooling coil: **1.** An evaporator. **2.** An oil cooler. **3.** An auxiliary cooler.

A B C D E F G H I J K L M N O P Q R S T U V W X Y Z

cooling-fan controller module: An electronic unit that will cycle the cooling fan ON and OFF, in response to signals from other engine sensors.

cooling method: The method used to remove excess heat from an engine or system, such as *air cooling* or *liquid cooling.*

cooling system: **1.** An air conditioner. **2.** A system of parts that circulates coolant through the engine to remove heat. **3.** A system to remove heat from a heat-generating mechanism.

cooling-system fan: An electrically- or mechanically-driven rotating device, having four to seven pitched blades to move air past a heat exchanger, such as the radiator.

Cool Pack: A trade name used to describe an early after-market hang-on or under-dash, after-market air-conditioning system.

copper *(Cu):* A reddish-colored metallic element, with great electrical properties; one of the oldest metals known.

copper-asbestos-copper gasket: A gasket made of copper-clad asbestos; now rare and not used due to the dangers associated with asbestos fibers.

copper gasket: A gasket made of copper and used in high-temperature and pressure conditions.

copper sulfate: A chemical used to detect chrome-plated cylinders and test *crankshafts* for *Tuftriding.*

cord: An inner material of textile, steel, fiberglass, or wire running through the plies of a tire that produce strength.

cord material: A material, such as rayon, nylon, fiberglass, polyester, or steel, used in tire construction to provide strength and maintain desired shape.

cord plies: The layers of rubber-impregnated cord material molded in the sidewalls of the tire casing.

core: **1.** The interior of a hollow casing. **2.** The coolant passages and fins of a radiator or heater found between the two header tanks. **3.** A used part or assembly that is to be returned to the vendor for rebuilding.

core hole: **1.** A hole provided to facilitate *sand casting.* **2.** A cavity in a casting caused by the shifting of the *core* during manufacturing.

core-hole plug: A plug used to provide a seal for a *core hole.*

core loss: The electric loss occurring in the *core* of an armature or transformer due to *eddy currents,* hysteresis, and like influences.

core plug: A metal, cup-shaped disc in a cast component, such as the engine block, to seal openings required by the casting tools that may pop out to protect it from freezing damage. Also known as *freeze plug.*

core shift: A condition where one side of a cylinder bore is thicker than the other side.

cornering light: Lamps in both sides of the vehicle body near the front that light up when the turn signals are activated and burn steady when the turn signal switch is in a turn position, providing additional illumination of the road in the direction of the turn.

Corporate Average Fuel Economy *(CAFE):* The fuel mileage standard for an automaker's line of vehicles set annually by the *Department of Transportation (DOT).*

corrosive flux: A flux composed of inorganic salts and acids, organic salts and acids, or activated rosin with a residue that chemically attacks the base metal.

corrosive rain: A form of pollution produced when sulfur and nitrogen are emitted into the air. Known as *acid rain.*

corrugated metal gasket: Thin sheets of metal used as a gasket that is shaped into parallel grooves and ridges for added strength.

corsa: Italian for course.

COSA: An abbreviation for *Carwash Owners and Suppliers Association, Inc.*

Cosmoline: A trade name for a heavy-grade lubricant used to protect machine surfaces.

cost of living allowance *(COLA):* An increase in stipends based on the average increase in the cost of living for any given period.

Cosworth: A British racing-engine manufacturer.

CO_2: The symbol for carbon dioxide.

cough: To damage or destroy an engine.

counter: A device, such as a register or computer-storage location, used to represent the number of occurrences of an event.

counterbalance: A weight added to a rotating shaft or wheel to offset vibrations.

counterbored ring: A compression piston ring with a counterbore on its inside diameter to promote cylinder sealing.

counterboring: A machining process, related to drilling, using an end-cutting tool to concentrically enlarge a hole to a limited depth.

counterclockwise *(ccw):* A rotation to the left; as opposed to the direction of the rotation of clock hands.

counterclockwise rotation: Rotating in the opposite direction of the hands on a clock.

counterelectromotive force: A term used for *back emf* or *counter emf.*

counter *emf:* Induced *voltage;* the *voltage* opposing the applied *voltage* and the *current* in a *coil,* caused by a flow of *current* in the *coil.*

countergear: An integral cluster of three or more various-sized *gears,* located in the lower *transmission* case, that revolve on a *countershaft* to provide the desired gear ratios, usually for second, low, and reverse.

countershaft: **1.** A shaft used in a *V-8 engine* to reduce the effects of imbalance. **2.** Two shafts used in an *I-4* engine to reduce the effects of imbalance. **3.** The shaft that supports the cluster-gear set in a manual *transmission* and rotates in the opposite direction of the *clutch* and *driveshaft.*

countersinking: A machining process, related to drilling, that bevels or recesses the material around the circumference of a hole.

counterweight: **1.** A weight that is cast opposite each offset connecting-rod journal to provide the necessary balance. **2.** A weight that is added to a rotating shaft or member to offset vibration by balancing the part.

coupe *(cpe):* A car with a close-coupled passenger compartment.

coupe de ville: A coupe with an open driver compartment and enclosed passenger compartment.

coupling: An attachment where one mechanism or part drives another mechanism or part, to follow the movements of the first.

coupling phase: The point in torque-converter operation where the turbine speed is 90% of the impeller speed, and there is no longer any torque multiplication.

coupling stage: A term referring to the torque-converter turbine speed, as it closely approaches the impeller speed; occurs during higher speeds under light loads.

coupling yoke: Two integral or separate Y-shaped components to which the universal-joint bearing cups are attached.

coupon: **1.** A piece of metal, of specified size, used for testing. **2.** A piece of metal from which a test specimen may be prepared.

courtesy light: Lamps that illuminate the vehicle interior and/or exterior when a door is opened, and are controlled from the headlight and door switches.

covered electrode: A composite filler metal consisting of a *core* of bare electrodes or a metal-cored electrode, with a covering sufficient to provide a slag layer on the weld metal.

cowl: That part of a vehicle between the passenger compartment and engine to which the windshield and dashboard are attached.

cowl air intake: The *inlet* at the base of the windshield that allows outside air to enter the heater/air-conditioning system, or driver/passenger compartment of the vehicle.

cP: An abbreviation for *centipoise.*

A B C D E F G H I J K L M N O P Q R S T U V W X Y Z

cpe: An abbreviation for *coupe.*

c-pillar: The structural support just behind the rear door that supports the *greenhouse.*

c-post: A term often used for *c-pillar.*

CPU: An abbreviation for *central processing unit.*

CR: An abbreviation for *compression ratio.*

cracked: **1.** Broken. **2.** A term often used for the mid-position of a two-way valve.

crank: **1.** A *crankshaft.* **2.** To start an engine. **3.** To go fast.

crankcase: The lower section of the engine that supports the *crankshaft,* confined by the lower block casting and the oil pan.

crankcase breather: A tube or vent that allows excessive *crankcase pressure* to escape.

crankcase dilution: The thinning of oil in the *crankcase,* caused by the condensation of *gasoline* due to blow-by, and by seepage past the *piston rings.*

crankcase emissions: Fumes that leave the *crankcase* by way of an open or closed ventilation system.

crankcase fumes: Vapors inside the *crankcase* that could contaminate the air, including unburned fuel vapors, water vapor, or blow-by gases. Also known as *crankcase vapors.*

crankcase pressure: The pressure produced in the crankcase from blow-by gases.

crankcase vapors: Fumes inside the *crankcase,* caused by blow-by, that could contaminate the air, including unburned fuel vapors, water vapor, or blow-by gases. Also known as *crankcase fumes.*

crankcase ventilation: A tube or vent that allows excessive *crankcase pressure* to escape.

cranking circuit: The starter and its associated circuit, including battery, relay (solenoid), ignition switch, neutral start switch, cables, and wires.

cranking motor: A special high-torque electric motor designed for the purpose of cranking the engine for starting.

crank kit: A reconditioned *crankshaft* with the appropriate rod and main bearings.

crank pin: The rod journal of a *crankshaft.*

crank *rpm:* A measurement that is required for an electronic engine-control system to determine when ignition should occur.

crankshaft: **1.** A revolving part mounted in the lower portion of the engine block. **2.** That part of a reciprocating compressor on which the wobble plate or connecting rods are attached.

crankshaft arm: A connector between the two rod journals and the main bearing on the crankshaft.

crankshaft counterbalance: A term used for *crankcase counterweight.*

crankshaft counterweight: A weight that is cast opposite each offset connecting-rod journal, to provide the necessary balance.

crankshaft end-play: A specified crankshaft endwise motion controlled by side flanges on one of the main bearings.

crankshaft gear: A *gear* or *sprocket* found on the front end of the *crankshaft* that is used, directly or indirectly, to drive the camshaft.

crankshaft main journal: That part of the *crankshaft,* ground round and polished smooth, around which the closely-fitted main bearings surround the journals and support the *crankshaft.*

crankshaft oil passage: Holes drilled through the *crankshaft* to permit oil to flow from the main bearings to the connecting-rod bearings.

crankshaft pulley: A pulley fitting on the front of the *crankshaft.*

crankshaft seal: **1.** A rubber-like seal around an engine *crankshaft* to prevent the leakage of oil. **2.** A rubber-like and/or ceramic seal surrounding the *compressor crankshaft* to prevent the leakage of oil and/or *refrigerant.*

crankshaft throw: One crankpin with two webs.

crankshaft thrust collar: A disc-shaped, machined surface between the main bearing and the two rod journals on the *crankshaft.*

crank signal: An electrical signal sent to the computer to tell it that the engine is cranking and to enrich the *air/fuel ratio* for easier starting.

crank start: To start an internal-combustion engine by hand cranking.

crank web: That part of a *crankshaft* that is between a crankpin and a main bearing.

crash box: A manual transmission with straight-cut, non-synchromesh gears.

crash sensor: Normally open, gold-plated, electrical switches that are designed to close when subjected to a predetermined impact.

crazing: Fine cracks that may extend in a network on or under the surface of a material, often occurring in the presence of an organic liquid or vapor.

cream hardener: An activating ingredient for a plastic-filler material, such as *Bondo.*

cream puff: A vehicle that is in especially fine condition.

creature comforts: Any options or amenities that improve vehicle driving or riding pleasures.

crescent: **1.** A concave and convex shape. **2.** A short roadway curved in a half circle.

critical pressure: The pressure at which a gas becomes unstable.

critical temperature: The temperature at which a flammable gas will ignite.

crocus cloth: A very fine-grit sandpaper that is used to clean or polish parts, such as a *crankcase.*

cross: The central component of the *U-joint* connecting the input and output yokes.

crossbolt main cap: A main-bearing cap that is secured with two down-facing bolts and two bolts that intersect at right angles.

cross-drilled crank: A *crankshaft* having two oil passages at right angles to each other in the main journal to provide extra lubrication for the main and *rod bearings.*

crossfire: The electromagnetic-induction spark that can be transmitted in another wire close to the wire carrying the current.

crossfire injection: A type of throttle-body injection system that uses two injectors mounted on the manifold. Each injector feeds a cylinder on the opposite side by using a crossover port.

crossflow head: A cylinder head having its intake ports on one side and the exhaust ports on the other side.

crossflow radiator: A *radiator* in which the coolant flow is from one side to the other.

cross-groove joint: A disc-shaped, inner, constant-velocity universal joint that bolts to a transaxle stub flange and uses balls and V-shaped grooves on the inner and outer races to accommodate the plunging motion of the half-shaft.

crosshatch: **1.** A honing pattern that is required in a *cylinder* to maintain the correct amount of oil retention and to facilitate ring rotation. **2.** A series of crisscrossing lines that indicate a specific area in a drawing or diagram.

cross heads: A T-shaped part on a diesel engine to open and close two valves at one time.

cross lock: A holding device used with automatic-throttle cable adjusters.

crossover network: An electrical circuit that routes different frequencies to the woofer, midrange, and tweeter in a multiple-speaker system.

crossover pipe: A pipe used to connect both sides of an exhaust system to equalize the pressure.

crossover tube: A tube that is used to transmit liquid or gases in or around an engine.

cross section: A section formed by a plane cutting through an object, generally at 90° to

A B C D E F G H I J K L M N O P Q R S T U V W X Y Z

the centerline to show interior details in a drawing.

cross up: To lose control of a vehicle, allowing it to spin or skid out of control.

crotch height: The height of a chain-link crotch above the pitch line of the link.

crowd: An accelerator action that maintains a constant manifold-vacuum reading, requiring a progressive opening of the throttle as the vehicle speed is increased.

crown: The top surface of a piston.

crown gear: The ring gear in a *differential.*

crown wheel: A term often used for *crown gear.*

CRS: An abbreviation for *cold-rolled steel.*

CRT: An abbreviation for *cathode-ray tube.*

crude oil: Petroleum in its natural state; as pumped from the ground before refinement.

cruise control: A control system that allows the vehicle to maintain a preset speed, though the driver's foot is off the accelerator.

cruising: **1.** A motorized version of a Saturday evening promenade around the town square. **2.** A social outing, generally by young people, driving a route repeatedly on the major city streets at a slow speed, often for hours at a time. **3.** A chance for young men, or ladies, to show off their wheels and flirt with the young ladies, or men.

crunch hat: A safety helmet.

crush: The amount of compression required to seat a rod or main bearing into its housing bore.

crush relief: The area at the edge of a rod, or main bearing, that allows the crush to occur.

crush sleeve: A sleeve that is used to position a pinion gear in the *differential;* designed to crush when torqued to specifications.

cSt: An abbreviation for *centistroke.*

C31: A computer-controlled, coil-ignition system that uses a computer to monitor, maintain, and adjust ignition timing.

CTO: An abbreviation for coolant temperature override.

C-train: A combination of two or more trailers in which the dolly is connected to the first trailer by means of two pintle hooks or coupler-drawbar connections, resulting in only one pivot point.

Cu: Symbol for copper.

cubes: A term for cubic-inch displacement or *CID.*

cubic feet per minute: The measure of a carburetor's air-flow capacity.

cubic inch *(ci):* An English measure for volume equal to 16.39 ccs.

cubic-inch displacement *(cid):* The English measure for the volume of space displaced by the piston as it moves from *BDC* to *TDC.*

cup expander: Metal discs formed to fit inside piston cups and to keep the lips of the cups in tight contact with the cylinder walls while the hydraulic system is not pressurized.

cup seal: The rubber seal inside a wheel cylinder.

cup shim: A cup-shaped spring adjuster used to reduce valve-spring bounce at high engine *rpm.*

cup wheel: A wheel shaped like a cup or bowl.

curb height: The measurement from the vehicle frame to the road surface.

curbside: The right-hand side of a vehicle; that nearest to the curb.

curb weight: The weight of a vehicle ready for the road, with a full complement of fuel, oil, and coolant, but without cargo, driver, or passengers.

curing: **1.** The hardening of a catalyzed compound, such as fiberglass. **2.** The drying of paint.

current: 1. At present. Now. 2. Rate of flow of electric charge through a *conductor.*

current-carrying capacity: The current, in *amperes,* a conductor can carry continuously, under the conditions of use, without exceeding its temperature rating.

current-draw test: Starting-system test that determines *amperes* consumed by the starter motor during operation.

cursor: A position indicator on a video display to indicate data or a command to be corrected, repositioned, or entered.

curtain area: An engineering term that relates to the efficiency of the flow of air and fuel entering the *combustion chamber.*

curved washer: A spring-type washer that exerts a relatively light thrust load and is used to absorb axial-end play.

custom: A car that has been restyled for a distinctive appearance.

customary measurements: United States customary measurements; the English measuring system of inches, feet, yards, miles, and so on.

customize: To restyle a vehicle.

custom system: 1. A proprietary system built to exact specifications. 2. An early model, deluxe, automotive air-conditioning system that used both inside and outside air.

cut: 1. To defeat or eliminate a competitor in a drag race. 2. To sever. 3. The direction and texture of the cutting teeth of a file.

cut a big one: To record a particularly high speed or fast time. Also referred to as *cut a fat one.*

cut a fat one: Same as *cut a big one.*

cutout: 1. To experience a momentary engine miss without a stall. 2. A fuse holder that may be used to isolate part of a circuit. 3. To bypass the exhaust system.

cutout relay: An electrical, protective ON and OFF switch between the generator and battery.

cutting attachment: A device for converting an oxyfuel-gas welding torch into an oxygen cutting torch.

cutting brake: A special type of master cylinder with two brake levers to control how much brake pressure is applied to either of the rear wheels, allowing an off-road vehicle to make a much sharper turn.

cutting fluid: Any fluid applied to a cutter or to work being cut, to aid in the cutting operation by cooling and lubricating.

cutting tip: The part of a cutting torch from which the gases are emitted.

CVCC: An abbreviation for *Compound Vortex Combustion Control.*

CV joint: A shortened version of *constant velocity joint.*

CVT: An abbreviation for *continuously variable transmission.*

cw: An abbreviation for *clockwise.*

CWR: An abbreviation for *cargo weight ratio.*

C_x: Symbol for *coefficient of drag.*

cycle: 1. The process of discharging and then recharging a battery. 2. A series of repeated events such as the intake, compression, power, and exhaust strokes of an engine. 3. A complete series of events. 4. Short for motorcycle.

cycle clutch time: The time from the moment the *clutch* engages until it disengages, then re-engages. Total time is equal to ON time, plus OFF time for one cycle.

cycle fenders: Individual fenders for each wheel of a car.

cycling clutch: An air-conditioning, electromagnetic clutch that is turned on and off to control cabin temperature.

cycling-clutch fixed orifice tube *(CCFOT)*: An air-conditioning system having a fixed-orifice tube in which the air temperature is controlled by starting and stopping the compressor with a thermostat or pressure control.

cycling-clutch orifice tube: A term often used for cycling-clutch, fixed-orifice tube.

cycling-clutch system: An air-conditioning system in which the cabin air temperature is controlled by starting and stopping the compressor with a thermostat or pressure control.

cylgastos: A gasket made of treated asbestos layers bonded to a metal plate; now obsolete due to personal and environmental hazards associated with *asbestos.*

cylinder: **1.** A storage tank for gases, such as refrigerant. **2.** The round hole(s) inside an engine block that provide space for the reciprocating piston(s).

cylinder arrangement: The way cylinders are placed in an engine, such as in-line, in a row; vee, in two banks or rows at an angle to each other; or flat, pancake.

cylinder block: The basic framework of an engine to which all other parts and assemblies are installed or attached.

cylinder bore: The diameter of a cylinder.

cylinder combustion pressure: The pressure in the cylinder from expanding gases immediately after the *air/fuel mixture* is ignited; about four times greater than compression pressure.

cylinder deglazing: The process of removing the glaze from cylinder walls after extended use.

cylinder head: That part of the engine that covers the cylinders and pistons.

cylinder-head gasket: The gasket used to seal the head to the block to promote compression and to ensure a leak-free bond.

cylinder liner: A replaceable cylinder wall.

cylinder numbering: The order in which the cylinders are numbered: cylinder one may be on either front side of a V engine and start with one at the front of in-line engines.

cylinder sleeve: A round, replaceable, cylindrical tube that fits into the cylinder bore.

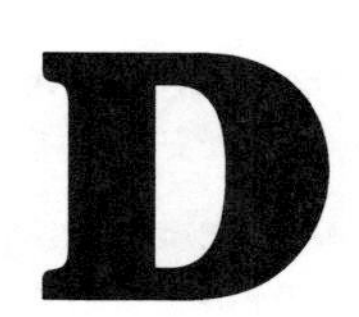

D: An abbreviation for *drive;* one of the forward gear positions in an automatic transmission.

DAC: An abbreviation for *digital to analog converter.*

dago axle: A dropped-beam axle used on older Fords to *dump* or lower the front end.

dam: The sealing provisions located between the radiator and condenser to ensure adequate ambient and ram air through both components.

damage appraiser: A term often used for *estimator.*

dampened pressure switch: An electrical pressure switch that opens the compressor clutch circuit when the low-side pressure is too low.

dampener:
A term often used incorrectly for *damper.*

dampening ball joint: A term used for *non-load-bearing ball joint.*

dampening effect: The effect shocks have on reducing suspension-spring oscillations.

damper: **1.** Friction or hydraulic shock absorbers or the equivalent electronic device. **2.** A device that reduces or eliminates vibration or *oscillation.* **3.** A balancer on the front of the *crankshaft* to reduce or eliminate *harmonic vibration.*

damper assembly: A device designed to decrease vibrations to the passenger compartment.

dark current: A small leakage current that flows through a photosensitive device in the absence of light.

dash: **1.** A short term for *dashboard.* **2.** A short race, usually about six laps.

dashboard: The section immediately behind the windshield that houses the instruments, accessory controls, and glove box.

dash components: Accessories, such as the air conditioner and/or heater, that are mounted on the firewall in the engine compartment.

dash control valve: Hand-operated valves located on the dash, such as parking-brake valves, tractor-protection valves, and differential locks.

dashpot: A device found on the *carburetor* that prevents a fast-closing throttle action.

Dash series: A *NASCAR* race series for four-cylinder, subcompact cars.

data link: A device in a computer system that sends and receives digital signals.

data processor: A device that is capable of performing data operations, such as a microcomputer.

datum point: The starting point for measuring an object.

daylight: To be ahead of a competitor by more than a car length in side-by-side racing, so daylight is visible between the two vehicles.

Daytona: The *Daytona International Speedway.*

Daytona International Speedway: A speedway in Daytona Beach, Florida. Home of the Daytona 500, Firecracker 400, and IMSA's 24-hour sport-car endurance race.

dB: A symbol for *decibel.*

DBC: An abbreviation for *dual-bed catalytic.*

DC: An abbreviation for *direct current.*

D-cam: The pattern used for grinding pistons in a cam shape with a 0.012 inch (0.3 mm) difference between the thrust face and the pinhole side.

dead axle: A shaft, such as the rear axle on a front-wheel drive vehicle, that connects wheels to either side of a vehicle but does not provide driving power.

A B C **D** E F G H I J K L M N O P Q R S T U V W X Y Z

dead player: A vehicle part or assembly that is inoperable.

dead spot: The momentary loss of power in an engine while increasing its speed.

dealership: A privately owned sales and service facility representing an automobile manufacturer.

deburr: To remove sharp edges from a part.

decarbonize: To remove carbon deposits. Also known as *decarburize.*

decarburize: Same as *decarbonize.*

decelerate: To decrease speed.

deceleration: **1.** A decrease in velocity or speed. **2.** To allow the vehicle to coast to idle speed from a high speed.

deceleration valve: A device used with the dual-diaphragm, vacuum-advance to advance engine timing under deceleration conditions.

decel valve: A device that reduces exhaust emissions during vehicle deceleration by keeping *rpm* up and vacuum down.

dechrome: **1.** To strip the vehicle of chrome for modification. **2.** To remove the chrome prior to body repair and/or painting.

decibel *(dB)*: A standard of measure for the relative loudness of a sound.

deck: **1.** The flat, mating surfaces of an engine block and head. **2.** The trunk of a passenger car. **3.** To machine an engine block deck flat. **4.** To dechrome the *deck lid.*

deck lid: The trunk door, or cover.

deck plate: A heavy metal plate that is bolted to an engine block during cylinder-honing operations.

declutching fan: An engine-cooling fan system, mounted on the water pump, having a temperature-sensitive device that governs or limits terminal speed.

decoke: A British term for *decarbonize.*

de Dion axle: A drive system with a differential or transaxle attached to the frame and exposed, universal-jointed half-shafts driving the wheels with a separate dead axle connecting the wheels that holds them upright and supports the springs.

deep cycling: A condition where a battery is completely discharged before it is recharged.

deep-dish wheel: A wheel with an extremely positive offset.

deep-rolled fillets: The transition radius between a journal and the cheeks of a *crankshaft.*

deep staging: A drag-racing technique where the driver uses minimum *roll out* to avoid *breakout.*

deep sump: An oil pan, used on some drag-race cars, that is deep enough to keep the oil away from the crankshaft at high *rpm* to slightly increase horsepower.

deflection: An axial or radial movement away from the normal or standard axis of a part.

deflection angle: The angle at which oil is deflected inside a torque converter during operation; the greater the angle, the greater the torque applied to the output shaft.

deflection rate: The number of pounds required to compress a spring one inch (25.4 mm).

defogger: That part of the heater system designed to clear the windshield of fog haze under certain conditions.

defrost door: A small door within the duct system that diverts a portion of the delivery air to the windshield.

defroster: That part of the heater system designed to clear heavy frost or light ice from the inside or outside of the windshield.

defroster door: A term often used for *defrost door.*

defrost switch: A thermostatic-type switch that senses evaporator temperature and turns the *compressor* off to prevent frosting or ice formation on the evaporator.

deg: An abbreviation for *degree.*

degreaser: A chemical that breaks oil and grease down.

degreasing: Cleaning parts in a solvent to remove oil and grease.

degree: Part of a circle; one degree is 1/360th of a circle.

degreeing a cam: Using instruments to determine the actual timing of valves opening and closing for a particular camshaft.

dehumidify: To remove water vapor from the air.

dehydrate: A process for *dehumidifying.*

dehydrator: **1.** A machine used to *dehydrate* a system or component. **2.** A term often used for *drier.*

dehydrator filter: A term used for *filter drier.*

de-ice switch: A switch used to control compressor operation to prevent evaporator freeze-up.

Delco Eye: A type of built-in battery tester that provides an immediate visual indication of battery conditions.

delta connection: The interconnection of three electrical-equipment windings, such as an alternator, in a triangular fashion.

DEMA: An abbreviation for *Diesel Engine Manufacturers Association.*

demagnetization: The process of removing magnetism from a magnetized substance.

demagnetize: To reduce or remove the magnetism from an object.

denatured alcohol: Ethyl alcohol, used to clean brake systems; contains methanol, rendering it unfit for human consumption.

Department of Environmental Regulation *(DER):* A department of the United States *Environmental Protection Agency* (EPA).

Department of Transportation *(DOT):* The United States Department of Transportation, a federal agency charged with the regulation and control of the shipment of all hazardous materials.

departure angle: The maximum angle, in degrees, of a line running rearward and upward from the rear tire contact point to the lowest obstruction under the rear of the vehicle.

depolarize: To remove or eliminate positive and negative poles from an item.

deposited metal: Any filler metal that may have been added during a welding process.

depressed park: The out-of-sight positioning, below the hood line, of windshield wiper blades of some wiper systems.

depressurize: To release or remove pressure.

depth of fusion: The penetration that a weld fusion extends into the base metal, or previous bead, from the surface.

DER: An abbreviation for *Department of Environmental Regulations.*

desiccant: **1.** A material, such as silica gel, that absorbs moisture from a *gas* or liquid. **2.** A drying agent used in refrigeration systems to remove excess moisture.

design working pressure: The maximum pressure under which a specific system or component is designed to work safely and without failing.

desmodromic valves: A valve system in which positive cam action, not spring action, is used to open and close the intake and exhaust valves.

detent: **1.** A recess to hold the gear selector in the gear range selected. **2.** A pin, stud, or lever which initiates or halts an action at a determined time or interval.

detergent: A chemical, added to engine oil, that possesses the ability to clean by preventing the accumulation of deposits.

detergent dispersant: A chemical component in motor oil that loosens dirt and varnish in an engine.

detergent oil: An oil with a detergent additive.

detonation: A phenomenon of internal combustion where the compressed air/fuel charge explodes violently instead of burning smoothly, usually due to the creation of a second flame front in the combustion chamber, away from the spark plug.

detonation-detection sensor: A term often used for *detonation sensor.*

detonation sensor: A device, mounted on an *engine block, cylinder head,* or *intake manifold,* that generates and sends a small voltage signal to the ECU, to retard timing, when encountering the vibration frequency associated with detonation.

Detroit locker: A specific brand of locking rear-end differential.

deuce: **1.** A 1932 Ford. **2.** A two-barrel *carburetor.* **3.** A 1962 through 1967 Chevrolet.

deuce and a half: A truck having a nominal payload capacity of 2-1/2 tons.

deuce and a quarter: The Buick Electra 225, particularly the 225-inch-long 1959 model.

Deutsche Institut für Normung *(DIN)*: A German standards authority.

Dexron II: A petroleum-based automatic transmission fluid developed by General Motors.

DG: An abbreviation for *double-groove valve stem.*

dia: An abbreviation for *diameter.*

diagnosis: A standard procedure that is followed to locate and identify the cause of a malfunction.

diagnostician: A person who determines the cause of problems when given all the signs and symptoms.

diagonal brake system: A dual-brake system with separate hydraulic circuits connecting diagonal wheels together; right front to left rear and left front to right rear.

diagonal cross check: In preparation for oval-track racing, the measure of weight distribution between the right-front and left-rear wheels and between the left-front and right-rear wheels.

dial: An instrument with an analog-gauge indicator.

dial in: **1.** To set the *Christmas tree* for *drag racing* events, with the interval between starting times for vehicles with different indexes. **2.** To set up a car with the right combination for maximum performance for any particular racing condition.

dial under: A practice allowed under *NASCAR* rules in handicap eliminations for Stock and Super Stock classes where the breakout rule is in effect.

diameter: The cross-section measurement of a round or circular object.

diaphragm: **1.** A flexible membrane in a speaker or microphone where electrical signals are converted to sound vibrations and vice versa. **2.** The flexible membrane found in a temperature- or pressure-control device that seals in an inert fluid from the atmosphere while allowing a mechanical movement.

diaphragm clutch: A *clutch* having a shallow, cone-shaped spring disc to provide pressure to the plate.

diaphragm spring: **1.** A spring shaped like a disk with tapered fingers pointed inward. **2.** A spring shaped like a wavy disk.

diaphragm-spring clutch: A clutch in which a diaphragm spring applies pressure against the *friction disk.*

dice: A tight contest between two cars on an oval track or road course generally battling for a specific position.

dichlorodifluoromethane: The proper chemical name for *Refrigerant-12.*

dielectric: **1.** An insulator. **2.** A term referring to the insulation between the plates of a *capacitor.*

dielectric strength: The ability of an insulator to insulate; to resist carrying current.

die out: To stall or stop running.

diesel cycle: An engine operating cycle where the air is compressed and the fuel is injected at the end of the compression stroke, causing ignition.

diesel engine: A *compression-ignition engine.*

Diesel Engine Manufacturers Association *(DEMA):* A professional association of diesel-engine manufacturers.

dieseling: A condition in which a *carbureted engine* continues to run after the *ignition* is shut off.

diff: An abbreviation for *differential.*

differential: The section of the rear-axle assembly that provides three functions: it allows the wheels to revolve at different speeds during turns, provides the final gear reduction, and changes the angle of drive 90 degrees.

differential action: An operational situation where one driving wheel can rotate at a slower speed than the opposite driving wheel.

differential assembly: The mechanism that relates to front- or rear-driving axles that permits unequal travel distances and speeds between the vehicle's driving wheels.

differential case: The housing of the *differential* that contains the side and *pinion gears* and the *pinion shaft,* and also serves as a mounting place for the *ring gear.*

differential drive gear: A large, circular helical gear that is driven by the *transaxle pinion gear and shaft* and that drives the *differential assembly.*

differential housing: Also known as rear-axle housing. A *cast iron* assembly that houses the *differential* unit and the drive axles.

differential lock: A toggle or push-pull type air switch that locks together the rear axles of a tractor so they pull as one for off-the-road operation.

differential pinion gears: Small, beveled gears located on the *differential pinion shaft.*

differential pinion shaft: A short shaft locked to the *differential case.* This shaft supports the *differential pinion gears.*

differential pressure valve: A device that maintains a constant pressure to fuel injectors.

differential side gear: The gears that are internally splined to the *axle shafts,* which are driven by the *pinion gears.*

digger: A *dragster.*

digital: **1.** Relating to the technology of computers and data communications where all information is encoded as bits of 1's or 0's that represent ON or OFF states. **2.** A numerical readout.

digital computer: A device that manipulates digital data and performs arithmetic and logic operations on such data.

digital electronic fuel injection: An early Cadillac *electronic fuel-injection system.*

digital instrument: A display that indicates an activity, such as *rpm, mph, voltage,* or fuel supply, with an electronic readout; used in most Indy cars, though some drivers still prefer *analog* instruments.

digital read out *(DRO):* A method of electronically reading conditions as opposed to an analog readout.

digital speech: Recorded speech broken into tiny units of sound, each having characteristics such as pitch and loudness that can be represented by numbers, becoming the digital code for speech.

digital-to-analog converter: A device, generally electronic, used to convert discrete digital numbers to continuous analog signals.

dig out: To accelerate suddenly, as in the start of a race.

digs: The drag races.

diluent: A liquid that extends a solution but definitely acts to weaken the solvent power of the active solution.

dilution: To make thinner or weaker. Oil is diluted by the unintentional addition of fuel and water droplets.

A B C D E F G H I J K L M N O P Q R S T U V W X Y Z

dimmer switch: A switch that allows the driver to select either high- or low-beam operation of the *headlights,* and to switch between the two.

dimple: The process of turning a hole under or down to allow the installation of a fastener.

dimpling: The distortion of an *oil-pan rail* around the bolt hole.

DIN: An abbreviation for *Deutsches Institut für Normung.*

ding: A small dent.

dipper: A small, metal scoop at the bottom of a connecting rod to scoop up oil from the pan to lubricate the bearing.

dipper trough: A trough aligned under the connecting rod that is fitted with the *dipper.*

dip stick: A thin, steel strip of metal used to measure the fluid level or quantity in the *engine, transmission,* or *compressor.*

direct-acting shock absorber: A double-action design controlling up and down suspension travel; it is mounted directly between the frame of the vehicle and the control arm or axle.

direct-acting thermostatic air cleaner: A component of heated air-intake systems that uses a thermostatic bulb connected to a rod to operate a flapper valve in the air-cleaner snorkel.

direct battery power: Power available to a circuit in the automobile directly from the battery, without an intermediate switch.

direct-bonded bearing: A bearing that is formed by pouring molten babbitt directly into the bearing housing and machining the cooled metal to the desired diameter.

direct-clutch solenoid: A solenoid that directs fluid flow to engage the direct clutch.

direct current: The current produced by a *battery.*

direct drive: A transmission mode in which the *driveshaft* and *engine crankshaft* are at the same speed.

direct ignition system: An ignition system where impulses are sent directly to the spark plugs by a modulator.

direct injection: A type of *fuel-injection system* that injects fuel directly into the cylinders.

directional signals: Lights at either side of both the front and rear of the vehicle that flash to indicate an intended turn.

directional tire: A tire having a tread pattern that must be mounted facing forward.

dirt dobber: A circle-track race driver that prefers a dirt track over an asphalt track.

dirt tracking: Driving on an asphalt track as if it were a dirt track.

disable: **1.** To remove or inhibit normal capability. **2.** A command that prevents further operation of a peripheral device.

disassemble: To take apart.

disc: **1.** An abbreviation for discount. **2.** A variant of *disk.*

disc brake: A type of brake that provides a means of slowing or stopping a vehicle using hydraulic pressure to apply pads against a rotor.

disc-brake fluid: A special fluid having a high boiling point, about 500°F (260°C). If exposed to air, it will attract moisture and be unsafe for use.

disc-brake pad: An assembly consisting of friction material and its steel backing.

discharge: **1.** Releasing some, or all, *refrigerant* from a system by opening a valve to permit the refrigerant to escape slowly into a recovery system. **2.** To purge air from a sealed system.

discharge air: Conditioned air as it passes through the outlets and enters the passenger compartment.

discharge line: The fluid line that connects the *compressor outlet* to the *condenser inlet.*

discharge pressure: The high-side pressure of the *refrigerant* as it is being discharged from the compressor. Also known as *high-side pressure*.

discharge pressure switch: A term used for *compressor-discharge pressure switch*.

discharge side: That portion of the refrigeration system under high pressure, extending from the *compressor outlet* to the *metering-device inlet*.

discharge valve: A term used for high-side service valve.

disc minimum thickness: The least thickness, usually stamped or cast into the disc, to which a brake disc can be machined or worn before it becomes unsafe.

disconnect: A switching device for disconnecting an electrical circuit or load from the power supply.

discrete components: An electrical component, such as a capacitor, that has been fabricated prior to its installation.

discrete device: An individual electrical component, such as a resistor.

dished piston: A piston having a depression in the crown.

disk: **1.** A magnetic device for storing information and programs accessible by a computer. **2.** The rotor, a revolving piece of metal, against which shoes are applied to provide a braking action. **3.** A two-dimensional figure defined by all points enclosed by a circle's diameter. **4.** Often referred to as *disc*.

disk drive: A device that stores, reads, or retrieves data from a magnetic disk and copies it into the computer's memory for use.

disk-operating system: An operating system in which the programs are stored on magnetic disks that keep track of, save, and retrieve files; allocate storage space and manage other control functions.

disk runout: The amount that a *brake disk* wobbles during rotation.

disperesant: A chemical added to oil that prevents impurities from clinging together and forming lumps that could clog the lubrication system galleys.

displacement: The volume within an engine's cylinders, usually expressed in cubic centimeters (cc).

displacement current: An expression for the effective current flow across a *capacitor*.

display terminal: An output device, such as a *CRT*, that produces a visual representation of graphic data.

disposable cylinder: A container for the one-time use of packaging, transporting, and dispensing of a fluid, such as *refrigerant*. It is a violation of Federal Law to refill these cylinders, commonly referred to as *"DOT-39s."*

disqualified: **1.** To render unfit. **2.** To be deprived of a right or privilege.

dist: An abbreviation for *distributor*.

distillation endpoint: The temperature at which a fuel is completely vaporized.

distortion: **1.** Inaccuracies in size and shape, as in an out-of-round cylinder. **2.** An unwanted change in purity of sound, as caused by a weak signal. **3.** A term used for *tread distortion*.

distributor: **1.** A jobber; one who buys from a manufacturer and sells to a wholesaler. **2.** A device used to direct electrical current to spark plugs. **3.** A device used to direct fuel to injectors.

distributor advance: A term used for *centrifugal advance, ignition advance,* or *vacuum advance*.

distributor cam: The four-, six-, or eight-sided lobe at the top end of the distributor shaft that rotates to cause the contact points to open and close.

distributor cap: A cover for the conventional ignition-system distributor, having a central terminal that receives secondary voltage from the coil and four, six, or eight peripheral terminals to send this voltage to the spark plugs.

A B C D E F G H I J K L M N O P Q R S T U V W X Y Z

distributor housing: A metal part that contains or provides a mounting for *distributor* components in a conventional ignition system.

distributor ignition: An ignition system that relies on a conventional distributor for proper operation.

distributorless ignition: A term used for *distributorless ignition system.*

distributorless ignition system: An ignition system that relies on a computer to distribute the electrical spark to the proper spark plug.

distributor pipe: A pipe to convey fuel from the fuel *distributor* to the *injector.*

distributor plate: The plate inside a *distributor* that is fastened to the housing and does not move.

distributor rail: A term used for *distributor pipe.*

distributor timing: A term used for *ignition timing.*

distributor vacuum advance: A term used for *vacuum advance.*

distributor vacuum-advance control valve: A term often used for *deceleration valve.*

div: An abbreviation for *division.*

dive: The tendency of the front of a vehicle to press down on the front springs during heavy braking.

divergent-convergent nozzle end: An exhaust-pipe nozzle that is expanded then reduced in size at the end.

divergent nozzle end: An exhaust-pipe nozzle end that expands from the pipe inlet to the end.

diverter valve: A vacuum-operated valve in an *air-injection system* that directs air-pump output to the atmosphere during high-vacuum deceleration to eliminate backfiring.

division: **1.** A branch or department of an organization. **2.** The mathematic operation to determine how many of one number is contained in another; for example, how many 10s are in 100.

DIY: An abbreviation for *do-it-yourself.*

DIYer: An abbreviation for a *do-it-yourselfer.*

dizzy: Slang for *distributor.*

D-Jetronic: An early *fuel-injection system* by Bosch.

DNF: An abbreviation for "did not finish."

dog: **1.** A poor performing vehicle. **2.** To follow another car very closely in oval-track racing. **3.** A pin or stub used to mate and/or drive a gear or assembly.

dog clutch: **1.** A simple splined clutch that cannot be slipped; generally found in dirt-track cars without transmissions. **2.** The mating collars, flanges, or lugs that can be moved as desired to engage or disengage similar collars, flanges, or lugs in order to transmit rotary motion.

dog house: **1.** The housing over an engine or transmission. **2.** The front fenders, *grille,* and hood assembly of a vehicle.

dog leg: A sharp, angular turn.

dog tracking: Off-center tracking of rear wheels as related to the front wheels.

DOHC: An abbreviation for *dual overhead camshaft.*

do-it-yourself *(DIY):* To repair one's vehicle.

do-it-yourselfer *(DIYer):* One known to repair their own vehicles.

dolly: To shape or form metal.

domed piston: A piston having a raised crown.

DOM tubing: An abbreviation for *drawn-over mandrel tubing,* a type of seamless tubing with precise and consistent inside and outside dimensions, used for race-car *chassis* construction.

D-1: The low forward-drive range of an *automatic transmission.*

donuts: 1. A 360-degree tire burnout. 2. Tires, especially big racing slicks.

dooley: A Chevrolet pick-up truck with dual rear wheels; generally a term given to any dual-wheel vehicle.

doorslammer: A full-bodied drag car with functioning doors.

doosy: A variation of *doozie.*

doozie: 1. Short for Duesenberg, considered the greatest racing car ever built in the United States. 2. Anything, such as a car, that is truly fine or outstanding.

doped solder: Solder containing a small amount of an added element to ensure retention of one or more characteristics of the materials on which it is used.

DOT: An abbreviation for *Department of Transportation.*

DOT 39: Trade jargon for a disposable refrigerant cylinder. Disposable cylinders should never be refilled or used as compressed air tanks.

double A-arm: A suspension system using two A-arms or A-frames to connect the *chassis* to the wheel spindle.

double-acting shock absorber: A shock that provides a dampening effect on both compression and rebound.

double-A frame: A term used for *double-A arm.*

double Cardian joint: A near-constant-velocity universal joint consisting of two *Cardian joints* connected by a coupling yoke.

double century: Two-hundred miles per hour.

double clutch: A driving technique to minimize gear clash when shifting gears with a manual transmission.

double filament lamp: A lamp designed for more than one function; used in the stoplight, tail light, and the turn-signal circuits combined.

double flare: A flare on the end of tubing, made by folding it over to form a double face.

double-groove valve stem: A valve stem having two *keeper* grooves.

double-hump heads: A term used for *camel-hump heads.*

double-J rim: The double-J shaped safety locks on a rim, used to hold the inner and outer tire beads securely.

double pivot control arm: A term used for *control arm.*

double pole double throw *(DPDT)*: A term used to identify a switch configuration having six terminals that connects one pair to either of the other two pairs of terminals.

double pole single throw *(DPST)*: A term used to identify a switch configuration having four terminals that connect or disconnect two pairs of terminals simultaneously.

double-reduction differential: A differential that contains extra gears to provide additional gear reduction.

double ton: Two-hundred miles per hour.

double wishbone: A term used for *double A-arm.*

double wrap: A flexible type of brake band designed with two segments that provide flexibility, which increases self-energizing action.

dowel: A pin inserted in an object or part to aid in the alignment of another object or part.

dowel pin: A round, solid, or hollow pin used to align two or more parts.

downdraft carburetor: A *carburetor* having a downward airflow.

downflow radiator: A *radiator* in which the coolant flow is from the top tank to the bottom tank.

downforce: The downward force of air on a speeding vehicle.

downshift: The automatic shift from a high gear ratio to a low gear ratio.

downstream blower: A blower arranged in the duct system so as to pull air through the heater and/or air-conditioner core(s).

downtime: **1.** The time a system or machine is not available for use, due to failure or routine maintenance. **2.** Time lost due to lack of service orders.

downward spring-and-wheel action: A term used for *rebound travel.*

DPDT: An abbreviation for *double pole double throw.*

D-port: An intake or exhaust port shaped like the letter D.

DPST: An abbreviation for *double pole single throw.*

DQ: An abbreviation for *disqualified.*

draft: To follow another vehicle close enough to take advantage of the slipstream or partial vacuum created behind it at very high speeds.

drafting: The technique of following another vehicle closely to save fuel.

draft tube: A vent to release engine-crankcase vapors.

drag: **1.** A performance term for a quarter-mile race from a standing start, against time or another car. **2.** An acceleration contest between two vehicles. **3.** The resistance of air against an object trying to pass through it.

drag coefficient: A term used for *coefficient of drag.*

drag link: The steering component that connects the Pitman arm to the *steering arm.*

dragster: A vehicle used for *drag racing.*

drag racing: An acceleration contest between two vehicles.

drag strip: A standard racing strip 60 feet (18.3 meters) wide by 4,000 feet (1,219 meters) long, where eighth- or quarter-mile-competition events occur.

drawbar: That part of a *converter dolly* that allows the *fifth wheel* to be repositioned on the *frame.*

drawbar capacity: The maximum, horizontal pulling force that can be safely applied to a coupling device.

drawn-over mandrel tubing: A type of seamless tubing with an accurate and precise inside and outside dimension; used in the construction of a race-car *chassis.*

dressing: A term used for *belt dressing.*

drier: **1.** A tubular device containing desiccant, usually placed in the liquid line to absorb moisture in an air-conditioning system. **2.** A *catalyst* added to paint to speed the time required for curing.

drift: A controlled slide through a turn involving all four wheels.

drift pin: A round, tapered, steel pin used to bring plates that are to be bolted or riveted into alignment to permit placing bolts or rivets easily.

drilled oil passage: Holes drilled in various parts of the engine for pressurized oil to flow through.

D-ring: A control in a drag-racing car's cockpit used to deploy the chute at the end of the run.

drip pan: A shallow pan under the evaporator core to catch the condensation that will be ejected by a drain hose fastened to the drip pan and extending to the outside.

driveability: The general operation of a vehicle, rated from good to poor, based on the characteristics of the average driver.

drive axle: An axle or axle shaft that transmits power to the drive wheels.

drive belt: Flexible belt or belts used to connect a *drive pulley* on the *crankshaft* to the coolant pump and accessories. Two basic types of drive belts are the serpentine or multiple-ribbed belt and the V-belt.

drive by wire: To use an electronic throttle control rather than mechanical linkage.

drive coil: A hollow field coil in a positive engaged starter that is used to attract the moveable pole shoe.

drive fit: A term used for *interface fit.*

drive gear: In a combination of two operating gears, the first is the *drive gear* and the other is the *driven gear.*

drive line: Assembly of various parts such as the *driveshaft, universal joints,* and *connecting yokes* that transmit *torque* from the transmission to the *differential.*

drive-line torque: The transfer of *torque* between the *transmission* and the driving axle assembly.

driveline windup: A reaction that takes place as a result of the transfer of engine *torque* through the *rear-wheel driveline.*

driven disk: The *friction disk* of a *clutch.*

driven gear: The gear that receives the driving action from the drive gear.

driven pinion: A rotating shaft in the *differential* that transmits *torque* to another *gear.*

drive pinion gear: One of the two main driving gears to multiply engine *torque* located within the *transaxle* or rear-driving axle housing.

driven plate: The friction plate in a *clutch* assembly.

driver: **1.** The operator of a vehicle. **2.** A car for everyday use.

driver-reaction distance: The distance traveled between the point at which the driver perceives a need to brake and the actual start of brake application.

drive shaft: An assembly of one or two universal joints connected to a shaft or tube used to transmit power from the *transmission* to the *differential.*

drivetrain: All of the components required to deliver engine power to the road surface.

driving axle: An axle capable of transmitting power by way of a *differential* or other *transmission* arrangement. Also a term used for drive axle.

driving lamps: Auxiliary lights to provide additional illumination for high-speed driving.

driving lights: A term used for *driving lamps.*

DRO: An abbreviation for *digital read out.*

drop: To lower a structural part of a vehicle.

drop-center rim: A wheel that has a smaller diameter center section, to facilitate tire replacement, designed to prevent the tubeless tire from rolling off the rim when the tire blows out.

drop-center safety rim: A wheel rim with a low center area and raised *flanges* designed to prevent the tire from accidentally coming off the rim.

drop-center wheel: A conventional wheel that has a space in the center for one bead to fit into while the other bead is being lifted over the rim flange.

drop head coupe: A British term for a two-door convertible.

drop the hammer: **1.** To engage the *clutch* and depress the *accelerator* suddenly. **2.** To suddenly depress the *accelerator* to the *floorboard.*

drop-throttle oversteer: A term used for *lift-throttle oversteer.*

dropped axle: A lowered beam axle used on older Fords to dump the front end.

drop through: The undesirable sagging or surface irregularity encountered when brazing or welding; caused by overheating with rapid diffusion between the filler metal and the base metal.

druid: A term used by a dissatisfied contestant, who has problems with the rules, when referring to a race official.

drum: The part of a brake that rotates with the axle hub and that the brake shoes press against to slow or stop the vehicle.

drum armature: A generator or motor armature having its coils wound longitudinally or parallel to its axis.

drum brake: A type of brake in which stopping friction is created by shoes pressing against the inside of the rotating drum.

drum-brake fade: The loss of braking efficiency due to excessive heat.

drum maximum diameter: The largest diameter, generally 0.060 inch (1.5 mm) over original, to which a brake drum can be machined or worn before it is considered unsafe.

dry: **1.** The process of changing from a liquid to a solid, as in paint. **2.** The process of evaporation, as in water.

dry-charged battery: A battery that is filled with electrolyte only when put into service.

dry deck: A condition in which the piston crown is at the level of the block deck.

dry-disk clutch: A *clutch* in which the friction faces of the friction disk are dry.

dry-film lubricant: A petroleum-based chemical used to lubricate operating parts during assembly, prior to engine start up, to provide lubrication until oil is circulated after start up.

dry friction: The friction between two dry solids.

dry fuel: A fuel with an additive to prevent water from collecting and freezing in the fuel system.

dry gas: Same as *dry fuel.*

drying agent: A term used for *desiccant.*

D-2: The high, forward-drive range for an *automatic transmission.*

dual-area diaphragm: An *automatic transmission shift control* that receives its vacuum signal from the *intake manifold* and the *EGR* port.

dual-bed catalytic: A term used for *dual-bed catalytic converter.*

dual-bed catalytic converter: A *catalytic converter* that passes different gases through an upper or front chamber coated with *platinum* and rhodium and a lower or back chamber coated with *platinum* and palladium.

dual-bed converter: A term used for *dual-bed catalytic converter.*

dual-brake system: A brake system that uses two hydraulic circuits; should one fail, the other remains operational.

dual camshaft: A type of engine that has two camshafts for opening and closing additional valves.

dual-diaphragm advance: A vacuum-advance mechanism with two diaphragms; one for normal ignition timing and the other to retard the spark during idling and part-throttle operation.

dual-diaphragm distributor: A distributor incorporating a dual-diaphragm advance.

dual exhaust: An exhaust system used on V-type engines with a separate muffler, exhaust, and tail pipe for each cylinder bank.

dual-fuel engine: **1.** An engine designed to run on two separate fuels, such as gasoline or propane. **2.** An engine designed to run on two fuels simultaneously, such as gasoline, alcohol, or a combination of the two.

dual-idler gear drive: A timing-gear system having two idler gears for the crankshaft to drive the camshaft.

dual-ignition system: An ignition system having two coils and two distributors, with two spark plugs for each cylinder.

duallie: A Dodge pick-up truck with dual wheels; popular reference for any dual-wheel vehicle.

dual master cylinder: A master cylinder, associated with a dual-brake system, having two sections for displacing fluid under pressure; generally one section for front brakes and one section for rear brakes or, in some applications, one for left brakes and the other for right brakes.

dual overhead camshaft: Two cams in a single-cylinder head.

dual-plane crankshaft: A *crankshaft* with two throws in two planes at right angles.

dual-plane manifold: An *intake manifold* with two air cavities to provide *air/fuel mixture* to the *cylinders.*

dual-point system: **1.** An ignition system with two sets of points in the distributor. **2.** A system that controls spark timing by electro-mechanical selection of separate advance and retard points.

dual quads: Two four-barrel *carburetors.*

duals: A term generally referring to two exhaust systems; two of anything, wheels, for example.

ducktail: An upswept rear end, often with a spoiler.

duct: A tube or passage used to provide a means to transfer air or liquid from one point or place to another.

ductility: The property of a material that permits it to be worked by drawing or stretching without rupture.

duct tape: A heavy duty, fiber-backed tape used for emergency repairs to body parts, in all forms of racing.

duesie: **1.** Anything truly fine or outstanding. **2.** Short for Duesenberg, touted by many to be the finest race car ever built. Also, *doozie.*

dummy: A mannequin, such as used in crash tests.

dummy load: An electrical test procedure to simulate actual operating conditions.

dummy shaft: A shaft used as an aid in the assembly or disassembly of a parts group.

dump: **1.** To lower the front end. **2.** To defeat a competitor in a drag race. **3.** To damage or destroy a component, such as a transmission.

dumped and tubbed: A vehicle that has been lowered in the front and fitted with oversize wheel wells.

dump station: A facility where RV's and others clean their holding tanks.

dump tubes: Straight through exhaust headers.

dune buggy: A small, light-weight, off-road vehicle with little or no body work.

duo-servo: A *drum brake* design with increased stopping power due to the servo or self-energizing effect of the brake.

durability: The quality of being useful, generally referring to the life of a catalyst or emission-control system.

Dura Spark ignition: An *ignition system* developed by Ford.

duration: The time, in crankshaft degrees, that a camshaft holds an *exhaust* or *intake valve* open.

dust: To overtake or defeat a competitor in a race.

dust off: A term often used for *dust.*

dust shield: **1.** The upper portion of a shock absorber that surrounds the lower twin-tube unit. **2.** A covering that protects an assembly from the elements.

dutched: A paint job finished with elaborate striping or flame painting.

duty cycle: The percentage of time that a power source, or one of its accessories, can be operated at rated output, without overheating.

duty solenoid: A solenoid on a feedback *carburetor* that cycles many times per second to control a metering rod, and therefore, the *air/fuel mixture.*

dwell: The degree of distributor-shaft rotation while the ignition breaker points are closed.

dwell angle: A term used for *dwell.*

dye-penetrant testing: A non-destructive, inexpensive method of testing metal for cracks.

Dyer drive: A type of starter-motor drive used in heavy-duty applications that provides mechanical meshing means and automatic demeshing.

dykum blue: A blue dye used to color metal.

Dykes ring: A compression piston ring having an "L" cross-sectional shape that provides sealing against the cylinder walls.

dynamically balanced: **1.** A term that often refers to a wheel and tire being balanced while spinning. **2.** In general terms, the balance of any object when it is in motion.

dynamo: An electric *DC-current generator.*

dynamometer: A machine on which a vehicle may be driven, simulating actual driving conditions for emissions and diagnostic purposes.

dyno: Short for *dynamometer.*

Dytel: A trade name for a red dye found in some *CFC* refrigerants.

Dzus fastener: A screw-like fastener that may be removed or installed with a quarter turn, ideal for race-car body panels that may have to be replaced quickly.

ea: An abbreviation for each.

early apex: Getting to the inside of a turn sooner than usual.

early fuel evaporation: A system that evaporates fuel in the *intake manifold* with the use of heat from the exhaust, or an electrical element to improve the engine's cold-running characteristics and to reduce emissions.

EAS: An abbreviation for *electronic air suspension.*

easy exit: A system that provides for easy entrance and exit of the vehicle by moving the seat all the way back and down. Some systems may also move the steering wheel up and to full retract.

EBCM: **1.** An abbreviation for *electronic brake control module.* **2.** A component of a General Motors electronically controlled brake system used on some of their car lines.

ECA: An abbreviation for *electronic control assembly.*

E-cam: A pattern used to grind pistons in an oval or cam shape.

ECC: An abbreviation for *electronic climate control.*

eccentric: An offset, such as a disc, used to convert rotary motion into reciprocation motion, as in a cam lobe.

eccentric bearing: A bearing that is thicker at the crown by a few ten thousandths of an inch.

eccentricity: A condition whereby two or more round parts or holes do not share the same central axis.

eccentric shaft: **1.** The *crankshaft* of a Wankle engine. **2.** A shaft with eccentric offsets, such as a *camshaft.*

ECCS: An abbreviation for: **1.** *Electronic computer control system.* **2.** Electronic concentrated control system. **3.** Electronic constant control system.

ECI: An abbreviation for *electronic controlled ignition.*

ECM: An abbreviation for *electronic control module.*

econobox: A small compact car.

economizer: A device in a *carburetor* that reduces fuel consumption, particularly under heavy throttle.

economy of motion: Minimum effort required to do a task.

econorail: A *dragster* with a single, normally aspirated gasoline engine.

ECS: An abbreviation for: **1.** *Electronic control system.* **2.** *Evaporative control system.*

ECT: An abbreviation for: **1.** *Electronic controlled transmission.* **2.** *Engine coolant temperature.*

ECTS: An abbreviation for *engine coolant temperature sensor.*

ECU: An abbreviation for *electronic control unit.*

eddy current: Currents induced in conducting materials by varying magnetic fields; considered undesirable because they represent loss of energy and cause overheating.

edge loading: A condition that exists when two parts rub together at their edges.

EDIS: An abbreviation for *electronic distributorless ignition system.*

edit: **1.** To check the correctness of data. **2.** To change, as necessary, the form of data by adding or deleting certain characters.

A B C D E F G H I J K L M N O P Q R S T U V W X Y Z

editor: A program permitting the user to create new data files or to alter existing data files.

EEC: An abbreviation for: **1.** *Evaporative emission control.* **2.** *Electronic engine control.*

EEC-I: An early Ford electronic engine control that only controlled ignition advance and retard.

EEC-II: An early Ford electronic engine control system that used a crank-mounted ignition sensor and electronic engine control unit.

EEC-III: An early 1980s Ford electronic microprocessor engine-control system.

EEC-IV: A Ford electronic engine-control system using sequential port fuel injectors.

EECS: An abbreviation for *evaporative emissions control system.*

E85: A fuel blend of 85% ethanol and 15% gasoline.

EFC: An abbreviation for **1.** *Electronic fuel control.* **2.** *Electronic feedback carburetor.*

EFI: An abbreviation for *electronic fuel injection.*

efficiency: The output of energy divided by the input of energy, expressed in percentage.

egg crate: A cross-hatch grille design.

EGR: An abbreviation for *exhaust gas recirculation.*

EGR system: An *EGR* valve, mounted on the *intake manifold,* that meters a small amount of *exhaust gas* into the *intake manifold* to dilute the *air/fuel mixture.* This keeps combustion temperatures below 2,500°F and reduces the formation of NO_X. The amount of exhaust gas recirculated into the engine is only a few percent.

EGR valve: An abbreviation for *exhaust gas recirculation valve.*

EGR valve-position sensor: A *potentiometer* that keeps the engine control computer informed relative to the *EGR* valve position.

EHMI: An abbreviation for the *Environmental Hazards Management Institute.*

EHS: An abbreviation for *extremely hazardous substance.*

EIN: An abbreviation for *engine identification number.*

EL: An abbreviation for *exposure limit.*

elapsed time *(ET)*: The time it takes a vehicle to cover a given distance, usually from a standing start, recorded to the thousandths of a second.

elastic limit: The maximum stress a metal can withstand without exhibiting a permanent deformation on release of the stress.

elastomeric seal: A seal made of rubber or a similar material.

ELC: An abbreviation for *electronic level control.*

eldo: Slang for Cadillac Eldorado.

elec: An abbreviation for electric.

electric assist choke: A choke containing a small electric heating element to warm the choke spring.

electric brakes: A brake system having an electro-magnet and armature at each wheel to provide the braking action.

electric car: A vehicle having an electric motor as the power source.

electric current: The movement of electrons through a conductor.

electric defrosting: Use of electric-resistance heating coils to melt ice and frost off: **1.** Evaporators. **2.** Rear windows.

electric-drive cooling fan: **1.** An engine-cooling fan driven by an electric motor. **2.** An electrically controlled fan that cycles ON and OFF with the air conditioner control, if predetermined system and/or ambient temperatures are exceeded.

electric engine-cooling fan: A 12-volt, motor-driven fan that is electrically controlled by either, or both, of two methods: an engine-coolant temperature switch (thermostat) and/or the air-conditioner select switch.

electric fuel pump: A device having either a reciprocating diaphragm or a revolving impeller operated by electricity to draw fuel from the tank to the fuel delivery system.

electric system: Any of the systems and sub systems that make up the automobile wiring harnesses, such as the lighting system or starting and charging system.

electric vehicle: A vehicle having electric motors like a power source driven by on-board rechargeable batteries.

electric welding: A term often used for *arc welding*.

electrochromic mirror: A mirror that automatically adjusts the amount of reflectance based on the intensity of glare.

electrocution: Death caused by electrical current through the heart, usually in excess of 50 ma.

electrode: **1.** A component of the electrical circuit that terminates at a gap across which current must arc. **2.** A rod used in welding.

electrohydraulic pressure actuator: A valve that will provide a continuous adjustment of fuel pressure in certain fuel-injection systems.

electrolyte: A substance in which the conduction of electricity is accompanied by chemical action.

electrolytic cell: A simple battery consisting of the container, two electrodes, and the *electrolyte*.

electromagnet: A device consisting of a ferrous-metal core and a coil that produces appreciable magnetic effects when an electric current exists in the coil.

electromotive force *(emf)*: Voltage.

electron: An element of matter that surrounds the nucleus and helps determine the chemical properties of an atom.

electronically controlled transmission *(ECT)*: A *transmission* that is electronically linked to the vehicle's electronic control system.

electronic air suspension *(EAS)*: A suspension system having provisions to adjust to road and/or load conditions to ensure a comfortable ride. May also include *automatic level control*.

electronic brake control module *(EBCM)*: A system having a monitor at each wheel to sense conditions and feed an electrical impulse into an *onboard computer* to reduce sideways skidding during rapid braking action.

electronic climate control *(ECC)*: A system used to regulate the temperature and humidity of a vehicle's cabin.

electronic computer control system *(ECCS)*: A term used for *electronic control assembly (ECA)*.

electronic control assembly *(ECA)*: A device that receives signals, processes them, makes decisions, and gives commands. More commonly referred to as a *computer*.

electronic controlled ignition *(ECI)*: A term that often applies to an *electronic ignition (EI)* system.

electronic controlled transmission *(ECT)*: A *transmission* that has electronic sensors to monitor *throttle* position, engine speed, *torque converter turbine* speed, and other drive-train operations that effect shifting, leading to fuel economy.

electronic control module *(ECM)*: An electronic device used to control some engine functions.

electronic control system *(ECS)*: An electronic device used to control certain electrical and engine operations.

electronic control unit *(ECU)*: A digital computer that controls engine and *transmission* functions based on data that it receives from sensors, relative to *engine* rpm and temperature, air temperature, *intake-manifold vacuum,* and *throttle* position.

electronic cycling-clutch switch: An electronic switch that prevents the evaporator from freezing by signaling various electronic control devices when the evaporator reaches a predetermined low temperature.

electronic engine control *(EEC)*: A system that regulates an engine's electrical functions.

electronic distributorless ignition system *(EDIS)*: An obsolete term for an ignition system that relies on a *computer* to time and route the electrical spark to the proper *spark plug* at the proper interval. Now known as *electronic ignition (EI)* system.

electronic feedback carburetor *(EFC)*: A *carburetor* that controls the *air/fuel mixture* according to commands from the engine control computer.

electronic fuel control *(EFC)*: A fuel system that uses electronic devices to monitor *engine* functions to ensure that the proper air/fuel ratio reaches the *combustion chamber* for optimum *engine* performance under any operating condition.

electronic fuel-injection system: A *fuel-injection* system that injects gasoline into a spark-ignition engine that includes an electronic control to time and meter the fuel flow.

electronic ignition *(EI)*: An ignition system where a solid state device has replaced mechanical breaker points.

electronic ignition system: An ignition system controlled by the use of small electrical signals and various semiconductor devices and circuits.

electronic leak detector: An electrically (AC or DC) powered leak detector that emits an audible and/or visual signal when its sensor is passed over a refrigerant leak.

electronic level control *(ELC)*: A device that automatically regulates the ride height of a vehicle under various load conditions.

electronic mail *(e-mail)*: The process of sending, receiving, storing, and forwarding messages in digital form over telecommunication facilities.

electron optics: Electronics that apply the behavior of moving electrons under the influence of electrostatic or electromagnetic forces to devices or equipment.

electronics: The branch of science pertaining to the study, control, and application of currents of free electrons, including the motion, emission, and behavior of the currents.

electronic spark control *(ESC)*: A system that controls and governs the vacuum signal to the distributor to assure proper distributor retard-advance under various engine-load conditions.

electronic spark timing *(EST)*: An electronic system that, based on input signals, provides the correct spark timing for a given *engine* condition.

electroplating: A process for depositing metal on a conductive surface that is made by the cathode in an electrolytic bath containing dissolved salts of the metal being deposited.

electrostatic shield: A shield that protects a device or circuit from electrostatic energy, but not necessarily from electromagnetic energy.

element: A substance that cannot be further divided; the smallest of matter.

elephant foot: A valve-adjusting screw having a ball that swivels when it contacts the valve stem.

elephant motor: The Chrysler Hemi 425 cid *engine.*

eliminations: A series of matches between two cars at a time, the winner advancing to the next race.

eliminator: The fastest drag car in its class.

Elky: The Chevrolet El Camino.

Elliot axle: A solid-bar front axle on which the ends span the steering knuckle.

elliptical spring: A term used for *leaf spring.*

elongation: The stretching of a material.

e-mail: A popular term for *electronic mail.*

embedability: The ability of a bearing to permit small dirt particles to sink into the bearing surface, thereby avoiding certain *crankshaft* scratches.

embossed gasket: A shim-type head gasket having a raised surface.

embrittlement: A condition whereby a part becomes hard and brittle due to excessive flexing.

emergency brake system: A brake system used for stopping a vehicle in the event of a malfunction in the means of operation and control of the service brake system.

emery paper: An abrasive-coated paper used for fine finishing.

emf: An abbreviation for *electromotive force.*

emission controls: The components that are directly or indirectly responsible for reducing air pollution, including *crankcase emissions, evaporative emissions,* and *tailpipe emissions.*

emissions: Unwanted, harmful chemicals and chemical compounds that are released into the atmosphere from a vehicle, especially from the *tailpipe, crankcase,* and *fuel tank* including unburned *hydrocarbons, carbon monoxide, oxides of nitrogen, particulates,* and sulfur.

emission standards: The federal-government-established emission and pollutant standards for all motor vehicles because of environmental damage in the past.

emissions warranty: A federally mandated 5-year/50,000-mile performance and defect warranty that covers all emissions-related components on all new vehicles built since 1981. In 1995 the emissions warranty was reduced to 2 years/24,000 miles on all emissions-related components except the *computer* and *catalytic converter,* which are extended to 8 years/80,000 miles.

emitter: The semiconductor material in a transistor that emits carriers into the base region when the emitter base junction is forward biased.

emulsifiable chemical: Any chemical that will mix with water.

emulsion: A fluid substance containing one liquid disbursed and suspended in another, rather than dissolved.

emulsion tube: A passage in a *carburetor* where air is mixed with fuel.

emissions control: A term used for *emission controls.*

enable: To switch a *computer* device or facility so it can operate.

enamel: A free-flowing clear or pigmented varnish, treated oil, or other form of organic coating that usually dries to a hard, glossy, or semiglossy finish.

enameled wire: Wire insulated with a thin, baked-on varnish enamel, used in coils to allow the maximum number of turns in a given space.

encoder: An electromagnetic transducer used to produce digital data (code) indicating angular or linear position.

end clearance: The distance a shaft moves longitudinally as the *crankshaft* in an engine. Also known as end play.

end gap: The distance between the ends of a *piston ring.*

end gas: The last part of an *air/fuel mixture* to burn during *combustion.*

endo: To flip a vehicle end over end.

end play: A term used for *end clearance.*

endurance limit: The maximum stress that a metal can withstand without failure, after a specified number of cycles of stress.

enduro: A race of 6 to 24 hours with emphasis on endurance and reliability rather than speed.

eng: An abbreviation for *engine.*

engage: The mechanical or automatic coupling of two members, like the driving *flywheel* and *pressure plate,* to rotate and drive the driven disc of a *clutch.*

engagement chatter: A shaking, shuddering action that takes place as the driven disc

A B C D **E** F G H I J K L M N O P Q R S T U V W X Y Z

makes contact with the driving members caused by a rapid grip and slip action.

engine: A device that burns fuel to produce mechanical power; to convert heat energy into mechanical energy.

engine bay: The area in a vehicle occupied by the *engine.*

engine block: A term used for *cylinder block.*

engine configuration: Relating to the style and type of engine, such as *V-8, pancake,* and so on.

engine-coolant temperature *(ECT):* The temperature of the *coolant* in an *engine.*

engine-coolant temperature sensor *(ECTS):* An electronic or electro-mechanical unit for sensing engine-coolant temperature.

engine cooling fan: A term used for cooling fan.

engine cooling system: A term used for *cooling system.*

engine displacement: The volume swept within an *engine* as its cylinders make one stroke, generally expressed in cubic inches, cubic centimeters, or liters.

engine dynamometer: Device to measure *engine* horsepower at the *flywheel.*

engine dyno: A term used for *engine dynamometer.*

engineering plastics: Thermoset and thermoplastic materials, such as polycarbonate, ABS, and nylon. Their characteristics and properties enable them to withstand mechanical loads, such as tension, impact, flexing, vibration, and friction, combined with temperature changes, making them suitable for application in structural and load-bearing product design elements.

engine fan: A bladed device found at the front of the *engine* used to draw air through the *radiator* and across the *engine.*

engine identification number *(EIN):* A code to identify the type of engine found stamped on the cylinder block.

engine idle compensator: A thermostatically controlled device on the *carburetor* that prevents stalling during prolonged hot weather periods, while the air conditioner is operated.

engine-management system: An electronic device that monitors, adjusts, and regulates the *ignition* and *fuel-injection systems* to maintain *engine* control under varying operating conditions.

engine overhaul: To perform more than minor repairs on a powerplant, though not as extensive as an *engine rebuild.*

engine paint: A paint specially formulated to withstand *engine* heat.

engine rebuild: To perform an extensive *engine* repair including machining, reboring, and honing to factory-stock specifications.

engine thermal switch: An electrical switch designed to delay the operation of a system in cool weather to allow time for the *engine coolant* to warm up.

engine thermostat: A temperature-sensitive mechanical device found at the *coolant* outlet of an *engine* that expands (opens) or contracts (closes) to control the amount of coolant allowed to leave the engine, based on its temperature.

engine torque: A rotating, twisting action developed by the *engine* that is measured in pounds, feet, or kilowatt hours.

engine tuneup: The inspection, testing, and adjusting of an engine; the replacement of any parts required to ensure maximum performance.

Englishtown *(E-Town):* A town in New Jersey; the site of the annual *NHRA* Summernationals.

Environmental Hazards Management Institute *(EHMI):* An educational organization that provides materials to teenagers and adults relative to responsible management and recycling of hazardous waste materials.

Environmental Protection Agency *(EPA):* An agency of the United States government charged with the responsibility of protecting the environment and enforcing the *Clean Air Act* of 1990.

EP: An abbreviation for *extreme pressure.*

EPA: An abbreviation for *Environmental Protection Agency.*

EP lube: A term used for extreme-pressure lubricant.

epoxy: A plastic compound that is often used to repair cracks and small holes in metal.

epoxy cement: A durable thermosetting adhesive.

EPR: An abbreviation for *evaporator pressure regulator.*

E-PROM: An abbreviation for *erasable programmable read-only memory.*

equalizer: **1.** A device in the brake-cable system that prevents one side from being applied before the other. **2.** The tube that balances pressure between two exhaust pipes.

equalizer line: A small-bore line used to provide a balance of pressure from one point to another, as in a thermostatic expansion valve.

equal length headers: An *exhaust system* having equal length tubes from each cylinder to the collector.

Equipment and Tools Institute *(ETI):* An organization of equipment and tool manufacturers, jobbers, and wholesalers.

erasable programmable read-only memory *(E-PROM):* A special *PROM* that can be erased under high-intensity ultraviolet light, then reprogrammed.

ergonomic hazards: Conditions that relate to one's physical body or to motion.

ergonomics: The science of adapting the workplace to the worker, as in the cockpit of a race car to the driver.

ESC: An abbreviation for *electronic spark control.*

esses: A series of s-shaped bends in a race course.

EST: An abbreviation for *electronic spark timing.*

estimator: The person who determines the cost value of the damage, plus the price required to repair the damaged vehicle, working with insurance adjusters or appraisers.

ESV: An abbreviation for *experimental safety vehicle.*

ET: An abbreviation for *elapsed time.*

etching: The localized attack of metal surfaces causing pitting and/or deterioration.

ethane: C_2H_6, a minor component of natural gas used as a fuel.

ethanol: A form of alcohol, C_2H_5OH, found in alcoholic beverages and also used as an additive in gasoline to produce gasohol.

Ethyl: The trade name for tetraethyl lead, a fuel additive no longer permitted due to environmental hazards.

ethyl alcohol: A term often used for *ethanol.*

ethylene glycol: The basic chemical used in automotive antifreeze; mixed with water for cooling-system protection and to increase the boiling point of the *coolant.*

ETI: An abbreviation for *Equipment and Tools Institute.*

E-Town: A popular term for *Englishtown.*

Eurostyle: A custom styling treatment to provide a European flair.

evacuate: **1.** To create a vacuum within a system to remove all traces of air and moisture. **2.** A service procedure using a vacuum pump to remove all atmospheric air and moisture from inside an air-conditioning system.

evap: An abbreviation for *evaporator.*

evaporation: The transformation of a liquid to the vapor state.

evaporation control system *(ECS):* A system that prevents fuel vapors from escaping to the atmosphere while the *engine* is off.

evaporative emissions: *Hydrocarbons* from fuel that evaporate from a vehicle's *fuel tank* and *carburetor;* eliminated by sealing the *fuel system* and using a charcoal canister to trap vapors from the *fuel tank* and *carburetor.*

evaporative-emissions control *(EEC):* A canister filled with activated charcoal to reduce raw fuel emissions.

evaporative-emissions control system *(EECS):* A system to reduce the amount of raw fuel vapors that are emitted into the ambient air from the *fuel tank* and *carburetor.*

evaporator: The heat exchanger of an air-conditioning system that removes heat from the air passing through it.

evaporator control valve: Any of the several types of suction pressure-control valves or devices used to regulate the evaporator temperature by controlling its pressure.

evaporator core: The tube and fin assembly located inside the evaporator housing, where *refrigerant* fluid picks up heat when it changes into a vapor.

evaporator equalizer valves-in-receiver: A term used for *valves-in-receiver.*

evaporator fan: A fan used to move air across the *evaporator* and heater core.

evaporator housing: The case that contains the *evaporator core,* diverter doors, duct outlets, and blower mounting arrangement.

evaporator pressure-control valve: A term used for *evaporator control valve.*

evaporator pressure regulator *(EPR):* A back-pressure regulated temperature-control device used on some early Chrysler car lines.

evaporator temperature regulator: A temperature-regulated device used by early Chrysler Air-Temp systems to control the *evaporator* pressure.

even fire: A *V–6 engine* with a 60- or 90-degree block having individual journals on each *crankshaft* throw staggered in relation to each other to provide even firing.

exc: An abbreviation for *except.*

except: A word referring to an omitted category or item.

excess flash: A term used for *flash.*

excessive crush: A condition that exists when a pair of bearings are installed in a bore that is too small, causing them to deform and leading to early failure.

exhaust analyzer: An automotive test and service device that uses a process involving infrared energy to determine and display the composition of an engine's *exhaust gases* such as the two-gas type to measure *hydrocarbons* and carbon monoxide or the four-gas type, which also measures oxygen and carbon dioxide.

exhaust back pressure: The resistance of the free flow of gas in the *exhaust system.*

exhaust emissions: Pollutants identified by clean-air legislation as being harmful or undesirable, including lead, unburned *hydrocarbons, carbon monoxide,* and *oxides of nitrogen.*

exhaust gas: The burned and unburned *gases* that remain after *combustion.*

exhaust-gas oxygen: The amount of oxygen present in the exhaust stream, as measured by an oxygen sensor and reported to the computer in closed-loop, feedback systems to aid in the control of the *air/fuel mixture.*

exhaust-gas recirculation *(EGR):* An emissions-control system that reduces an engine's production of *oxides of nitrogen* by diluting the *air/fuel mixture* with *exhaust gas* so that peak *combustion* temperatures in the cylinders are lowered.

exhaust-gas recirculation valve *(EGR valve):* A valve, generally vacuum operated, to regulate the *exhaust gas* flow into the *intake manifold.*

exhaust-gas speed: The speed at which *exhaust gases* pass through the header pipes, usually at about 200–300 feet (61–91 meters) per second.

exhaust headers: **1.** A term used for *exhaust manifold.* **2.** A special *exhaust manifold* with tubes of equal length from the exhaust ports to the header of the *exhaust system.*

exhaust manifold: A component, generally of *cast iron,* with passages of unequal lengths that carry the *exhaust gases* from the *exhaust ports* to the header of the *exhaust system.*

exhaust pipe: A pipe that connects the *exhaust manifold,* the *muffler,* or *catalytic converter.*

exhaust port: In a *rotary combustion engine,* peripheral opening in the rotor housing that allows the burned gases to leave the *engine.*

exhaust stroke: The upward motion of the *piston* forcing burned *gases* out the open *exhaust valve.*

exhaust system: The tubing, *mufflers,* and *catalytic converters* that direct *exhaust gases* from the *engine* to the atmosphere.

exhaust valve: Valve that, upon opening, allows the burned *gases* to leave the *combustion chamber* during the *exhaust stroke.*

exothermic: A chemical reaction, such as *combustion,* that gives off heat.

expansion control: The grinding of a *piston* in a slightly oval shape so that it becomes round when it expands with heat.

expansion plug: A term used for *core plug.*

expansion tank: **1.** A coolant-recovery tank. **2.** An auxiliary tank, usually connected to the inlet tank or a *radiator,* that provides additional storage space for heated *coolant.*

expansion tube: A metering device used at the inlet of some *evaporators* to control the flow of liquid *refrigerant* into the *evaporator core.*

expansion valve: A term used for *thermostatic expansion valve.*

experimental safety vehicle *(ESV)*: A prototype vehicle used to test and evaluate specific safety features.

explosion-proof cabinet: Cabinet used in the automotive shop to store gasoline and other flammable liquids.

exposure limit *(EL)*: A term often used for *permissible exposure limit (PEL).*

extension housing: An aluminum or iron casting that encloses the transmission output shaft and supporting bearings.

external combustion engine: An engine that burns the *air-fuel mixture* in a chamber outside the engine cylinder, such as a *steam engine* or a *Stirling engine.*

external equalizer: A term used for *equalizer line.*

external in-line filter: **1.** *Filter* placed in the transmission-cooler return line outside the transmission housing. **2.** A supplemental filter placed in the air-conditioning system to prevent system damage after repairs.

externally-tabbed clutch plate: *Clutch plates* having tabs around the outside periphery to fit into grooves in a housing or *drum.*

extra capacity: **1.** A coupling device that has strength capability greater than standard. **2.** A fluid reserve system.

extremely hazardous substance *(EHS)*: Any substance or the by-product of any substance that is poisonous or contains *carcinogens* considered unsafe, in any concentration, for humans, animals, or plant life.

extreme pressure *(EP)*: A much higher-than-average pressure condition.

extreme pressure lubricant: An *API*-rated lubricant for heavy loads.

eyeball: To make a visual estimation or determination.

eyes: The light beam that operates an electronic, race-timing system.

eyewash fountain: An emergency water fountain that directs water to the eyes for flushing and cleaning.

A B C D **E** F G H I J K L M N O P Q R S T U V W X Y Z

F: Abbreviation for *Fahrenheit.*

fabric cord: A cord-like material used for strength and reinforcement in belts and tires.

fab shop: A shop that makes or fabricates parts.

face angle: The angle of the valve face.

face feed: The application of a filler metal to a joint during brazing and soldering.

face shield: A clear-plastic shield that protects the entire face from outside elements.

factor: The mathematical evaluation used by *NHRA* to estimate true horsepower and weight, and assign production cars to specific stock classes.

factory experimental: A type of drag-racing car developed by Dodge/Plymouth and Ford/Mercury in the 1960s using their largest *engines* in light vehicles, a forerunner of the *pro-stock* class.

factory recall: A vehicle manufacturer's effort to correct a defect discovered after the vehicle has been delivered to the customer, often several years later.

factory specs: Factory specifications; the manufacturer's dimensions, clearances, and tolerances.

factory team: A racing team sponsored and supported by a vehicle manufacturer.

factory tool: A tool designed by a vehicle manufacturer for a specific application.

fad car: A *T-bucket* with a bobbed pickup bed.

fade: A term often used when referring to *brake fade.*

fadeaway fenders: A front-fender design that flowed back and blended into the front doors of the vehicle; a popular style of the 1940s.

fad T: A term used for *fad car.*

Fahrenheit *(F):* The English scale for the measure of temperature.

failure: The total cessation of function of a system or device, such as *coil failure.*

failure code: A term used for *trouble code.*

false air: Air in a *fuel-injection system* in excess of that required for *combustion.*

false guide: A valve guide used to replace a worn integral-valve guide.

family buggy: **1.** The family car. **2.** A vehicle more than ten years old.

fan: **1.** A device having two or more blades attached to the shaft of a motor. **2.** A device mounted in the heater/air-conditioner duct that causes air to pass over the heater core and *evaporator.* **3.** A device having four or more blades, mounted on the water pump, that causes air to pass through the *radiator* and *condenser.* **4.** The spray pattern of a paint spray gun. **5.** The incorrect technique of applying paint by waving a spray gun back and forth with the wrist.

fan belt: A flexible V-, or flat poly-groove-type drive belt that transfers power from the *crankshaft pulley* to the *water pump* and/or accessories, such as the *alternator.*

fan blade: In an engine cooling-system fan, four to six wings on the fan, usually spaced unevenly to reduce vibration and noise.

fan clutch: A device installed between the water-pump pulley and fan of an engine-driven fan that is sensitive to *engine* speed and underhood temperature.

fan hub: The mounting surface for the *fan.*

fanning: The use of pressurized air through a spray gun to facilitate drying, a practice not recommended.

fanning the brakes: A term used for *brake fanning.*

fan shroud: Plastic or metal housing inside which the fan rotates; on certain vehicles, this allows the *fan* to pull more air past *radiator* finned tubing and prevents air recirculation.

FAQT: An abbreviation for *Federation of Automotive Qualified Technicians.*

fastback: An autobody style having a roof line that extends in a single, simple curve from the windshield to the rear bumper.

fast flush: The use of a special machine to clean the cooling system by circulating a cleaning solution.

fast idle: The higher speed, 1,100 to 1,500 rpm, at which an engine idles during warm-up, when first started.

fast-idle cam: A cam-shaped lever on the *carburetor* that provides fast-idle action when the *engine* is cold.

fast-idle screw: A screw in the *carburetor* linkage to adjust fast-idle speed.

fast-idle solenoid: An electro-mechanical device on the *carburetor* for adjustment of the fast-idle speed.

fast overdrive: A planetary gear set operating with the planetary carrier as input, the *sun gear* as output, and the *ring gear* held.

fast reverse: A planetary gear set operating with the planetary carrier held; the *ring gear* is input, and the *sun gear* rotates in the opposite direction.

fat fenders: Bulbous fenders, as on vehicles in the 1930s and 1940s.

fatigue: The tendency of a material to break under conditions of repeated stressing considerably below its tensile strength.

fatigue failure: Metal failure due to repeated stress so that the character of the metal is altered and it cracks. This is a condition that frequently causes *engine* bearing failure due to excessive *engine* idling or slow engine-idling speed.

fatigue strength: The measure of a material's resistance to fatigue.

faying surface: The mating surface of a member that is in contact or in close proximity with another member to which it is to be joined.

Fe: The chemical symbol for *iron.*

featheredge: The technique of blending the repair of a damaged area into the undamaged area, maintaining the original surface texture and sheen.

feathering: The technique of modulating the throttle lightly and smoothly for precise, controlled changes in engine speed.

featherweight leaf spring: A fiber composite spring.

feature car: A vehicle displayed at a car show for appearance money and not trophy competition.

federal version: A car that meets the United States' exhaust emission standards, but not California's standards.

Federation Internationale de l'Automobile *(FIA):* An international association of national automobile clubs that sanction and regulate major international auto racing series, such as *Formula One.*

Federation Internationale du Sport Automobile *(FISA):* A division of *FIA* that sanctions and regulates major international auto racing series.

Federation of Automotive Qualified Technicians *(FAQT):* A professional association that provides life, health, and disability insurance.

feedback: **1.** A principle of fuel-system design wherein a signal from an oxygen sensor in the *exhaust system* is used to give a computer the input it needs to properly regulate the *carburetor* or *fuel-injection system* in order to maintain a nearly perfect *air/fuel ratio.* **2.** A signal to a *computer* that reports on the position of a component, as an *EGR* valve.

feedback carburetor: A carburetor that controls the *air/fuel mixture* according to

A B C D E F G H I J K L M N O P Q R S T U V W X Y Z

commands from the engine control computer, typically through the operation of a duty solenoid.

feed holes: The holes to supply *coolant* or oil to an *engine.*

felt: Natural felts are produced by compressed wool, hair, wool/hair, or synthetic fibers, yielding a wide range of densities and permeabilities of consistent density, pore size, and mesh geometry so that performance is reasonably predictable.

felt dust seal: **1.** An *engine* seal made of felt, usually used on the front crankshaft pulley. **2.** A compressor-shaft seal made of felt, usually found between the seal face and armature of the *clutch.*

female: The universal designation of a part into which a mating (male) part fits.

fender cover: A protective cover placed on the fender when a mechanic works on an *engine,* preventing damage to the finish.

fenderside: A narrow-bed pickup truck.

Ferguson Formula *(FF):* A four-wheel drive system developed in the 1960s, a forerunner of the *AWD* and *4WD* systems in high-performance cars today.

ferrite: An iron compound that has not been combined with *carbon* in pig iron or steel.

ferrous metal: A metal containing iron, such as steel.

ferrous wheels: Vehicle wheels made of iron or steel alloy.

FF: An abbreviation for: **1.** Ferguson Formula. **2.** Formula Ford.

F-head: An *engine* design having the *intake valves* in the head and the *exhaust valves* in the block.

F-head engine: An *engine* with some of its valves in the head and some in the cylinder block, giving an F-shaped appearance.

FI: An abbreviation for *fuel injection.*

FIA: An abbreviation for *Federation Internationale de l'Automobile.*

fiber-composite spring: A term used for *fiber-composite leaf spring.*

fiber-composite leaf spring: Leaf springs made of fiberglass, laminated and bonded together by tough polyester resins; incredibly lightweight, they possess some unique ride-control characteristics.

Fiberglas: The trade name for Owens Corning *fiberglass-reinforced plastic.*

fiberglass: **1.** The generic spelling of *Fiberglas.* **2.** A product used to fabricate or mold durable lightweight parts and auto body panels.

fiberglass-reinforced plastic *(FRP):* A plastic structure or panel that is reinforced with *fiberglass.*

fiberglass spring: A term used for *fiber-composite spring.*

fiber optics: A data-transmission medium made of tiny threads of glass or plastic that transmit huge amounts of data via light waves, at the speed of light.

fiber timing gears: *Camshaft* timing gears made of fiber composition material that reduce gear noise.

field: **1.** Electrical force in the space around electrically charged particles. **2.** The lines of force in a natural or man-made magnet between its north and south poles.

field coil: The coil, or winding, around the field magnets or pole pieces of a motor or *generator.* Also known as field winding.

field density: The density of the magnetic field, measured in the number of lines of force per unit area, dependent upon the strength of the field element, the number of turns of wire, and the size and characteristics of the pole piece.

field-frame assembly: The round, soft iron frame of a *generator* or motor into which the field coils are assembled.

field relay: A relay that connects the alternator to the battery when alternator voltage is greater than battery voltage and disconnects it when battery voltage is greater than alternator voltage.

field terminal: The electrical terminal connecting the field coil to the voltage regulator.

field winding: A term often used for *field coil.*

15-year rule: A regulation by *NHRA* banning vehicles more than 15 years old from participating in the association's major drag events.

fifth wheel: A coupling device mounted on a truck, and used to connect a semi-trailer. It acts as a hinge point to allow changes in direction of travel between the tractor and semi-trailer.

fifth-wheel height: **1.** The distance from the ground to the top of the fifth wheel when it is level and parallel with the ground. **2.** The height from the tractor frame to the top of the fifth wheel, as applies to data given in fifth-wheel literature.

fifth-wheel hitch: A hitch having a slotted wheel-shaped plate on a tow vehicle into which the kingpin of a fifth-wheel trailer locks to connect the trailer to the vehicle.

fifth-wheel top plate: The portion of the fifth-wheel assembly that contacts the trailer bolster plate and houses the locking mechanism that connects to the kingpin.

file: **1.** A hand tool with fine teeth for removing small amounts of material. **2.** A collection of records treated as a basic unit of storage.

filled: The seams in a welded body panel that have been covered with lead, putty, or plastic.

filled axle: An I-beam axle with concave portions that are reinforced with metal *gussets.*

filler: A material used to fill dents and repair damaged body parts.

filler cap: Manually removable lid or seal on the filler neck of a fuel tank, radiator, or other reservoir.

filler metal: The metal or alloy added for bonding in making a welded, brazed, or soldered joint.

filler-neck restrictor: A restriction plate located in the inlet of the fuel tank to prevent leaded fuel from being put into the tanks of cars that require unleaded fuel.

fillet: The radius connecting the journal to the *crankshaft* cheek.

fillet weld: A nearly triangular cross-section weld joining two surfaces at approximately right angles to each other in a lap joint, T-joint, or corner joint.

filter: **1.** A system designed to remove solid particles. **2.** A device used with the drier, or as a separate unit, to remove foreign material from *refrigerant.* **3.** A device used to clean the air as it enters the engine.

filter bypass: A spring-loaded valve built into or next to most *oil filters* that allows oil to pass around the element if it becomes clogged.

filter drier: A device having a *filter* to remove foreign material and a desiccant to remove moisture from *refrigerant.*

filter plate: An optical material that protects the eyes against excessive ultraviolet, infrared, and visible radiation.

final drive: The *pinion, ring,* and *differential* gears that provide power to the drive wheels.

final drive ratio: The ratio between the *drive pinion* and *ring gear.*

fins: Thin metal strips in an *evaporator, condenser,* or *radiator,* found around the tubes to aid in *heat transfer.*

fire: Cause to burn; a flame.

fire bottle: A term used for *fire extinguisher.*

fire extinguisher: A device used to put out *fires* with the use of chemicals.

fire point: The lowest temperature at which a flammable liquid will flash ignite at its surface, and continue to burn.

fire suit: An aluminized fire-resistant driving suit.

fire (class A): A *fire* resulting from the burning of wood, paper, textiles, and clothing.

fire (class B): A *fire* resulting from the burning of gasoline, greases, oils, and other flammable liquids.

fire (class C): A *fire* resulting from the burning of electrical equipment, motors, and switches.

fire (class D): A *fire* resulting from the burning of combustible metals, such as *magnesium.*

fire wall: The partition between the engine and passenger compartment.

firing order: The order in which the cylinders deliver power strokes.

first-aid box: A kit made of various first-aid bandages, creams, and wraps for the emergency treating of minor injuries.

first law of motion: A body in motion tends to remain in motion and a body at rest tends to remain at rest.

FISA: An abbreviation for *Federation Internationale du Sport Automobile.*

fish: To pull wire or cable through a conduit, raceway, or other confined space.

fish eye: A crater-like opening in a newly painted surface, caused by water, oil, or a silicone-based material.

fish-eye mirror: A wide-angle mirror providing a broader view than a standard mirror.

fishtail: To lose traction with the rear wheels, allowing them to uncontrollably slither from side to side.

fit: The range of tightness or looseness that results from the application of a specific combination of tolerances in mating parts.

fitting: **1.** An accessory such as a locknut, bushing, or other part of a wiring system that is intended primarily to perform a mechanical, rather than an electrical function. **2.** A hose end that is designed to mate with a mechanical part.

five-point seat belt: A safety-belt system with two lap belts and two shoulder belts with a single buckle.

5-6-7: 1955, 1956, and 1957 Chevrolets.

five-window coupe: A coupe body having five windows; two in the doors, two in the rear quarter panels, and one in the rear.

fixed caliper: A disc-brake caliper that does not float or slide, rigidly mounted to the steering knuckle, spindle, or control arm, having one or two *pistons* on each side of the disc.

fixed-caliper disc brake: A brake system with the caliper attached to the mounting bracket and pads that adjust themselves to rotor position and thickness.

fixed-orifice tube: A *refrigerant* metering device, used at the inlet of *evaporators,* to control the flow of liquid refrigerant allowed to enter the evaporator.

fixed-orifice-tube, cycling-clutch system: An air-conditioning system having a fixed-orifice tube as a metering device and a thermostat-controlled cycling clutch as a means of temperature control.

fixed tappet: A solid valve lifter or cam follower.

fixed-type constant velocity joint: A joint, found on the outer ends of the drive shafts of *FWD* vehicles, that cannot telescope or plunge to compensate for suspension travel.

flame cutting: Using a welding torch fitted with a cutting head to cut metal.

flame front: The leading edge of an *air/fuel mixture* during *combustion.*

flame harden: To heat-treat metal to increase its surface hardness and wear resistance.

flame kernel: The initial shape of a freshly ignited *air/fuel mixture* during the first few milliseconds of *combustion.*

flame out: An *ignition* failure while a vehicle is in motion.

flame propagation: The expansion of the *air/fuel mixture* in the chamber as *combustion* is completed.

flame spraying: A thermal spraying process in which an oxyfuel-gas flame is the source of heat for melting the surfacing material.

flame suit: An aluminized, fire-resistant, driving suit.

flame travel: Distance across the *combustion chamber* that the flame of the ignited mixture travels.

flange: A projecting rim or edge of a part, usually narrow and of approximately constant width for stiffening or fastening.

flanged bearing: **1.** A bearing having a *flange* to affix its position in a bore or on a shaft. **2.** A main bearing having a *flange* to control the *end play* of a *crankshaft.*

flanged sleeve: A cylinder sleeve having a *flange* at the top to allow it to be set at a specific depth in the block.

flange gasket: A gasket that seals the mating surface of a flanged part and base surface.

flange nut: A fastener that incorporates a *flange* or washer thrust surface.

flange weld: A weld made on the edges of two or more joint members, at least one of which is flanged.

flank: The flat part of a *camshaft lobe.*

flapper valve: A ball valve that operates with a vacuum, pressure diaphragm, or motor.

flare: A cone-shaped *flange* end applied to a piece of tubing to provide a means of fastening to a fitting.

flared: Wheel wells emphasized with raised edges.

flash: **1.** Material that is expelled from a flash weld prior to the upset portion of the welding cycle. **2.** Excess material found along the parting edges of a cast or forged part. **3.** A condition that occurs when the first coat of paint appears to be dull, prior to final drying.

flashback: A recession of the flame back into the mixing chamber of the oxyfuel gas torch or flame-spraying gun.

flashback arrester: A device to limit damage from a flashback by preventing propagation of the flame front beyond the location of the arrester.

flash chrome: Thin chrome plating on certain engine parts to provide good wear characteristics.

flasher: An automatic-reset, circuit-breaker-type switch used in directional signal and emergency signal circuits.

flash gas: *Gas* resulting from the instantaneous *evaporation* of *refrigerant* in a pressure-reducing device, such as an *expansion valve.*

flash point: The lowest temperature at which a flammable liquid produces sufficient vapor to flash near its surface, but without continuing to burn.

flash to pass: A steering-column-mounted dimmer switch having an additional feature that illuminates the high-beam headlights even with the headlight switch in the OFF or PARK position.

flat four: An opposed four-cylinder engine.

flathead: An *engine* having intake and *exhaust* valves in the block.

flat motor: A Ford flathead *V–8.* Also known as a *flatty.*

flat out: Full throttle.

flat rate: A method of charging for services based on the time normally required to perform a particular service.

flat six: An opposed six-cylinder engine.

flat spot: The momentary loss of power as *engine* speed is increased.

flat tappet: A mechanical valve lifter.

flat tow: To tow a vehicle on all four wheels.

flat track: A dirt oval track without banked turns.

flatty: A Ford *flathead V–8.* Also known as a *flat motor.*

fleet service: Service given to a fleet of vehicles owned by a particular company.

flex drive plate: A disc-shaped, slightly flexible, steel part transferring power from the *crankshaft* to the *torque converter.*

A B C D E F G H I J K L M N O P Q R S T U V W X Y Z

flexible flyer: A dragster chassis with a light, flexible structure.

flexible hose: A fluid or vapor hose having the ability to be routed around various components without creating a restriction or blockage.

flexible radiator hose: A large-diameter universal hose connecting the radiator to the *engine cooling system.*

flex joint: A flexible connection such as between the steering column and steering gear.

flexplate: A lightweight *flywheel* with a starter *ring gear* around its outside diameter, used on *engines* equipped with an *automatic transmission;* also serves as the attachment point for the *torque converter.*

flip: To overturn a vehicle.

flip top: A convertible with a retractable hardtop.

float: **1.** Part that floats in the fuel bowl to assist in controlling the gasoline level in the *carburetor* by operating a float-needle valve. **2.** A cruising drive mode in which the throttle setting matches *engine* speed to road speed, neither accelerating nor decelerating.

float bowl: **1.** A section of the *carburetor* main body that acts as a fuel reservoir. **2.** The reservoir from which fuel is metered into the passing air.

float chamber: The fuel reservoir at the bottom of a *carburetor.*

float circuit: A circuit that maintains the correct fuel level in the *carburetor float bowl.*

floater: A term used for *full-floating axle.*

floating axle: A term used for *full-* or *semi-floating axle.*

floating caliper: A disc-brake caliper that has piston(s) on only one side of the disc. The *caliper bore* moves away from the rotor in order to press the pad on the other side against the disc when the brakes are applied.

floating-caliper disc brakes: A brake system in which only one of the two pads are energized and move the caliper so that it is caught between both pads.

floating drum: A brake *drum* that is not secured to a hub.

floating piston: A piston having a *floating piston pin.*

floating piston pin: A *piston pin* that rotates freely within the bore in the *connecting rod.*

float level: The float position when the needle valve is against its seat, cutting off the fuel supply.

float system: The system that controls fuel into the *carburetor* and the fuel level in the *float bowl.*

flog: To abuse a vehicle by pushing it too hard.

flood: A condition whereby more fuel is in the *combustion chamber* than can be ignited.

flooded: A condition whereby the *air/fuel mixture* in a cylinder is too rich to burn.

flooding: A condition caused by: **1.** Too much liquid *refrigerant* being metered into the *evaporator* for *evaporation.* **2.** Too much gasoline metered into an *engine* for *combustion.*

floorboard: The slanted section of the floor pan immediately behind the firewall.

floorboard it: A term used for *floored* or *floor it.*

floored: To run at full throttle. Also *floorboard it* and *floor it.*

floor it: To have the accelerator pushed to the floor. Also *floorboard it* and *floored.*

floor pan: The panel forming the floor of the interior of the vehicle.

flopper: A funny car with a fiberglass body that flops up in the front to provide access to the *engine, chassis,* and driver compartment.

floppy disk: A flexible 5.25- or 3.5-inch disk used widely with microcomputers and minicomputers, providing electronic media storage at a relatively low cost.

flowability: 1. The ability of molten filler metal to flow. 2. The ability of a fluid or vapor to flow.

flow coating: A method of applying paint by passing parts on a conveyor through a chamber where several nozzles direct a shower of coating material over the parts.

fluid: Any liquid or *gas.*

fluid aeration: Air bubbles formed in a fluid, giving the appearance of foam.

fluid cooler: 1. Small heat-exchanger component in a hydraulic line near the pump to reduce power-steering-fluid temperature. 2. A device inside the *radiator* to provide cooling for *transmission* fluid. 3. Any *heat exchanger* designed to reduce the temperature of a fluid.

fluid coupling: A device in the power train containing two rotating members, one of which transmits power to the other via fluid.

fluid pressure: The pressure of a fluid that is invariable and uniform in all directions.

fluorocarbon: Any of a group of *chlorofluorocarbon refrigerants,* such as *R-12.*

flush: 1. To use a fluid to remove solid particles such as metal flakes or dirt. 2. To purge *refrigerant* passages with a clean, dry *gas,* such as *nitrogen* (N).

flush bucket: An oversized *carburetor* that feeds more *air/fuel mixture* than the *engine* can handle.

flushing: A term used for various acts of cleaning a system, such as brake flushing.

flushing agent: An approved liquid or *gas* used to flush a system.

flushing hydraulic system: The procedure for replacing old brake fluid with new fluid.

flux: A material to dissolve and prevent the formation of oxides in molten metal and solid metal surfaces; a wetting agent that facilitates the bonding of a filler metal.

flyboy: A *drag* competitor who races only occasionally as a hobby.

flying kilometer: The international standard for attempting to set a speed record by entering the measured kilometer after attaining the highest speed possible.

flying mile: The international standard for attempting to set a speed record by entering the measured mile after attaining the highest speed possible.

flyweight governor: Governor assembly, sensitive to *centrifugal force,* whose action is controlled by primary and secondary weights.

flywheel: 1. A heavy metal wheel with starter *ring gear* that is mounted at the rear of the *crankshaft.* It absorbs energy on power stroke(s), returns energy on other stroke(s), and transfers power to the *clutch* or *torque converter.* 2. Front-most part of a clutch assembly that is bolted to the *engine crankshaft* with a rear surface to provide a smooth friction area for the disc-front facing to contact during clutch engagement.

flywheel ring gear: A gear, fitted around the flywheel, that is engaged by teeth on the starting-motor drive to crank the engine.

foaming: 1. A condition caused by the churning of oil or other fluids. 2. A term used for *shock foaming.*

fog lamp: Auxiliary lamps, often amber, mounted in front of a vehicle to aid visibility during snow dust or fog conditions.

fold: To bend a material, usually to 180 degrees.

follower: A term used for *lifter.*

following ball joint: A term used for *non-load-carrying ball joint.*

FoloThru drive: An inertia-starting motor drive, similar to a *Bendix Folo-Thru drive.*

foot *(ft)*: An English measure equal to 12 inches.

foot in it: A driver accelerating rapidly and/or refusing to yield during an attempted pass.

foot in the carburetor: The accelerator pushed to the floor.

foot-pounds *(ft-lb)*: An English measure for *torque.*

footprint: **1.** The portion of the contact area of a loaded tire with the ground. **2.** The bolt pattern of a device.

foot valve: The foot-operated brake valve that controls air pressure to the service chambers.

footwell: The area for passengers' feet in a vehicle.

force: Any push or pull exerted on an object.

forced induction: An intake system that provides a means for the *air/fuel mixture* to enter the *combustion chamber* at greater than atmospheric pressure.

force fit: A term used for *press fit.*

Fordor: An early designation for a Ford four-door vehicle.

forge: To form metal into a desired shape.

forging flash: Remnants of metal on a forged part, usually that which is squeezed out of a mold.

forked eight: A *V–8* engine.

forked six: A *V–6* engine.

fork lift: A power-driven truck or tractor that carries pallets or platform-loaded cargo on forks ahead of the machine.

fork truck: A term used for *fork lift.*

formed-in-place gasket: A gel-like material that forms a gasket when clamped between two surfaces.

forming: Making any change in the shape of a metal piece that does not intentionally reduce the metal thickness and produces a useful shape.

forming charge: The initial charge applied to a new battery before use.

formula car: A single-seal, open-wheeled race car built to a particular set of specifications.

Formula Ford *(FF)*: An entry level, highly competitive class of racing car with a stock 1,600 cc engine.

formula libre: A type of road racing open to all without any limitation or restrictions on *engine* size, bodywork, or other design features.

Formula One: A single-seat, open-wheeled car for international Grand Prix racing.

49-state car: A vehicle that meets federal vehicle-exhaust standards for all states except California.

forward clutch: A *clutch* that is engaged whenever the vehicle moves forward, controlled by the valve-body forward circuit.

forward rake: A car having a lower front than rear, to create an extra down force at high speeds.

forward sensor: A sensor used in *air-bag restraint systems.*

forward shoe: A term used for *leading shoe.*

fossil fuels: Fuels formed underground from animal and plant matter by chemical and physical change, such as coal, *petroleum,* and *natural gas.*

foul: To leave before the green light in *drag racing* and be disqualified.

four banger: **1.** A four-cylinder engine. **2.** A vehicle equipped with a four-cylinder engine.

four-barrel carburetor: A *carburetor* having four venturis.

four-bolt mains: A main bearing cap that is held in place with four bolts.

four by: A term used for *4×4.*

4×4: A four-wheeled vehicle with *four-wheel drive.*

4×2: A four-wheeled vehicle with *two-wheel drive.*

four cammer: A V-type engine with dual overhead *camshafts* on each cylinder bank.

four cycle: A term used for *four-stroke cycle.*

four-gas analyzer: An exhaust-gas analyzer able to detect and measure exact amounts of *hydrocarbons, carbon monoxide, carbon dioxide,* and *oxygen.*

four on the floor: A four-speed manual transmission with a floor-mounted shift lever.

four-point seat belt: A safety-belt system with two shoulder straps, two lap straps, and a single buckle.

four speed: A manual *transmission* with four forward gears.

four-speed transmission: A *transmission* providing four forward-gear ratios, one reverse ratio, and neutral, permitting closer matching of *engine* speed to load requirements than a three-speed transmission.

four-stroke cycle: **1.** A cycle of *engine* operation whereby the *combustion* occurs in every cylinder on every second revolution of the *crankshaft.* **2.** A complete cycle includes *intake* or *induction, compression, combustion* or expansion, and exhaust. **3.** Also known as *four cycle.*

four-valve head: A head design having four valves per cylinder.

four-wheel drift: A controlled four-wheel slide on a paved surface.

4WD: A term used for *four-wheel drive.*

four-wheel drive: A vehicle having a driving axle in front and rear, so all four wheels are driven. Also *4WD.*

four wheeling: The off-highway travel in a four-wheel vehicle.

4WS: A term used for *four-wheel steering.*

four-wheel steering *(4WS)*: A system whereby all four wheels are used to steer a vehicle.

fractionation: A condition when one or more *refrigerants* of the same blend leak at a faster rate than other refrigerant in that same blend.

frame: The substructure of a vehicle supported by the suspension system that supports the bodywork, *engine,* and power train.

frame alignment: The measurement of a vehicle frame to ensure that it is within a manufacturer's specifications.

franchised dealer: A dealer that has a contract with the vehicle manufacturer to sell and service its vehicles.

Frantz oil filter: The tradename of a popular filter that uses a roll of toilet tissue as its filtration medium.

Freedom of Information Act: A federal law giving everyone the right to certain information.

free-flow exhaust: An *exhaust system* with reduced back pressure.

free-standing deck: An *engine* block construction in which the cylinders are cast in place without being tied by the *deck* to the outer walls.

freewheeling: A mechanical device in which a driving member imparts motion to a driven member in one direction, but not the other.

freeze crack: A crack in the *engine* caused by expansion due to *coolant* freezing.

freeze plug: A term used for *core plug.*

freeze protection: **1.** The controlling of *evaporator* temperature so that moisture on its surface does not freeze and block the airflow. **2.** An additive added to *coolant* to prevent freeze-up by lowering its freezing temperature.

freeze-up: **1.** Failure of a unit to operate properly due to the formation of ice at the metering device. **2.** A term used for *seize.*

frenched: A body part, normally separate, having been molded together with another body part.

french seam: A fabric seam in which the edge of the material is tucked under and sewn on the inner side.

Freon: Registered trademark of E.I. duPont, for a group of its *refrigerants.*

Freon 12: The trade term for *refrigerant-12* by E.I. duPont.

A B C D E F G H I J K L M N O P Q R S T U V W X Y Z

frequency: The number of complete cycles an alternating electric current, sound wave, or vibrating or rotating object undergoes in a given time.

frequency valve: A valve that is used to stabilize the *air/fuel mixture* on a fuel-injected engine.

friction: The resistance to motion of two items in contact with each other.

friction ball joint: A term used for *non-load-carrying ball joint.*

friction bearing: A bearing in which there is a sliding contact between the moving surfaces, such as a connecting rod bearing.

friction disk: A flat disc surfaced with a friction material on one or both sides, such as a clutch disk.

friction facing: A hard-molded or woven material that is riveted or bonded to the clutch-driven disc.

friction horsepower: Engine horsepower losses due to friction from such sources as the engine, transmission, and drive train.

friction material: One of several types of material used for friction surfacing, such as *metallic friction material* or *organic friction material.*

friction-modified fluid: Automatic transmission fluid that provides smooth automatic shifts; designed to slip.

frontal area: The area, in square feet, of the vehicle's cross section, as viewed from the front.

front and rear suspension systems: The suspension system, with the frame, supplies steering control under all road conditions and maintains proper tracking and directional stability as well as providing proper wheel alignment to minimize tire wear.

front-axle limiting valve: A valve that reduces pressure to the front service chambers, thus eliminating front wheel lockup on wet or icy pavements.

front-body structural components: In a perimeter frame design, the front body section is made up of the radiator support, front fender, and front fender apron. These are installed with bolts and form an easily disassembled structure.

front chute: The front straightaway on a circle race track.

front clip: The complete replacement of the front bodywork back to the cowl or A-pillar.

front control arm: Horizontal arms that connect the front wheels to the car and that support the weight of the front of the car.

front drive: A drive system that transmits power through the front wheels.

front-end drive: A vehicle having its drive wheels located on the front axle.

front-end geometry: The angular relationships involving the front suspension, steering system, and tires.

front engined: A vehicle with the engine in front, ahead of the passenger compartment.

front idler: A pulley used as a means of tightening the drive belt.

front-of-dash components: The heating and air-conditioning components that are mounted on the firewall side in the engine compartment.

front pump: A pump, located at the front of the transmission, driven by the engine through two dogs on the torque-converter housing, to supply fluid whenever the engine is running.

front-roll center: The center, determined by the front suspension geometry, around which the forward part of a vehicle tends to roll.

front seating: Closing off the line, leaving the air-conditioner compressor open to the service port fitting, allowing service to the compressor without purging the entire system.

front steer: A steering gear mounted ahead of the front wheel centerline.

front straight: A straight area of a race track, such as between turns four and one at the *Indianapolis Speedway.*

front suspension: To support the weight of the front of the vehicle.

front-suspension system: Components that provide support of the vehicle front section, allow wheels to move vertically, and provide adjustments for front wheel alignment. The common parts include upper and lower ball joints; control arms, shaft bushings and shims; sway bar and bushings; strut rod and bushings; coil springs; stabilizers; shock absorbers; and steering knuckle and spindle.

front to rear brake bias: The difference in balance of brake pressure between the front and rear brake cylinders or calipers; higher in front due to weight transfer during heavy breaking.

front-wheel drive: A drive system that transmits power through the front wheels.

fronty: In early racing, the Chevrolet Frontenac and the Fronty Ford Model T.

frosting back: The appearance of frost on the air-conditioning suction line extending back as far as the compressor.

FRP: An abbreviation for *fiberglass-reinforced plastic.*

fruit cupper: An amateur driver.

fry: Overheat.

ft: Abbreviation for *foot.*

ft-lb: Abbreviation for *foot-pound.*

fudge factor: An allowance made for extra time and/or material when working on a vehicle.

fuel: **1.** Any combustible liquid, such as gasoline, that can be used to fuel an engine. **2.** As slang, any fuel other than gasoline that is used to fuel an engine.

fuel block: A *manifold* used to distribute fuel to multiple *carburetors.*

fuel burned: The type of fuel used; usually gasoline, diesel fuel oil, or liquefied petroleum gas.

fuel cell: A special fuel tank designed for a race car.

fuel consumption: The amount of fuel that is consumed or used by the vehicle. Also known as *gasoline consumption.*

fuel decel valve: A device that supplies additional air/fuel to the *intake manifold* during *deceleration* to help control *hydrocarbon emissions.*

fuel distributor: A mechanical or electro-mechanical device used to route fuel to the injectors.

fueler: A race car running on fuel other than gasoline.

fuel filter: A device located in the fuel line to remove impurities from the fuel before it enters the carburetor or injector system.

fuel-fouled plug: A term used for *carbon-fouled plugs.*

fuel gas: A gas, such as acetylene, natural gas, or hydrogen, normally used with oxygen in an oxyfuel process, and for heating.

fuel gauge: A gauge that indicates the amount of fuel remaining in the tank.

fuel hose: A term used for *fuel line.*

fuelie: **1.** An early production car with fuel injection rather than carburetion. **2.** A term for the fuel categories in drag racing, such as *Top Fuel* and *Funny Car.*

fuel injection *(FI)*: A term often used for *fuel-injection system.*

fuel-injection system: A system that sprays fuel under pressure into the *intake manifold* or directly into the cylinder intake ports, allowing more precise control of the *air/fuel mixture* for improved performance, fuel economy, and reduced exhaust emissions.

fuel injector: A mechanical or electro-mechanical device that meters fuel into an engine.

fuel line: Rubber or metal lines that: **1.** Carry fuel from the *fuel tank* to the *carburetor* or *injector.* **2.** Return fuel not used to cool the *carburetor* and/or *injectors.*

fuel log: A manifold that delivers fuel to two or more *carburetors.*

fuel map: A chart to show the relationship of engine *rpm,* fuel flow, and ignition for a particular engine.

fuel nozzle: A tube in the carburetor through which fuel passes from the *float bowl* into the passing air.

fuel pump: A mechanical or electrical device used to move fuel from the fuel tank to the *carburetor* or *injectors.*

fuel pump eccentric: An engine part, usually bolted to the front of the *camshaft,* that is used to operate a mechanical fuel pump.

fuel-pump inertia switch: A normally closed, manually reset switch that opens if the vehicle is involved in an impact over 5 mph, or rolls over, turning off power to the fuel pump. This safety feature prevents fuel from being pumped onto the ground or hot engine components if the fuel line is ruptured or the engine dies.

fuel rail: A conduit to deliver fuel from the distributor to the injectors of an *FI* system.

fuel separator: A device that separates fuel and water.

fuel system: The system that delivers fuel to the cylinders, consisting of a fuel tank and lines, *gauge, fuel pump, carburetor* or *injectors,* and *intake manifold* or *fuel rail.*

fuel tank: A storage tank for fuel in a vehicle.

fuel-vapor recovery system: An evaporative emission-control system that recovers gasoline vapors escaping from the *fuel tank* and carburetor *float bowl.*

fuel wash: A condition that sometimes occurs during initial start-up of a rebuilt engine wherein fuel washes away the protective assembly oils to allow raw metal to metal contact.

fulcrum: A support, often wedge-shaped, on which a lever pivots when it lifts an object.

full bore: Full throttle. Also known as *full chat.*

full chat: A term used for *full bore.*

full coil suspension: A vehicle suspension system in which all four wheels have their own coil spring.

full floater: A term used for *full-floating axle.* Also known as *floater.*

full floating: Any part that moves and rotates within another part, such as a floating piston pin.

full-floating axle: An axle that performs only one function: to transfer torque to drive the vehicle; a type of axle popular on trucks.

full-floating pin: A *piston pin,* held in position by *snap rings* fitted into grooves in the *piston boss,* that is allowed to move in both the *piston* and the *connecting rod.*

full flow: **1.** A flow without restrictions. **2.** A type of oil filter, having no bypass, through which all of the oil from the oil pump flows.

full-flow filter: A type of oil filter designed so that all of the oil from the oil pump flows through it.

full house: An engine that has had every normal hot-rodding modification that may be set up for racing or for street use.

full load: An engine delivering its maximum output.

full-load enrichment: The injection of additional fuel into an engine during full-load conditions.

full-metallic brake lining: Brake linings made of metal particles that have been fused together into a solid material.

full race: An engine that is built for maximum racing performance.

full-race cam: A camshaft that is ground for maximum performance.

full throttle: **1.** The wide open throttle position. **2.** The accelerator pressed to the floorboard.

full trailer: A trailer that employs a towbar coupled to a swiveling or steerable running gear assembly at the front and does not transfer any of the load to the towing vehicle.

fully oscillating fifth wheel: A fifth-wheel type with fore-aft and side-to-side articulation.

fume: The airborne dispersion of minute particles, a byproduct of heating a solid, that may produce an oxide of the solid.

fundamental tone: The tone produced by the lowest frequency component of an audio or radio frequency (RF) signal.

funny car: Any unusual or unorthodox racing vehicle.

Funny Car: A drag-racing vehicle covered with a lightweight plastic replica of a passenger car body.

furnace brazing: A welding process used to repair complex cast-iron castings.

fuse: **1.** A protection device that opens a circuit when the fusible element is severed by heating, due to overcurrent passing through. **2.** To join two pieces of metal by bonding them together.

fuse block: A box-like enclosure that contains the fuses and circuit breakers for the electrical circuits of a vehicle.

fuse link: A term used for *fusible link.*

fuse wire: **1.** Wire made of an alloy that melts at a low temperature. **2.** A term used for *fusible link.*

fusible link: A type of fuse in which a special wire melts to open a circuit when the current is excessive.

fusion: The melting together of filler metal and base metal, or of base metal only, to produce a weld.

fusion welding: Any welding process that uses fusion of the base metal to make the weld.

G

***g*:** Symbol for gravity; unlike most other single-letter abbreviations, usually lower case and always italicized.

g: Green.

gage: **1.** A less popular way of spelling *gauge.* **2.** An element used to contain and space the rollers in a one-way roller clutch.

gallery: A passageway inside a wall or casing, such as for oil circulation.

galling: **1.** The friction-induced roughness of two metal surfaces in direct sliding contact. **2.** The welding up and tearing down of metals due to extreme high temperature and/or inadequate lubrication.

galloping hinge: A connecting rod.

galvanic action: The action of electropositive and electronegative metals due to the wasting away of the positive metal. This is most noted when iron (Fe) and brass, copper (Cu), or copper (Cu) and zinc (Zn), are in contact in the presence of acid-diluted water.

galvanic reaction: An electrical reaction caused by heat and water passing through metal pipes or parts at high speeds.

galvanize: An electromechanical process of coating zinc (Zn) to ferrous metals to reduce or prevent rust.

galvanometer: An instrument used for measuring a small electrical current.

gap: The space between two adjacent parts or surfaces.

gas: **1.** A vapor. **2.** A short term for *gasoline.*

gas bypass line: A term used for *hot-gas bypass line.*

gas bypass valve: A term used for *hot-gas bypass valve.*

gas cylinder: A container used for the transportation and storage of compressed gas.

gas defrosting: The use of hot, high-pressure gas in the evaporator to remove frost or ice buildup.

gas-filled shock absorber: A shock that uses nitrogen gas, at 25 times atmospheric pressure, to pressurize the fluid in the shock to reduce or prevent aeration or foaming. Also known as *gas shock.*

gas guzzler: A vehicle having excessive fuel consumption.

gas hog: A term often used for *gas guzzler.*

gasifier section: The part of a gas-turbine engine that draws in air, mixes it with fuel, and burns the mixture.

gasket: **1.** A metal ring used on some spark plugs to seal pressure between the plug shell and the cylinder head. **2.** A piece of thin compressible material such as cork, rubber, or soft metal, placed between two mating surfaces to form a seal.

gasket cement: A liquid gel or paste adhesive material used to bond gaskets to their mating surfaces.

gasket shellac: A liquid form of *gasket cement.*

gas-metal arc cutting *(GMAC)*: An arc-cutting process that uses a continuous, consumable electrode and a shielding gas.

gas-metal arc welding *(GMAW)*: An arc-welding process that uses an arc between a continuous-filler metal electrode and the weld pool.

gasohol: A mixture of 85–90% gasoline and 10–15% alcohol, usually ethanol.

gasoline: A liquid blend of hydrocarbons used as automotive fuel and processed from crude oil.

Gasoline Alley: The infield-garage area at the *Indianapolis speedway.*

gasoline consumption: A term used for *fuel consumption.*

gasoline container: Specially approved *OSHA* containers used to safely hold and store gasoline.

gas ports: A series of holes drilled from the crown of a piston to the top ring groove, providing combustion pressure to force the ring against the cylinder wall, ensuring a more effective seal.

gas regulator: A device used to control the delivery of gas at a substantially constant pressure.

gas shock: A term used for *gas-filled shock absorber.*

gassing: The conversion of battery water into hydrogen and oxygen, and oxygen gas.

gas-transfer velocity: The speed at which the *air-fuel mixture* spreads out in the *combustion chamber* during the power phase.

gas-tungsten arc cutting *(GTAC):* An arc-cutting process that uses a single tungsten electrode with gas shielding.

gas-tungsten arc welding *(GTAW):* An arc-welding process that uses an arc between a non-consumable tungsten electrode and the weld pool.

gas-turbine engine: An internal-combustion engine in which the shaft is spun by combustion gases flowing against curved turbine blades located around the shaft.

gauge: **1.** An instrument used for measuring, such as for compression. **2.** A dashboard-mounted component used for visual indication of engine and system conditions, such as oil pressure.

gauge pressure: A pressure indicated on a scale with atmospheric pressure as 0 *psi* or *kPa.*

gauge set: **1.** Two or more instruments attached to a manifold and used for measuring or testing pressure. **2.** Two or more instruments used to indicate engine and system conditions, such as oil pressure, coolant temperature, and charging-system indicators.

gauging point: The starting point for measuring a part.

gate: **1.** The money that is collected for admission. **2.** The starting position for a drag race. **3.** To take the lead in a drag race right from the start.

GAWR: An abbreviation for *gross-axle weight rating.*

GCW: An abbreviation for *gross combined weight.*

gear: **1.** A cogged device that mates or meshes with another. **2.** One's personal belongings and/or equipment.

gear backlash: The measurable gap or slack between gears.

gear carton: A transmission housing.

gear drive: A system of two or more gears, such as one that transmits power from the crankshaft or camshaft.

geared speed: A theoretical vehicle speed based on engine *rpm,* transmission-gear ratio, rear-axle ratio, and tire size, not accounting for slippage.

gear lubricant: A type of oil or grease especially formulated to lubricate gears.

gear oil: A thick lubricant, generally with an *SAE* number of 80 or above, used in standard transmissions or differentials. These often contain additives, such as an EP additive, to guard against being squeezed out from between gear teeth.

gear oil, limited-slip: A lubricant specified for use in certain limited-slip differentials to prevent chattering during turns and/or abnormal wear to the parts.

gear pitch: The number of teeth in a given unit of pitch diameter.

A B C D E F **G** H I J K L M N O P Q R S T U V W X Y Z

gear ratio: The speed relationship that exists between a driving (input) and a driven (output) gear. For example, a driving gear that revolves twice for each driven-gear revolution has a 2 to 1 (2:1) ratio.

gear rotor pump: A type of pump that uses sun gears to move liquid.

gears: Mechanical devices containing teeth that mesh that transmit power, or turning force, from one shaft to another.

gear shift: **1.** A floor- or steering-wheel-mounted lever used to manually change gears in the transmission. **2.** A linkage-type mechanism by which the gears in a transmission are engaged.

gear-type pump: A pump that uses two rotating gears to draw in fluid that is carried around the outer pump body in cavities between gear teeth and dispensed under pressure as the gear teeth mesh together.

gear whine: A high-pitched sound developed by some types of meshing gears.

gel coat: The first layer that is applied to a female mold for a fiberglass layup, before the mat or cloth layers.

general engine specifications: Specifications that are used to identify a particular style and type of engine.

general over-the-road use: A fifth wheel truck designed for pulling multiple standard duty highway equipment such as trailers, flat beds, tankers, and so on.

general purpose *(GP)*: A military designation during World War II, for the Willys *4×4;* later known as a Jeep.

generator: An ac or dc electrical-generating device that converts mechanical energy to electrical energy.

GEN III system: A fuel and engine ignition-management system developed in the late 1980s by General Motors.

gerotor: A term used for *gear rotor pump.*

get off it: To release the accelerator and slow down. Also known as *get out of it.*

get out of it: To release the accelerator and slow down. Also known as *get off it.*

get with the program: To perform a task properly.

g-force: The force that is exerted on a vehicle during acceleration, deceleration, or cornering, expressed in units of gravity (*g*'s).

gilhooley: To spin out on a dirt track.

Gilmer belt drive: An accessory drive system using a cogged belt to ensure positive engagement, such as with a *supercharger.*

gingerbread: Nonfunctioning visual ornamentation, such as *chrome.*

girdle: A heavy-duty, main-bearing support for a racing engine.

G.I. spacer: A device used to space out a piston ring in a piston groove that has been intentionally machined oversize because of wear.

gladhand: Connectors between tractor and trailer air lines.

glass: **1.** A term used for fiberglass-reinforced plastic. **2.** A hard, brittle, transparent substance composed of silicates mixed with potash or suds and lime; used for windows in a vehicle.

glass-bead cleaning: The cleaning of parts and panels using glass beads propelled with the use of compressed air.

glass beader: An apparatus for using air-propelled glass beads to clean parts.

glass-bead test: A method of determining the absolute rating of a filter and its efficiency by introducing a measured quantity of glass beads of varying, but known, diameter, as a contaminant into the fluid, which is then filtered through the element.

glass wrapped: A term used to refer to a vehicle with a fiberglass body.

glaze: A smooth, glossy surface.

glazed: A very glossy, thin, smooth surface.

gloss: An appearance characteristic that gives the perception of the brightness and luster of a smooth, polished surface.

glove box: A small, enclosed storage area in the front passenger compartment of a vehicle.

glove compartment: A term used for *glove box.*

glow plug: An electrical plug used to preheat the combustion chamber to aid in starting a cold diesel engine.

glycol: A term used for *ethylene glycol.*

Glyptal: A specific brand of *Class F insulating enamel.*

GMAC: An abbreviation for *gas-metal arc cutting.*

GMAW: An abbreviation for *gas-metal arc welding.*

GMC supercharger: A positive displacement, mechanically-driven blower developed by General Motors for two-stroke, diesel-truck engines later adapted for gasoline racing engines.

GN: An abbreviation for *Grand National.*

goat: A term used to identify a Pontiac GTO.

gobble: To run extremely fast.

go button: The accelerator.

gofer: **1.** Go for. **2.** A person who runs errands.

Go Kart: A term for a specific brand of racing kart, often used as a generic term for any kart.

gold: A trophy.

goodies: **1.** Prizes and awards for race-car winners or show-car participants. **2.** High performance equipment on an engine.

goosed moose: A car having a severe *forward rake.*

gourd guard: A helmet.

governor: **1.** A device that controls another device, usually on the basis of *speed* or *rpm.* **2.** A speed-sensitive mechanical assembly in the *automatic transmission* driven by the output shaft, to supply primary control of when shifting is to occur.

governor assembly: A vehicle speed-sensing device that produces governor pressure to force the transmission upshift and permit the downshift.

governor pressure: The transmission's hydraulic pressure, used to control shift points, which is directly related to *output shaft speed.*

governor valve: A device attached to the output shaft and used to sense vehicle speed.

gow job: A term used in the early 1940s to identify a *hot rod.* Also *gow wagon.*

gow wagon: A term used in the early 1940s to identify a *hot rod.* Also *gow job.*

gow out: To accelerate quickly in a *gow job.*

GP: **1.** An abbreviation for *general purpose.* **2.** An abbreviation for *Grand Prix.*

GPH: An abbreviation for gallons per hour.

GPM: An abbreviation for gallons per minute.

gr: Grey.

grade: **1.** A hill; generally a steep hill. **2.** Sort by size or quality.

gradeability: The ability of a truck to negotiate a given grade at a specified *GCW* or *GVW.*

grade labeling: Tire ratings that may be used as a guide in buying, enabling the buyer to compare, at a glance, the apparent value of one tire with that of another.

grams-per-mile: A measurement of the amount of *exhaust emissions* a vehicle produces.

Grand National: A premier series of *NASCAR* stock car races.

Grand Prix *(GP)*: **1.** A French term that means "grand prize." **2.** An international series of Formula One racing events. **3.** The

24-hour race at LeMans. **4.** An annual *USAC* midget-car race once held on Thanksgiving in Southern California.

grand touring: A sports coupe with enclosed, rather than open, bodywork.

granny gear: The combination of low gear in the transmission and low range in the transfer case.

gran turismo *(GT)*: Italian for "grand touring."

gran turismo omologato *(GTO)*: Italian term for GT cars that have been qualified for racing as a production vehicle.

graphics: Any computer-generated image or illustration produced on a screen, paper, or film.

graphite: A crystalline form of carbon (C) found in natural deposits or formed by heating black carbon (C).

gravity bleeding: A method of purging air from a system by allowing the fluid to force air out of an opened bleeder valve by its own weight.

gravity casting: A process that relies on gravity to pull the material to the bottom of the mold during casting.

gravity energized: A *one-way roller clutch* that has no mechanical means for holding the roller in contact with the cam and race.

gray market: The individual importing of a foreign car into the United States.

gray-market parts: Any part that is not covered by a manufacturer's warranty, generally because it was not sold through an authorized dealer.

gray water: Used wash water from the kitchen and bathroom of a mobile home or recreational vehicle.

grease: Lubricant consisting of a stable mixture of oil, soap thickeners (usually lithium, sodium, or calcium), and other ingredients for the desired physical or operating characteristics.

grease fitting: The orifice, having a check valve, that a grease gun fits on during the greasing process.

grease plug: A small plug in a lubrication hole where there is a danger of over lubrication by high-pressure equipment. It is removed for lubrication and then reinstalled.

greasy friction: The friction between two surfaces coated with a thin layer of oil.

green: **1.** A color. **2.** A term used for *green flag.*

green flag: The flag used to signal the start of a race.

green goo: A term used for *green Loctite.*

greenhouse: The upper part of a vehicle body, including the windows, pillars, and roof.

green Loctite: A high-temperature anaerobic compound used for securing parts together.

grenaded motor: An engine that has blown up.

grenade motor: An engine that is expected to deliver a very high horsepower for a short time, before it blows up.

grid: **1.** A wire mesh, such as that used for a grille. **2.** The starting position of race cars based on their qualifying order.

grid growth: A condition where the battery grid grows little metallic fingers, extending through the separators, shorting out the plates.

grille: An ornamental opening in the front of a vehicle through which air is delivered to the radiator.

grind: **1.** To use an abrasive wheel to remove metal. **2.** The specific contour of a camshaft lobe, such as a quarter-race grind. **3.** A long race with emphasis on endurance rather than performance.

grip: A term used for *traction.*

grit: Fine dust or dirt.

grocery getter: A car used for everyday transportation.

grommet: **1.** A reinforced-metal eyelet through which a fastener is attached. **2.** A rubber or plastic eyelet inserted in a hole to protect wires that pass through it.

groove: **1.** The recess in a part to hold a ring or snap ring. **2.** Space between two adjacent tread ribs or lugs; the lowered section of the tire tread. **3.** The specific path through a turn in an oval track or in road racing. **4.** A rut.

groove weld: A weld made in a groove between the work pieces.

gross axle-weight rating *(GAWR):* The maximum allowable fully loaded weight of a given axle.

gross combination weight *(GCW):* The total weight of a fully equipped vehicle including payload, fuel, and driver.

gross horsepower: The maximum engine output as measured on a *dynamometer.* Also known as *gross torque.*

gross torque: A term used for *gross horsepower.*

gross trailer weight *(GTW):* The sum of the weight of the trailer(s) and the payload.

gross valve lift: The valve lift, including the running-valve clearance.

gross vehicle-weight rating *(GVWR):* The maximum that a vehicle should weigh when fully loaded with all passengers, fuel, and cargo.

ground: **1.** A large conducting body, such as the Earth, used as a common return for an electric circuit. **2.** The metal part of the vehicle structure in a single-wire system. **3.** A term sometimes used for *short circuit.*

grounded: Connected to Earth or to some conducting body that serves in place of the Earth.

grounded circuit: A condition that allows current to return to ground before it has reached the intended load component.

ground effect: The reduced air pressure under a vehicle that allows the normal pressure above the vehicle to push the vehicle downward, providing a better grip on the road surface.

ground electrode: A part of the lower end of the spark-plug shell making up the electrode gap, providing a path for current flow while jumping the gap.

ground-end spring: A spring that has flattened ends, usually by grinding, to better fit the perches or retainers.

grounding: The connecting of an electrical unit to the vehicle frame to provide a complete path for electrical current.

ground-return system: A common system of electrical wiring whereby the chassis or frame serves as a part of the electrical circuit.

ground wave: An electromagnetic wave that travels along the Earth's surface; usually emitted by a radio transmitter.

groupie: One, especially a female, who follows the racing circuits hoping to be accepted by its members.

group injection: A method of injecting fuel into the manifold areas of several cylinders of an electronic fuel-injection system while at the same time, actually entering each cylinder as its *intake valve* is opened.

GT: An abbreviation for *gran turismo,* Italian for "grand touring."

GTAC: An abbreviation for *gas-tungsten arc cutting.*

GTAW: An abbreviation for *gas-tungsten arc welding.*

GTO: An abbreviation for *gran turisimo omologato.*

GTW: An abbreviation for *gross trailer weight.*

guard: A barrier, such as a shroud, that physically prevents entry of the operator's hands or fingers into the point of operation.

guide: **1.** A part used to align an assembly or another part. **2.** The support for a valve stem in the head.

guide block: A metal or hard-rubber device used to maintain timing-chain tension.

guide insert: A valve-guide insert that is not an original part of the cylinder head; used to replace a worn integral valve guide.

guide liner: A thin, bronze sleeve used to repair a worn-out valve guide.

Guide-Matic: An electronic headlamp control that automatically shifts between upper and lower beams as required.

guide-mounted seal: A seal that mounts directly onto the valve guide.

guide plate: A metal plate on the cylinder head that keeps the push rod aligned with the rocker arm.

guide rail: A device that keeps the timing chain aligned and in the correct position.

guide seal: A seal that prevents oil from being drawn into the valve guide.

guide shoe: A device used to maintain tension on the timing chain of an overhead-cam engine.

guide sleeve: A tubular sleeve that is placed on a connecting rod bolt when it is removed, to prevent the bolt threads from scratching the crank pin.

gullwing: A coupe-body style with the doors hinged so they open upward.

gulp valve: A valve that opens in an air-injection system to admit extra air into the *intake manifold* upon deceleration, thus "leaning out" the mixture to prevent backfiring.

gum: Residue that remains when gasoline is allowed to sit for a period of time.

gumballs: **1.** Drag racing tires made of especially soft, sticky compound. **2.** The rotating light in a hemispherical housing on top of a police car or emergency vehicle.

gun-drilled oil holes: Holes that are drilled to allow oil to be fed to the piston pin.

gun it: To rev up an engine suddenly.

gurgle method: A method of adding refrigerant from small cans to a system without running the engine.

gusset: A triangular piece of metal used to add strength to a corner.

gut: To remove non-functional parts of a vehicle in order to reduce its weight.

guts: **1.** The internal structure of a vehicle. **2.** The essential working parts of a device.

GVWR: An abbreviation for *gross vehicle-weight rating.*

gymkhana: An individually clocked maneuverability contest for sports cars over an extremely tight course.

H: The chemical symbol for *hydrogen.*

hack: **1.** A taxicab. **2.** A passenger car for everyday driving. **3.** To cut out a section of body work.

hairpin: **1.** A turn greater than 90 degrees on a road or race course. **2.** A *crankshaft.*

half-moon key: A term often used for *Woodruff key.*

half shaft: The axle shafts on either side of the differential that transmit power to the drive wheels.

halide leak detector: A device consisting of a tank of acetylene gas, a stove, a chimney, and a search hose used to detect *CFC* refrigerant leaks by visual means.

halogen lamp: A high-output, white-light lamp used as a headlight. Also known as *halogen light.*

halogen leak detector: An electronic device used to detect leaks in a refrigerant system.

halogen light: A high-output, white-light lamp used as a headlight. Also known as *halogen lamp.*

halogens: Refers to any of five nonmetallic and electronegative chemical elements: astatine (At), bromine (Br), chlorine (Cl), fluorine (F), and iodine (I), such as those that may be found in refrigerants.

hammer forming: To form or shape metal by hitting it with a hammer over, on, or into a base form.

handicap: A system that allows two cars from different classes in a drag race to compete with each other.

handler: **1.** A driver who is able to get the most out of his or her car. **2.** A car that the driver can get the most out of.

hand shield: A hand-held protective device that has a filter plate and is used in arc welding, arc cutting, and thermal spraying for shielding the eyes, face, and neck.

hand tight: To assemble plain or threaded parts as much as can be done by hand, without the aid of a wrench or tool.

Hand Tools Institute *(HTI):* A trade association that represents automotive hand-tool and tool-box manufacturers.

hand valve: A valve mounted on the steering column or dash, used by the driver to apply trailer brakes independently of the tractor brakes.

hang a left: To make a left turn.

hang a right: To make a right turn.

hang it out: To deliberately throw the rear end into a slide during a turn.

hang on unit: An under-dash, aftermarket air conditioner. May also refer to other under-dash devices, such as a CB radio.

hangover: Undesired speaker-cone oscillation caused by a transient signal.

hard chrome: A heavy-chrome plating applied to metal to increase its durability and resistance to wear.

hard copy: A printed copy of machine output in readable form, such as reports, listings, or graphic images.

hard edge: A sharp corner where two surfaces meet.

hardener: A chemical component of an epoxy resin that starts a catalytic reaction and causes the mixture to harden.

hard face: The application of a hard material, such as chrome, to a surface to increase its durability and wear resistance.

hardness test: The technique used to determine the hardness of a material.

hard parts: The internal engine parts required to rebuild an engine.

hard pedal: **1.** A loss in braking efficiency so that an excessive amount of pressure is needed to actuate brakes. **2.** A condition whereby the load literally overrides the brakes.

hard solder: Any solder requiring more than 1000°F (538°C) to melt.

hard spots: Areas in the friction surface of a brake drum or rotor that have become harder than the surrounding metal.

hard throttle: A term used when the accelerator is pressed to the floor.

hardtop: A two- or four-door hardtop convertible body type with front and rear seats, generally characterized by the lack of door or "B" pillars.

hardware: **1.** The components that make up a computer system, such as a keyboard, a floppy disk drive, and a visual-display terminal. **2.** The nuts, bolts, and accessories required to assemble an add-on component.

hardwired: The physical connection of two pieces of electronic equipment by means of a cable.

harmonic: The rhythmic vibration of a moving part or assembly.

harmonic balancer: A balance shaft or wheel that reduces or eliminates harmonics.

harmonic damper: A term used for *harmonic balancer.*

harmonic vibration: A rhythmic vibration of a moving part or assembly at a specific speed.

harness: A group of electrical conductors laced or bundled in a given configuration, usually with several breakouts.

harness clip: A metal or plastic clip used to secure a wiring harness to the car frame or body at various points, providing safe and neat routing.

harness connector: An electrical connector at the end of a wire or harness used to connect the conductor to a device or system.

hat: A pipe-like housing for the injector on a constant-flow fuel-injection system.

hatchback: A passenger-car body style where a full opening hatch, including the rear window, lifts for access to the cargo storage area.

hat trick: Three successive victories in the same annual event.

hauler: **1.** A top-performing car. **2.** A top-performing driver.

hauling the mail: Performance at peak efficiency.

hazardous waste: **1.** Any product used in a system that is considered harmful to people and/or the environment. **2.** Any poisonous substance, the byproduct of a poisonous substance, or one that contains carcinogens and is considered unsafe for animal or plant life by the EPA.

hazard system: A driver-controlled system of flashing front- and rear-marker lights to warn approaching motorists of a potential hazard.

H-beam rod: A connecting rod having an H-beam-shaped cross section.

HC: An abbreviation for *hydrocarbon.*

HC-CO meter: A device using an infrared sensor to measure the amount of hydrocarbons, in parts per million, and carbon monoxide, in percent, in the vehicle exhaust.

HD: An abbreviation for *heavy duty.*

HDMA: An abbreviation for *Heavy Duty Manufacturers Association.*

HDRA: **1.** An abbreviation for *Heavy Duty Representatives Association.* **2.** An abbreviation for *High Desert Racing Association.*

head: **1.** That part of an engine that covers the top of the cylinders and pistons and usually contains the combustion chambers and valve train. **2.** That part of a compressor that covers the valve plates and separates the high side from the low side of an air-conditioning system.

headache rack: A rack that is provided at the rear of a truck-cab interior to prevent a load from crashing through the rear window.

head bolts: The bolts that fasten the cylinder heads to the engine block.

header: **1.** A vehicle crash. **2.** A free-flowing exhaust manifold used on high-performance cars.

header tank: The top and bottom tanks or side tanks of a radiator in which coolant accumulates or is received.

head gasket: A gasket used to seal the cylinder head to the engine block.

head-gasket notch: A series of one to four notches used to measure the thickness of a gasket to control the piston-to-head clearance on a diesel engine.

headlamps: The lights at the front of a vehicle used to illuminate the road ahead. Also known as *headlights.*

head-land ring: A top compression ring with an L-shaped cross section.

headlight delay: A system that turns off the headlamps automatically after the operator leaves the vehicle, though the lights were left on.

headlights: A term used for *headlamps.*

headliner: Part of the automobile-interior overhead or the covering of the roof inside.

head pressure: Pressure of the refrigerant from the discharge valve through the lines and condenser to the metering-device inlet.

head-saver shims: Shims that are used to compensate for material removed during head resurfacing to maintain correct overhead-cam drive-chain tension.

heads-up display *(HUD):* The images of a vehicle's instrument panel, displayed on the inside surface of the windshield in front of the driver where they can be seen without taking the eyes off the road.

heads-up racer: A race driver that is very good.

heads-up racing: Direct, no handicap, competition.

heat: A form of energy released by the burning of fuel.

heat-control valve: A thermostatically controlled valve in the exhaust manifold to divert heat to the *intake manifold* before the engine reaches normal operating temperature.

heat dam: The narrow groove cut into the top of the piston. It restricts the flow of heat down into the piston.

heated-air-intake system: A vacuum operated power door in the air-cleaner snorkel that maintains intake air at a near constant temperature by blending ambient air with heated air picked up from a shroud over the exhaust manifold. Also known as *heated air system.*

heated-air system: A term used for *heated air-intake system.*

heated-exhaust gas-oxygen sensor *(HEGOS):* An oxygen sensor that is heated electrically and by engine exhaust so it warms up to normal operating temperature more quickly, allowing the engine to enter closed-loop operation sooner than with a non-heated sensor.

heater: A device used to warm the vehicle interior.

heater core: A small water-to-air radiator-like heat exchanger that provides heat for the passenger compartment.

heater valve: A manual or automatic valve in the heater hose used for opening or closing, providing coolant flow control to the heater core.

heat exchanger: An apparatus in which heat is transferred from one medium to another on the principle that heat moves to an object with less heat.

heating: The use of an apparatus that produces a relatively high degree of warmth under controlled conditions, by natural or

A B C D E F G H I J K L M N O P Q R S T U V W X Y Z

mechanical means, as an aid to ensure personal safety and comfort.

heating value: The measure of heat released when a fuel is burned.

heat intensity: The measurement of heat concentration with a thermometer.

heat load: The "load" imposed on an air conditioner due to ambient temperature, humidity, and all other factors that produce unwanted heat.

heat of compression: An increase of temperature brought by the compression of gas, air, or an *air/fuel mixture.*

heat of fusion: The amount of latent heat, in *Btu,* required to cause a change of state from a solid to a liquid.

heat of respiration: The heat given off by ripening vegetables or fruits in the conversion of starches and sugars.

heat quantity: **1.** The amount of heat as measured on a thermometer. **2.** The amount of heat contained in an object, measured in *Btu's* or *calories.*

heat radiation: The transmission of heat from one substance to another while passing through, but not heating, intervening substances.

heat range: The measure of a spark plug's ability to transfer heat from the tip of the insulator into the cylinder head.

heat riser: A channel in an *intake manifold* through which exhaust gas flows in order to heat the *manifold,* thus aiding in fuel vaporization.

heat-riser valve: A control valve between the *exhaust manifold* and exhaust pipe that restricts the flow of exhaust, causing it to flow back through the heat-riser channel under the *intake manifold.* It acts as an aid to fuel evaporation and to speed engine warm-up.

heat sink: **1.** A finned, aluminum housing that dissipates heat and serves as a base on which electrical components, such as semiconductor devices, are mounted. **2.** A heat dissipating section of the alternator that holds the insulated diodes or all the diodes, and carries heat away from the devices.

heat tab: A small aluminum disk attached to the cylinder head or block having a lead center designed to melt at 260°F (127°C) to indicate if the engine has been overheated.

heat transfer: The movement of heat from a hotter to a colder area by conduction, convection, or radiation.

heat treatable: A metal alloy that can be hardened by heat treatment after being shaped.

heat treatment: A sequence of controlled heating and cooling operations applied to a solid metal to provide the desired properties.

heavy duty *(HD):* A device or product that is designed for heavy use.

heavy-duty coil spring: Coil springs having larger wire diameter, and 3 to 5 percent greater load-carrying capability than a conventional spring.

Heavy Duty Manufacturers Association *(HDMA):* A trade association of original equipment and aftermarket manufacturers of parts, tools, and equipment for servicing heavy trucks.

Heavy Duty Representatives Association *(HDRA):* An association of independent heavy-duty manufacturers representatives.

heavy-duty shock absorber: *Shock absorbers* having improved seals, a single tube to reduce heat, and a rising rate valve for precise spring control.

heavy-duty truck: A truck having a gross vehicle weight of over 26,000 pounds (11,794 kilograms).

heavy-duty vehicle: Any motor vehicle having a *gross vehicle-weight rating* over 8,500 pounds (3,856 kilograms), that has a vehicle curb weight over 6,000 pounds (2,722 kilograms), or has a frontal area exceeding 45 square feet (4.2 square meters).

heel: **1.** The innermost part of the gear tooth on a ring and pinion gear set. **2.** The bottommost portion of a cam on a camshaft.

heel and toe: The act of applying one's right toes to the brake pedal while applying the right heel to the accelerator to slow the vehicle as it enters a turn while keeping the engine revved up.

heel to toe clearance: The space between the brake drum and the *heel and toe* of a *brake shoe.*

HEGO: An abbreviation for *heated exhaust gas-oxygen sensor.*

HEI: An abbreviation for *high-energy ignition.*

Heim joint: A spherical-rod end joint that allows precise adjustment of wheel position.

Heliarc: The trade name of a tungsten inert-gas form of welding developed by Linde Welding and Cutting Systems.

Heliarc welding: The act of electrically joining two pieces of aluminum or stainless steel using a high-frequency electric weld and an inert gas, such as argon, that is fed around the weld to prevent oxidation.

helical: A device, such as a spring, shaped like a helix or spiral.

helical compression spring: An open-pitch spring, in a variety of configurations, used to resist applied compression forces.

helical extension spring: A spring, usually made of round wire, that exerts a pulling force with various types of end hooks or loops by which they are attached to the load.

helical gear: A gear having teeth cut at an angle across its face, or diagonally to the rotational axis.

helical spring: Springs made of bar stock or wire, coiled into a helical form, including compression, extension, and torsion springs.

Heli-Coil: A trade name for a screw-thread insert used to repair damaged internal threads.

helmet: **1.** A device worn on the head to protect eyes, face, and neck from arc radiation, radiated heat, and spatter or other harmful matter expelled during *arc welding, arc cutting,* and thermal spraying. **2.** Protective head gear.

helper spring: An additional spring device that permits a greater load on an axle.

hemi: Half of a sphere.

hemi cammer: An engine having both hemispherical combustion chambers and an overhead camshaft.

hemi head: A cylinder head with hemispherical combustion chambers providing room for extra large valves or multi-valve arrangements and providing an ideal shape for smooth, even combustion.

hemispherical combustion chamber: A type of combustion chamber that is shaped like half of a circle. This combustion chamber has the valves on either side with the *spark plug* in the center.

HEPA: An abbreviation for *high-efficiency particulate air filter.*

heptane: A hydrocarbon (C_7H_{16}) that serves as a primary reference fuel with an index of 0.

hermaphrodite caliper: A combination tool having a caliper leg and a leg from a divider. Used in testing centered work and laying off distances from the edge of a piece of work.

herringbone gear: A gear cut in the V-shaped pattern of the small bony extension of a herring's spine.

hertz: A unit of frequency of any regularly occurring event.

hesitation: A momentary lack of throttle response or pause in the rate of acceleration.

HFC-134a: A hydrofluorocarbon gas that is not damaging to the environment and ozone layer and that can be used in automotive air-conditioning systems and to replace *CFC-12.*

Hg: The chemical symbol for the element mercury.

HHV: An abbreviation for *higher heating value.*

highboy: A *hot rod* that has not been chopped, channeled, or sectioned.

high CG load: Any application in which the load center of gravity of the trailer exceeds 40 inches (102 centimeters) above the top of the *fifth wheel.*

High Desert Racing Association *(HDRA):* A professional off-road racing association that stages desert races in Nevada and adjacent states.

high-energy ignition *(HEI):* A General Motors high-voltage ignition system.

high-efficiency particulate air filter *(HEPA):* A filter system that is 99.97% efficient in trapping particles.

higher heating value *(HHV):* The measure of the heat that is released when a fuel is burned.

high gear: The condition achieved from a planetary gear set when any two members, sun gear and planet gears or planet gears and internal gears, are locked together.

high gearing: The drive ratio that applies maximum engine output at a relatively high road speed.

high head: A term used when the high-side pressure of a system is excessive.

high heat load: Refers to the maximum amount of heat that can be absorbed by a refrigerant as it passes through the evaporator.

high-lift cam: A camshaft that provides increased valve lift.

high-lift rocker arms: Rocker arms that provide increased valve lift.

high-load condition: Those instances when the air conditioner must operate continuously, at maximum capacity, to provide the cool air required.

High Mobility Multipurpose Wheeled Vehicle *(HMMWV):* A GM diesel powered, multipurpose four-wheel-drive vehicle developed by AM General and built in a variety of configurations for civilian and military use.

high modulus belt: Those belts designed with special reinforcing members with a high degree of elasticity such as steel, fiberglass, or aramid fiber.

high-mounted coil spring suspension: A type of suspension having a coil spring located above the upper control arm, with the top end of the spring contacting the car body inside a spring tower.

high output *(hp):* A replacement device having a higher output than the original equipment device.

high pedal: **1.** The position of a brake pedal when it is adjusted near the top of its upward gravel. **2.** The position of the clutch pedal when engaged in high gear in the transmission of a *Model T.*

high-performance speed rating: A rating system used for tires designed for high-performance cars such as SR, HR, and VR.

high-performance tire: A tire having a *high-performance speed rating,* such as SR, HR, or VR.

high-pressure control: A term used for *high-pressure cut off switch.*

high-pressure cutoff switch: An electrical switch activated by a predetermined high pressure to open a circuit during certain high-pressure periods.

high-pressure line: **1.** The lines or hoses from the compressor outlet to the metering device inlet of an air conditioner. **2.** Any line or hose that carries high-pressure liquid or gas, such as a power-steering hose.

high-pressure relief valve: A mechanical device designed so that it releases the extreme high pressures of the system to the atmosphere.

high-pressure side: **1.** That part of an air-conditioning system, from the compressor outlet to the evaporator inlet, that is under high pressure. **2.** Any part of a system, such as fuel injection, that is under high pressure.

high-pressure switch: An electrical switch activated by a predetermined high pressure.

high-pressure vapor line: A line or hose running from the compressor outlet to the condenser inlet.

high riser: **1.** The raised cylinder heads of the 427 cid Ford engine of the 1960s. **2.** The term given to a *manifold* that places the *carburetor* above its original design location.

high side: A term used for *discharge side.*

high-side pressure: A term used for *discharge pressure.*

high-side service valve: A device located on the discharge side of the compressor, permitting the service technician to check the high-side pressures and perform other necessary operations.

high-speed steel *(HSS):* An extremely hard alloy.

high-speed system: A carburetor main-metering system that supplies fuel to the engine at speeds above 25 mph (40 km/h).

high spot: The outward curvature of a normally flat or smooth surface.

high suction: A term used when the low-side pressure of a system is higher than normal due to a malfunction of the system.

high-swirl combustion *(HSC):* Ford's name for a cylinder head and valve design that promotes turbulence in the combustion chamber during the power stroke, contributing to complete and efficient burning of the *air/fuel charge.*

high-swirl port technology: A cylinder-head design with high-speed air flow to pack more *air/fuel mixture* into the combustion chamber.

high tech: A term for developments in technology resulting from new, or new uses of materials and machinery that produce products fully comprehensible only to experts in that particular technology.

high-tension cables: A term used for *spark plug* wires.

high-tension wire: The wire that transmits high voltage from the *coil* to the *distributor.*

high vacuum: A vacuum below 500 microns (0.009 *psia* or 0.66 *kPa*).

high-vacuum pump: A two-stage vacuum pump that has the capability of pulling below 500 microns (0.009 *psia* or 0.66 *kPa*).

high-voltage cables: The secondary cables that carry high voltage from the *distributor* to the *spark plugs.*

Highway Loss Data Institute *(HLDI):* A research organization sponsored by the property-casualty insurance industry.

Highway Users Federation *(HUF):* A transportation coalition dedicated to preserving freedom of choice in transportation and mobility.

hi-po: High performance.

hi-rev kit: A set of auxiliary springs to keep roller lifters in contact with the camshaft lobes.

history: Something that is damaged or destroyed beyond repair.

hitch: A device that is attached to the rear of a vehicle to allow it to pull a trailer.

hitch ball: A steel sphere mounted on top of the hitch to receive the cup-like fitting of a trailer.

hitch pin: A term used for *hitch receiver.*

hitch receiver: A heavy square tube into which the removable shank supporting the hitch ball is inserted.

hitch weight: The load imposed to the tow vehicle's hitch by the trailer tongue, not to exceed 10–15% of the gross trailer weight.

hi-test: Gasoline that has an octane number near 90–95.

HLDI: An abbreviation for *Highway Loss Data Institute.*

HMMWV: An abbreviation for *High Mobility Multipurpose-Wheeled Vehicle.*

HO: An abbreviation for *high output.*

hog: **1.** Among motorcyclists, a Harley Davidson. **2.** A large vehicle such as an older Lincoln or Cadillac.

hog out: To enlarge the openings and passages in an engine.

hold-down spring: A spring tension mechanism that holds a brake shoe against a backing plate.

hold the road: To steer and handle well at high speeds.

hole: **1.** The starting position for a drag race. **2.** A cylinder.

hole shot: Beating a drag-racing competitor right from the start.

hollow: **1.** The concave area of a body panel. **2.** The inside of a pipe or tube.

hollow-domed piston: A piston with a compression dome with equal thickness across the top.

hollow flank: A concave or indented flank on a camshaft lobe.

hollow pushrod: A tubular pushrod that is hollow inside.

homologated: A term used when a sufficient number of a particular car have been built to qualify them for racing as a stock-production vehicle.

honcho: Someone in charge.

Honda system: An efficient combustion-chamber design that uses a small auxiliary combustion chamber (containing the spark plug) that receives a rich mixture to get an overall lean mixture in the cylinder to fire dependably.

honeycomb: A converter internal-substrate structure, usually made of ceramic, used to support the catalyst material, such as platinum, that allows maximum catalyst surface exposure to exhaust gases.

honk: To beat a competitor, especially in a *drag race.*

honker: A top-performing car.

hood: An opening body panel in front of the cowl covering the engine in a front-engine vehicle.

hood scoop: An air duct in the hood to allow air to flow directly across the engine.

hooks: Open loops or ends of extension springs.

hook up: **1.** To enter a draft in oval-track racing. **2.** To set up a car for the best possible traction off the starting line of a *drag race.*

hooligan: A consolation event in dirt-track racing featuring cars that failed to make the show for the main event.

hop up: To modify an engine for increased performance.

horizontal position: **1.** Level or parallel to the horizon. **2.** The position in which a fillet weld is performed on the upper side of an approximately horizontal surface and against an approximately vertical surface. **3.** The position of a groove weld in which the weld axis lies in an approximately horizontal plane and the weld face lies in an approximately vertical plane.

horn: An electrical noise making device.

horn brush/slip-ring: An electrical contact ring used in the horn circuit, located in the steering wheel, and used to maintain electrical contact when the steering wheel is turned.

horn relay: A relay used to carry the heavy-horn current requiring light current to actuate.

horsecar: A Ford Mustang.

horsepower *(HP):* A measurement of mechanical power or the rate at which work is done.

hose-and-line tape: A term used for *insulation tape.*

hose clamp: A device used to attach hoses to the engine, heater core, radiator, and water pump. A popular type of replacement clamp is the high-torque, worm-gear clamp with a carbon steel screw and a stainless-steel band.

host computer: A main computer that monitors and controls other computers.

Hotchkiss drive: A popular type of drive-line system that features an open or visible drive shaft.

hot dog: A top performer in almost all forms of racing.

hot drawn: A process of forming metal by heating it and pulling it through rollers or dies.

hot gas: The condition of refrigerant as it leaves the compressor until it gives up its heat and condenses.

hot-gas bypass line: The line that connects the hot-gas bypass valve outlet to the evaporator outlet.

hot-gas bypass valve: A device used to meter hot gas back to the evaporator, through the bypass line, to prevent condensate from freezing on the core.

hot-gas defrosting: The use of high-pressure gas in the evaporator to remove frost.

hot-idle compensator: A temperature-sensitive carburetor valve that opens when the inlet air temperature exceeds a certain level to allow additional air to enter the *intake manifold,* preventing an overly rich *air/fuel ratio.*

hot iron: An early term for a *hot rod.*

hot patch: A patch applied to a tube or tire with the use of heat to vulcanize it onto the damaged surface.

hot rod: A passenger car that has been modified and rebuilt for high performance and a distinctive appearance used primarily for straightaway speed or acceleration racing.

hot-rolled steel *(HRS):* Steel that is formed while hot.

hot setup: A combination of modifications and components that enhances maximum performance.

hot shoe: A top race driver.

hot soak: A condition that may occur when an engine is stopped for a prolonged period of time after a hot, hard run causing the gasoline to evaporate out of the carburetor.

hot spot: **1.** A localized area where excessive heat can build up, such as around a spark plug, that could cause preignition. **2.** An area on an auto body where corrosion is most likely to occur.

hot tank: A large vat containing a hot, caustic, chemical solution that is used to clean the engine block and parts.

hot-tank gloves: Long-sleeved, thick rubber gloves for use in the hot tank.

hot wire: **1.** Wire that carries current from the ungrounded terminal of the battery to an electric load. **2.** To bypass the *ignition switch* to start the vehicle.

hot-wire sensor: An electrical device inside the mass-airflow meter that measures air flow and density.

hot working: The mechanical working of metal at a temperature above its recrystallization point and high enough to prevent strain hardening.

hourglass frame: A term used for *X-frame.*

housekeeping: The type of safety in the shop that keeps floors, walls, and windows clean, lighting proper, and containers and tool storage correct.

housing bore: The inside diameter of a bearing housing.

housing breather: A venting device that allows air to enter or leave the axle housing.

HP: An abbreviation for *horsepower.*

HPV: An abbreviation for *human-powered vehicle.*

HRS: An abbreviation for *hot-rolled steel.*

HSC: An abbreviation for *high-swirl combustion.*

H-shift pattern: An arrangement of shift positions in the form of an H, for a manual transmission.

HSS: An abbreviation for *high-speed steel.*

HTI: An abbreviation for *Hand Tools Institute.*

H_2O: The chemical symbol for water.

hub: The center of a wheel.

hubbed drum: A brake drum that is mounted on a hub.

hub cap: A wheel covering.

HUD: An abbreviation for *heads-up display.*

A B C D E F G **H** I J K L M N O P Q R S T U V W X Y Z

HUF: An abbreviation for *Highway Users Federation.*

huffer: A supercharger.

hum: Interference from ac power, normally audible and of low frequency.

human-powered vehicle *(HPV):* Any vehicle powered by the operator, such as a bicycle.

Hummer: A term used for *Humvee.*

Humvee: An abbreviation for *High Mobility Multipurpose-Wheeled Vehicle.*

hunting: The tendency of a system to oscillate around its normal position.

hunting shift: Tendency of an automatic transmission to upshift, downshift, and upshift rapidly.

H-valve: An air-conditioning expansion valve used on some Chrysler and Ford car lines.

hybrid: **1.** A vehicle having two separate propulsion systems, such as a gasoline engine and an electric motor. **2.** A vehicle having its major components from more than one source, such as a Ford engine in a Chevrolet chassis.

hybrid battery: A battery that combines the advantages of both the low maintenance and the maintenance-free battery.

hydraulic brakes: A braking system using hydraulic pressure to press the *brake shoes* against the *brake drums.*

hydraulic clutch: A clutch that is actuated by hydraulic pressure.

hydraulic lifter: Valve lifter located between the camshaft and pushrod that uses internal oil pressure to cause the lifter to expand lengthwise.

hydraulic pressure: A pressure exerted through the medium of a liquid.

hydraulic principles: The use of a liquid under pressure to transfer force or motion, or to increase an applied force.

hydraulics: Hydraulic ram units, installed in a low rider at each suspension point, to lower or raise the ride height. Also referred to as *hydro.*

hydraulic valve lifter: A device using oil pressure to adjust its length and maintain zero valve clearance so that valve noise is at a minimum. Also referred to as *hydro.*

hydrazine: A highly explosive jet fuel sometimes used as an additive fuel for drag racing.

hydro: **1.** A term used for *hydraulics.* **2.** A term used for *hydraulic-valve lifter.*

Hydro: A short term for hydromatic, a transmission built by General Motors.

hydro-boost: A type of brake power booster that uses hydraulic pressure provided by the power-steering pump to reduce pedal effort.

hydro cam: A camshaft especially designed to operate with hydraulic valve lifters.

hydrocarbon *(HC):* A compound containing hydrogen (H) and carbon(C), such as gasoline.

hydrocarbon emissions: A term used for *exhaust emissions* or *unburned hydrocarbons.*

hydrocarbon reactivity: A measure of the smog-forming potential of a hydrocarbon.

hydrochloric acid: A corrosive acid produced when water and R-12 are mixed as within an automotive air-conditioning system.

hydrochlorofluorocarbon: A group of refrigerants that contain the chlorine (Cl) atom and the hydrogen (H) atom, which causes the chlorine atom to dissipate more rapidly in the atmosphere.

hydrofluorocarbon: A man-made compound used in refrigerants, such as R-134a.

hydrogen: A colorless, odorless flammable gas.

hydrolyzing action: The corrosive action within the air-conditioning system induced by a weak solution of hydrochloric acid formed by excessive moisture chemically reacting with the refrigerant.

hydrophilic: Attraction to water.

hydrophilic tread: A tire tread that is attracted to water.

hydroplaning: A term used for *aquaplaning*.

hydrostatic lock: A condition in an engine when coolant leaks into and fills a cylinder to block movement of the piston.

hydrostatic testing: Testing the integrity of a tank or other such containers by using water under high pressure.

hypalon wire: A special dielectric wire covering used for high-voltage wires, such as spark-plug wires.

hypoid gears: A spiral-bevel gear that allows the *pinion* to be placed below the center of the ring gear in a final drive assembly.

hypoid gear set: Two gears that transmit power at a 90 degree angle or other angles having the driving gear below the driven gear centerline to allow for lowering the drive shaft.

I

I-beam axle: A beam axle having an I-shaped cross section.

I-beam rod: A connecting rod having an I-shaped cross section.

IC: An abbreviation used for *internal-combustion engine.*

ICE: An acronym for *internal-combustion engine.*

ice-melting capacity: A refrigeration effect equal to the latent heat of fusion of a stated weight of ice, at 144 *Btu* per pound.

i.d.: An abbreviation for internal diameter or *inside diameter.*

ideal air/fuel ratio: A term used for *Stoichiometric air/fuel ratio.*

ideal humidity: Most people are comfortable at a relative humidity of 45% to 50%.

ideal temperature: Most people are comfortable at a temperature from 68° to 72°F (20° to 22.2°C).

IDI: An abbreviation for *indirect injection.*

idiot box: An *automatic transmission.*

idiot light: The system-malfunction warning lights on a dashboard.

idiot proof: A product that is designed to be as simple as possible to operate.

idle: The engine speed with no load and the accelerator pedal fully released.

idle air-control valve: A General Motors electrically operated valve that, on commands from the engine-control computer, varies the size of an air passage that bypasses the throttle plate of an electronic fuel-injection system, thus controlling *idle speed.*

idle circuit: The system within a *carburetor* that meters fuel when the engine is running at low *rpm.*

idle compensator: A term used for *engine-idle compensator* or *hot-idle compensator.*

idle limiter: A device that controls the maximum richness of the idle *air/fuel mixture* in the *carburetor.*

idle-limiter cap: A plastic device pressed over a carburetor's idle-mixture screw that limits the amount of adjustment during service to help eliminate excessive air pollution that is caused if the mixture is too rich.

idle mixture: The *air/fuel mixture* supplied to the engine while idling.

idle-mixture adjusting screw: An adjustment screw on some carburetors that is turned in or out to lean or enrich the idle *air/fuel mixture.*

idle-mixture adjustment: An adjustment made on carbureted engines by removing the antitampering plugs and turning the idle-mixture screws until the proper idle mixture is achieved. Setting the idle mixture on an emissions-controlled engine may require monitoring exhaust CO or using a special procedure called *propane enrichment.*

idle-mixture screw: A screw on the carburetor used to adjust engine-idle speed.

idle port: An opening into the carburetor throttle body through which the idle system discharges fuel.

idler arm: A pivoting component that supports the right side of the steering relay rod in much the same manner as the pitman arm supports the left side.

idler eccentric: A device used in a belt-drive system using the idler pulley as a means of tightening the belt.

idler gear: A gear connecting two other gears in a manner so that they will turn in the same direction.

idler pulley: A pulley that is used to adjust the belt in a belt-drive system.

idle speed: The speed at which an engine idles in revolutions per minute (*rpm*), usually between 600 and 850 *rpm*. The idle speed is specified on the under-hood emissions decal.

idle-speed adjustment: The idle speed on carbureted engines without computer idle-speed control is set by turning a screw that opens or closes the throttle plates. On fuel-injected engines without computer idle-speed control, idle speed is set by turning an idle-air bypass screw that allows air to bypass the throttle plates. *Idle speed* is not adjustable on engines with computer idle-speed control.

idle-stop solenoid: An electromagnetic device mounted on the carburetor linkage that maintains the proper throttle opening for specified idle speed while the ignition is on, but allows the throttle to close further when the ignition is switched off, reducing the amount of air that can enter the engine and the likelihood of dieseling.

idle system: The passages of a carburetor through which fuel is fed when the engine is idling.

idle vent: An opening in an enclosed chamber through which air can pass under idle conditions.

I-4: An abbreviation for an inline four-cylinder engine.

IFS: An abbreviation for *independent front suspension.*

if you can't find 'em, grind 'em: An admonition to a driver who misses a shift.

ignition: The firing of a spark plug to ignite the *air/fuel mixture* in the *combustion chamber.*

ignition advance: The moving forward, in time, of the ignition spark relative to the piston position.

ignition coil: A transformer containing a primary and secondary winding that acts to boost the battery voltage of 12 volts to as much as 30,000 volts to fire the spark plugs.

ignition distributor: A term used for *distributor.*

ignition map: A chart showing the precise advance and retard of the ignition on an electronic control-equipped engine.

ignition points: A term used for *points.*

ignition reserve: The difference between the minimum available and maximum required voltage for proper operation.

ignition resistor: A resistance element in series with the primary circuit to reduce the voltage supplied to the coil during engine operation.

ignition retard: The moving back in time of the ignition spark relative to the position of the piston.

ignition switch: A five-position switch that is the power distribution point for most of the vehicle's primary electrical systems. The spring-loaded START position provides momentary contact and automatically moves to the RUN position when the key is released. The other switch detent positions are ACCESSORIES, LOCK, and OFF.

ignition system: The major components, such as the battery, coil, ignition switch, distributor, high-tension wiring, and spark plugs, that provide the right spark at the right time to ignite the *air/fuel mixture.*

ignition temperature: The lowest temperature at which a combustible material will ignite and continue to burn independent of the heat source.

ignition timing: The timing of the spark, expressed in crankshaft degrees, in relation to *top dead center.*

ignition timing-retard sensor: A term used for knock sensor.

I-head: An overhead valve engine having both intake and exhaust valves directly over the piston.

I-head engine: An overhead valve engine with the valves in the head.

IHP: An abbreviation for *indicated horsepower.*

IHRA: An abbreviation for the *International Hot Rod Association.*

IKF: An abbreviation for the *International Karting Federation.*

I/M: An abbreviation for *inspection maintenance.*

IMACA: An abbreviation for *International Mobile Air Conditioning Association.*

IMCA: An abbreviation for *International Motor Contest Association.*

IMEP: An abbreviation for *indicated mean effective pressure.*

impact failure: The failure of a part or assembly due to repeated impact or off-square seating.

impact hardness: The hardening of a part caused by repeated impacts.

impact particle cleaning: A method of cleaning parts using air-propelled glass beads.

impeller: A rotor that transmits motion such as a centrifugal pump, supercharger, turbine, or fluid coupling.

IMS: An abbreviation for the *Indianapolis Motor Speedway.*

IMSA: An abbreviation for the *International Motor Sports Association.*

I/M 240: A 240-second emissions-inspection program that uses loaded-mode testing that simulates the federal urban test cycle for certifying new vehicle-emissions performance.

inactive spring coil: Inactive coils located at the end of a spring introduce force into the spring when a wheel strikes a road irregularity.

inadequate crush: An insufficient crush on a pair of bearing shells allowing excessive clearance between the bearings and the shaft.

inadequate joint penetration: Joint weld penetration that is sufficient to properly bond the base metals.

in and out box: A racing transmission having a neutral and a single direct drive forward gear.

inboard: Toward the centerline of the vehicle.

incandescence: Glowing due to heat.

incandescent: That which gives light or glows at a white heat.

incandescent lamp: An electric lamp or bulb containing a thin wire or filament of infusible conducting material.

in-car sensor: A term used for *in-car temperature sensor.*

in-car temperature sensor: A thermistor used in automatic temperature-control systems for sensing the in-car temperature.

inches of mercury (in-Hg): A term used to designate a vacuum on the English scale.

inches of water column (in-H_2O): An engineering term used to designate pressure such as inches of mercury or kilo Pascals.

inch-pound: An English measure of *torque.*

inclined surface: **1.** A slope or slanted surface. **2.** A valve rotator.

included angle: **1.** Sum of the kingpin or ball-joint inclination angle and the camber angle of a vehicle's front wheel. **2.** The sum of the camber angle added to the steering axis-inclination angle. **3.** The angle included between the outer surfaces of the link-plate contours of a silent chain that meshes with the sprocket tooth form.

incomplete joint penetration: Joint penetration that is unintentionally less than the thickness of the weld joint.

independent front suspension *(IFS):* A suspension-system method of supporting the chassis on the wheels without the use of rigid axles where the movements of the two front wheels are not interdependent; one wheel does not force the other wheel to change its plane of rotation.

independent publishers: Publishers that provide service and repair information on a variety of subjects, such as automobiles. An example is this manual by Delmar Publishers.

independent rear suspension *(IRS):* **1.** A suspension system in which both rear wheels are free to move independently of the other, providing a reduction of unsprung weight as well as overall vehicle-weight reduction. **2.** An *independent suspension* system at the rear of the vehicle.

independent service: General or specialty service provided by independent garages on all types and makes of vehicles.

independent suspension: A suspension system by which a wheel on one side of a vehicle can move vertically without affecting the wheel on the other side, and wheel jounce or rebound travel of one wheel does not directly affect the movement of the opposite wheel.

index: **1.** To examine and adjust rod journal spacing and alignment. **2.** The elapsed time assigned to various classes in drag racing used to calculate starting-time differences for handicap or bracket racing. **3.** Installing a *camshaft* relative to the *crankshaft* so the valves open and close at precisely the right time.

Indianapolis Motor Speedway *(IMS):* A very popular race course, not in Indianapolis, but in Speedway, Indiana.

Indianapolis Raceway Park *(IRP):* Better known as *Indy.* The site of the U.S. Nationals on Labor Day weekend.

indicated horsepower *(IHP):* A measurement of power developed by the combustion processes in the cylinders, not considering friction.

indicated mean effective pressure *(IMEP):* The average of the pressures developed in the cylinders during the four-stroke cycle, *intake, combustion, exhaust,* and *compression.*

indicated speed: The vehicle speed as indicated on a *speedometer.*

indicated torque: *Torque* as calculated from the *indicated mean effective pressure,* which is an actual indication of output developed within the cylinders.

indirect injection *(IDI):* A system in which fuel is injected into a prechamber, where it is ignited before entering the main combustion chamber.

induction: **1.** A synonym for *intake.* **2.** A process whereby an electromagnet or electrically charged object transmits magnetism or an electric current to a nearby object without making physical contact.

induction coil: An electrical device that receives low battery voltage in its primary winding and delivers pulsating high voltage in its secondary winding as the primary field is energized and collapsed by use of a set of points.

induction hardening: A process of hardening the surface of a part by placing electric coils next to it.

in-duct sensor: A thermistor used in automatic temperature control units for sensing the in-duct return air temperature.

Indy: **1.** Short for *Indianapolis Motor Speedway.* **2.** Short for *Indianapolis Raceway Park.*

Indy car: An open-wheeled vehicle built especially for *Indy* and *CART* racing.

Indy 500: A race held every Labor Day weekend at *Indianapolis Motor Speedway.* It is an American classic, the single largest spectator-attended sporting event in the world.

Indy Lights: A junior series of *Indy* racing sponsored by *CART.*

inert gas: A gas that normally does not combine chemically with materials.

inertia: The property of a moving object that causes it to resist any change in speed or direction.

inertia lock retractor: A passive seat-belt system that uses a pendulum mechanism to lock the belt tightly during a sudden change in movement.

inertial scavenging: An exhaust system that uses the momentum of flowing gases to pull more gases through.

inertia weight: A small, steel flywheel used to dampen drive-line vibrations on certain applications; generally located on the *slip yoke,* drive pinion yoke, or near an end of the *drive shaft.*

infield: An area enclosed by an oval track or a road course.

inflation pressure: The amount of air pressure required to inflate a tire, measured in *psi* or *kPa.*

information center: The electronic display in some luxury vehicles that shows such data as the date, inside and outside temperature, trip distance traveled and distance remaining, time of departure and estimated time of arrival, average and current speed, average and instantaneous fuel economy, fuel used and remaining, and, often, an appointment reminder.

information sheet: A form that is used to record pertinent information about a vehicle.

infrared exhaust analyzer: An instrument able to detect and measure *hydrocarbons, carbon monoxide, carbon dioxide,* and *oxygen levels.*

in-Hg: An abbreviation for *inches of mercury.*

inhibitor: A substance used to reduce the rate of a chemical or electrochemical reaction.

in-H_2O: An abbreviation for *inches of water column.*

injected: An engine equipped with a *fuel injector.*

injection cylinder: A term used for *oil-injection cylinder.*

injection valve: A term used for *fuel injector.*

injector: **1.** The tube or nozzle through which fuel is injected into the intake airstream of the combustion chamber. **2.** A term used for *fuel injector.*

injector body: The housing for a *fuel-injection system.*

injector fuel: The fuel that is stored inside the injector prior to use.

injector tube: The copper or brass tube used in diesel cylinder heads to cool fuel injectors and seal combustion pressures.

in-lb: An English measure of *torque.*

inlet: A synonym for *intake.*

in-line engine: An engine having all its cylinders in one line or row.

in-line steering gear: A type of integral steering system.

innards: The internal parts of an assembly.

inner-cam clutch: A *one-way roller clutch* that has the cam profile surfaces on the inner element.

inner liner: A synthetic, gum-rubber layer molded to the inner surface of a tire for sealing purposes.

inner shaft: A shaft upon which the inner end of the upper control arm pivots in a front suspension system.

inner tube: An inflatable rubber bladder mounted inside some tires to contain the air.

in-phase: The in-line relationship between the forward coupling-shaft yoke and the drive-shaft slip yoke of a two-piece drive line.

input retarder: A paddle-wheel-type unit located between the torque converter housing and the main housing in the retarder housing; primarily for over-the-road operations.

input shaft: The transmission shaft that receives power from the engine.

insert valve guides: Replaceable valve guides.

insert valve seats: Replaceable valve seats.

inside diameter *(i.d.)*: The diameter of the inner walls of a pipe or tube.

inside-vehicle lubrication services: Work performed inside the vehicle during a chassis lubrication such as changing the mileage sticker, lubricating the ignition-lock cylinder and the glove-compartment hinges and lock, and inspecting safety-related items.

inspect: To examine a component or system for any audible or visual signs of a defect.

inspection: The process of measuring, examining, testing, gauging, or otherwise comparing the unit with the applicable requirements or specifications.

inspection maintenance *(I/M):* The periodic and systematic inspection and maintenance of a vehicle's ignition, fuel, and emissions-control systems.

install: To set up an accessory part or kit for use in a vehicle.

installed height: The distance from the valve-spring end to the valve-spring seat.

instant center: An imaginary center point around which a wheel appears to pivot.

insulated return system: The ground wire of a two-wire electrical system.

insulation: Any material that blocks the flow of electricity.

insulation tape: Tape (usually cork) used to wrap refrigeration hoses and lines to prevent condensate drip.

insulator: A material with high electrical resistance that supports or separates conductors to prevent undesired flow of current to other objects.

insulator ribs: In a *spark plug,* unevenly spaced circular ridges on the upper section of the insulator to reduce or prevent flashover of the high voltage from the terminal to the lower shell.

Insurance Information Institute: A public information and communications association.

Insurance Institute for Highway Safety: A highway and vehicle safety research organization.

intake: The point at which a fluid or gas enters a pipe or channel Z. The act of taking in.

intake manifold: A metal component used to duct: **1.** The *air/fuel mixture* from the *carburetor* to the *intake ports.* **2.** Air in an injected engine to the *intake ports.*

intake over exhaust *(IOE):* An engine design having intake valves in the cylinder head and exhaust valves in the block.

intake ports: **1.** Passages in the cylinder head that direct *air/fuel mixture* from the *intake manifold* to the *intake valves.* **2.** Passages in an L-head engine block that direct the *air/fuel mixture* from the *intake manifold* to the *intake valves.* **3.** Passages in the head or block of an injected engine that direct air to the *intake valves* from the *intake manifold.*

intake stroke: A downward stroke of a *piston* that draws the *air/fuel mixture* into the cylinder.

intake valve: A valve that opens to admit the *air/fuel mixture.*

integral: Internal or built-in as a part of the whole.

integral-type power steering: A steering system with a built-in power assist.

integral valve guide: A valve guide machined into the cylinder head.

integral valve seat: A valve seat machined into the cylinder head.

inter-axle differential: A third differential between the front and rear differentials that permits differences in speed between the front and rear driving axles.

inter-changeability: The basis of mass production; the production of parts that are consistent in size, tolerance, and specifications.

intercooler: A radiator-like heat exchanger between the supercharger or turbocharger and the engine to cool the *air/fuel mixture.*

interface fit: A term used for *interference fit.*

interference angle: The difference between the valve face angle and the valve seat angle.

interference fit: A fit having limits of size, 0.0005 to 0.0007 inch (0.0127 to 0.0178 millimeters) so prescribed that an interference always results when mating parts are assembled by pressing them together. Also known as *drive fit* and *interface fit.*

interference valve angle: The difference between the valve face angle and the valve seat angle, usually one or two degrees.

Inter Industry Conference on Auto Collision Repair: An international, collision-repair training organization.

A B C D E F G H **I** J K L M N O P Q R S T U V W X Y Z

interior lights: The inside vehicle-lighting system, including courtesy, dome, map, and instrument-panel lights.

interlock: An electrical-circuit arrangement that prevents a second operation from taking place until the first operation is completed.

intermediate gear: A small torque increase obtained from a planetary gear set when the sun gear is held and power is applied to the internal gear.

intermittent duty: A service requirement that demands electrical or mechanical operation for alternate intervals of load and no load; load and rest; or load, no load, and rest.

internal-combustion engine *(IC)*: An engine that burns the *air/fuel mixture* in a chamber inside the engine cylinder, like a conventional reciprocating engine or a rotary engine.

internal gear: A gear with teeth pointing inward toward the center of the orifice.

International Carwash Association/National Carwash Association: A trade association consisting of carwash owners, operators, manufacturers, distributors, and suppliers.

International Franchise Association: A trade association representing franchisors, franchisees, suppliers, educational institutions, and franchise associations from other countries.

International Hot Rod Association *(IHRA)*: A drag-racing sanctioning body.

International Karting Federation *(IKF)*: A major kart-race sanctioning body.

International Mobile Air Conditioning Association *(IMACA)*: A trade association of manufacturers and distributors of automotive air conditioners and parts.

International Motor Contest Association *(IMCA)*: An automobile race-sanctioning body of drivers, owners, and mechanics created to establish rules, regulations, and specifications.

International Motor Sports Association *(IMSA)*: A road-racing sanctioning body.

International Race of Champions *(IROC)*: An annual series for top drivers from various forms of oval-track and road racing who are invited to compete, driving identical *spec cars.*

International Show Car Association *(ISCA)*: A promoter of International Championship Custom Car Shows.

International Truck Parts Association *(ITPA)*: A trade association of companies that buy and sell used and remanufactured heavy-duty truck parts.

interpole: A small field pole placed between the main field poles and electrically connected in series with the armature of an electric rotating machine.

in the chute: Staged for a drag race.

intrinsically safe: A system that is incapable of releasing sufficient electrical or thermal energy under any condition to cause ignition of a specific hazardous mixture in the atmosphere, even in its most ignitable concentration.

in-vehicle sensor: A term used for *in-car sensor.*

in-vehicle temperature sensor: A term used for *in-car sensor* or *in-car temperature sensor.*

inversion valve: A normally open air valve used primarily in parking and emergency brake systems.

inverted start: When cars are started in the reverse order of their qualifying times, in oval-track racing.

investigation table: A term used for *component location table.*

IOE: An abbreviation for *intake over exhaust.*

ionic bond: A primary bond arising from the electrostatic attraction between two oppositely charged ions, such as electrostatic painting.

IROC: An abbreviation for *International Race of Champions.*

iron: A basic element with the chemical symbol *Fe.*

Irontite: A tradename for the process of filling cracks in cast-iron cylinder heads and blocks using tapered plugs.

IRP: An abbreviation for *Indianapolis Raceway Park.*

irs: An abbreviation for Internal Revenue Service.

IRS: **1.** An abbreviation for Internal Revenue Service. **2.** An abbreviation for *independent rear suspension.*

ISCA: An abbreviation for *International Show Car Association.*

I-6: An in-line, six-cylinder engine.

Isky: The trade name of a popular brand of camshaft and valve-train components.

islands of steel: A term used for *hard spots.*

isoblock: A hard-rubber insulator block used for motor mounts.

isolate: A technique where a component or system may be separated from the rest of the component or system without the loss of fluids or pressures.

isolator: A rubber or synthetic device used to separate two parts to reduce noise or vibration.

isomers: Chemical compounds having the same molecular weight and atomic structure, but a different molecular structure.

iso-octane: A hydrocarbon with an octane rating of 100.

ITPA: An abbreviation for *International Truck Parts Association.*

A B C D E F G H I **J** K L M N O P Q R S T U V W X Y Z

J

jacketed gasket: A gasket having metal grommets around bolt and water holes.

jacking: Modifying the suspension to raise or lower one corner of an oval-track race car in order to provide better handling characteristics.

jackrabbit start: The sudden acceleration from a standing start.

jackshaft: A shaft found in most overhead cam engines that is used to drive the distributor, fuel pump, and oil pump.

Jack the Bear: A top-performing driver.

Jacobs brake: A device more commonly known as *Jake brake.*

Jacob's ladder: A triangular control linkage to center the rear axle assembly found on some rear-wheel-drive race cars.

Jake brake: A hydraulically operated *compression braking system* used to slow the truck by alternating the engine's exhaust valve opening time; named for its inventor, Jacobs.

jalopy: **1.** A low-budget, dirt-track racing car rebuilt from an older vehicle. **2.** An old automobile.

JAMA: An abbreviation for *Japan Automobile Manufacturers Association.*

jam nut: A second nut tightened against a primary nut to prevent it from working loose; used on inner and outer tie-rod adjustment nuts and on many pinion-bearing adjustment nuts.

Japan Automobile Manufacturers Association *(JAMA):* An international trade association of Japanese car, truck, motorcycle, and bus manufacturers.

Japanese Industrial Standard *(JIS):* Japan's equivalent of the *DIN* in Germany or the *SAE* in the United States.

jeeping: Driving a *4WD* vehicle, generally off the road.

jeep trail: A road or trail suitable only for a *4WD* vehicle.

jet: A calibrated passage in the carburetor used to meter fuel flow.

jet car: A race car powered by a jet aircraft engine for open course competition.

jet valve: An intake valve, located in a pre-combustion chamber, that admits a highly concentrated *air/fuel mixture* into a stratified charge engine.

jig: **1.** A fixture for holding work. **2.** A device used as an aid to straighten the frame of a vehicle.

JIS: An abbreviation for *Japanese Industrial Standard.*

Johnson rod: A mythical automotive component that is blamed for any problem that cannot otherwise be explained.

joint: A junction of members that are to be joined or have been joined.

joint penetration: The depth a filler metal extends from its face into a joint, exclusive of weld reinforcement.

jounce: The inward reaction of the *spring* and *shock absorber* when a wheel hits an obstruction.

journal: That part of a rotating shaft that turns in a bearing.

journal size: The finished diameter of a crankshaft journal.

J-rim: A groove around the outer edge of a wheel rim to provide a lock for the tire bead.

judder: A low-frequency vibration of the clutch or brakes.

jug: A carburetor.

juice: **1.** The current in an electrical system. **2.** A special racing-fuel formula. **3.** Any fluid, such as brake fluid.

jump: **1.** To begin early rapid acceleration in a rolling-start race before the green flag. **2.** To make an early, sudden start off the line in drag racing before the green light. **3.** To bypass the ignition switch to start a vehicle. **4.** To *jump start* a vehicle. **5.** An obstacle that causes an off-road vehicle to become airborne.

jumper cable: Two heavy-duty cables used to connect two batteries, as for starting a vehicle.

jumpers: A term used for *jumper cable* or *jumper wire.*

jumper wire: A wire used to bypass electrical circuits or components for testing purposes.

jump out: A condition when a fully engaged gear and sliding clutch are forced out of engagement.

jump start: Using battery jumper cables to connect an able battery to a disabled battery to start the vehicle.

junction block: A device on which two or more junctions may be found.

Junior fueler: An old drag-racing term for a lower-class fuel dragster.

Junior stocker: An old drag-racing term for a lower-class fuel stocker.

junk: Anything that is so badly worn or deteriorated that it is of no further use.

junk box: **1.** A vehicle that has been rebuilt from well-worn parts. **2.** A vehicle that is ready to be scrapped.

junker: A vehicle that is ready for the junk box.

junky: Anyone in the vehicle salvage business.

junkyard: A vehicle salvage business that sells used parts.

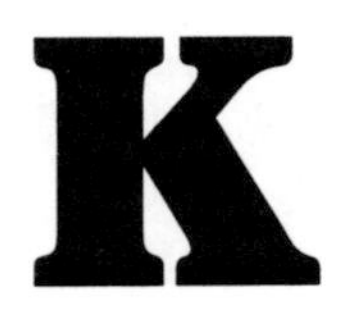

k: An abbreviation for "kilo," a metric term.

kandy apple: A rich, red finish popular on *kustom* cars.

kart: A small, open, four-wheeled vehicle with a single cylinder, two- or four-cycle gasoline engine.

K-car: A compact front-wheel drive vehicle introduced by Chrysler in the early 1980s.

KD set: An abbreviation for *knocked down set.*

Keenserts: A trade name for a thread-repair system using spiral inserts to repair damaged threads.

keeper grooves: The grooved area on a valve stem to accommodate the *keepers.*

keepers: Key-like, tapered-metal locking devices used to hold valve retainers in place on the valve stem.

KE-Jetronics: A continuous electronic fuel-injection system by Bosh that has been modified using a *lambda* oxygen sensor.

kemp: A *kustom car* or *lead sled.*

keyboard: An input device used to key programs and data into the computer's storage.

keyless entry: A system using a coded keypad that allows the operator to unlock the doors or the trunk from outside the vehicle without the use of a key.

key-off loads: A term used for *parasitic loads.*

keystone ring: Compression piston ring, double tapered and shaped like a keystone.

keyway: A groove milled or machined onto a shaft or into a bore to accept a square, half moon, or round piece of metal.

kg: An abbreviation for the metric *kilogram.*

kickdown: A downshift to the next lower gear in an automatic transmission when the driver applies full throttle, as in overtaking and passing another vehicle.

kickdown pressure: The pressure developed to downshift the transmission from a high gear to a lower gear.

kickdown valve: A valve located in the valve body that develops *kickdown pressure.*

kick out of gear: To shift to neutral.

kickpad: The area along the inside bottom of a car door.

kickup: A section of a chassis frame that is raised to clear the axles or suspension.

killed: **1.** To bring to a stop, such as to kill an engine. **2.** To overcome completely or with irresistible effect.

kill switch: A switch used to disconnect the electrical system in an emergency.

kilogram *(kg)*: A metric unit of measure for weight, as in English pounds.

kilometer *(km)*: A metric measure for distance, as in English miles.

kilo Pascal *(kPa)*: A metric unit of measure equal to 6.895 psi.

kinetic energy: The energy of motion, such as that of a flywheel.

King Kong: A hemi-engined Dodge or Plymouth stock car racer during the 1960s and 1970s.

kingpin: **1.** A pin or shaft on which the steering spindle assembly rotates. **2.** The pin, mounted through the center of the trailer upper-bolster plate, that mates with the *fifth wheel* and locks to secure the trailer to the *fifth wheel.*

kingpin angle: A term used for *kingpin inclination.*

kingpin axis: The inward tilt of the steering axis from the vertical.

kingpin inclination: A major factor in a vehicle's directional control and stability; the angle of a line through the center of the kingpin in relation to the true vertical centerline of the tire viewed from the front of the vehicle.

kit car: A knocked-down vehicle designed to be built by the owner.

K-Jetronic: A continuous fuel-injection system; a forerunner of the *KE-Jetronic system.*

km: An abbreviation for the metric *kilometer.*

knee bolster: An energy-absorbing pad used on a passive-restraint system to cushion the forward motion of the driver during an accident by restricting leg movement.

knock: A noise within an engine generally caused by *detonation* or *preignition.*

knocked-down set *(KD set)*: A package of automotive parts, assemblies, and subassemblies packaged at one location to be assembled at another location.

knock off: **1.** A counterfeit. **2.** A cheap auto part packaged to represent a popular brand.

knock-off hub: A large, single, two- or three-eared wing nut used to retain a wheel; illegal for street use.

knock sensor: A sensor that signals the engine-control computer when detonation is detected, momentarily retarding ignition timing until detonation ceases.

knuckle: A term used for *steering knuckle.*

knuckle arm: The arm that extends backward from steering knuckle to provide attachment for the tie rods.

knurl: A series of ridges formed on the outer surface of a piston or the inner surface of a valve guide to help reduce clearance and hold oil for added lubrication.

KOEO: An abbreviation for "key on engine off."

KOER: An abbreviation for "key on engine running."

kPa: An abbreviation for the metric kiloPascal, as in the English pounds per square inch.

kustom: A term George Barris used for "custom," now in common use.

kustom car: A "custom" car, specifically one built by George Barris, a noted California customizer.

L

L: **1.** An abbreviation for Low, one of the forward gear positions of a transmission. **2.** Liter, a unit of metric liquid measure.

labeled: An identifying mark or trademark of a nationally recognized testing lab that is attached to signify that the item has been tested and meets appropriate standards.

lacing: A method used to mend cracks in cylinder heads and blocks using threaded repair plugs side-by-side and overlapping each other.

lacquer: A type of paint that dries by solvent evaporation that must be rubbed out to produce a gloss.

ladder bars: A device attached to the chassis or frame at one point and to the rear axle at two points on a drag-race car to reduce rear-axle windup.

ladder chassis: A conventional frame design that consists of two side rails, not necessarily parallel, connected to each other by a series of cross members like a ladder.

ladder frame: A term used for *ladder chassis.*

lag: **1.** The incorrect operation of a *shock absorber* because of aeration due to the mixing of air with oils, causing the *shock absorber* to produce a poor ride. **2.** To fail to keep up; to fall behind.

lakes: A dry lake used for high-speed performance trials.

lakes pipes: Straight exhaust pipes with no muffler.

lakester: A *hot rod* with minimal frontal area and fully exposed wheels designed for lakes competition.

LA Kit: A reconditioned crankshaft that is supplied with the appropriate bearings and an installation kit.

LAL: An abbreviation for *lowest achievable level.*

lambda: **1.** The Greek letter L. **2.** A term used by engineers to represent the *air/fuel ratio.* **3.** A European auto maker's term for the oxygen sensor.

lambda control: Using a lambda, sensor-controlled computer to adjust the *air/fuel mixture.*

lambda sensor: A European auto maker's term for the oxygen sensor.

Lambda Sond: The first closed-loop, fuel-injection system to appear in production, developed jointly by Robert Bosch and SAAB.

laminar airflow: **1.** The smooth, continuous movement of one layer of gas or liquid over another. **2.** The movement of a body of *air/fuel mixture* through the *intake manifold* and ports as it flows over the *boundary layer.*

laminate: **1.** A structure consisting of two or more layers of material, such as fiberglass. **2.** To fabricate a structure of two or more layers of material, such as fiberglass.

laminated case: A container, such as a safety fuel cell, with walls of two or more layers of material.

laminated core: An assembly of steel sheets for use as an element of a magnetic circuit having the property of reducing eddy-current losses.

lamp: A device to convert electrical energy to radiant energy, which is normally visible light.

land: That part of a piston between the ring grooves that separates and supports the rings.

landau: A semi-convertible car with a folding top over the rear passenger compartment.

landau bar: Ornamental *ogee* or S-shaped trim sometimes applied to *C-pillars* or *sail panels.*

landau iron: A term used for *landau bar.*

landau top: A passenger-car top partially covered with vinyl to give it a convertible-like appearance.

landing gear: Retractable supports for a semi-trailer to keep the trailer level when the tractor/truck is detached from it.

land speed record *(LSR):* The maximum speed obtained by: **1.** A wheel-driven, internal-combustion engine vehicle. **2.** A thrust-driven jet or rocket engine vehicle.

lap: **1.** A complete trip around an oval track or road course. **2.** To gain a full lap over the second place competitor.

lap joint: A joint between two overlapping members in parallel planes.

lapping: A valve-grinding process using a paste-like grit on the face of the valve.

lapping compound: A paste-like grit used for lapping valves.

lap weld: A welding process between two overlapping members in parallel planes.

laser: **1.** An acronym for light amplification by stimulated emissions of radiation. **2.** A device that produces a concentrated, coherent light beam by stimulated electronic or molecular transitions to lower energy levels.

lash: The clearance between two parts.

lash-pad adjusters: Small round metal pieces of various thickness used to adjust valve clearance.

late apex: Getting to the inside of a turn later than usual due to a late entry into the corner.

latent heat: The amount of heat required to cause a change of state of a substance without changing its temperature.

latent heat of condensation: The quantity of heat given off while changing a substance from a vapor to a liquid.

latent heat of evaporation: The quantity of heat required to change a liquid into a vapor without raising the temperature of the vapor above that of the original liquid. Also known as *latent heat of vaporization.*

latent heat of fusion: The amount of heat that must be removed from a liquid to cause it to change to a solid without causing a change of temperature.

latent heat of vaporization: A term used for *latent heat of evaporation.*

lateral acceleration: The centrifugal force that tends to push a vehicle sideways, toward the outside of a turn.

lateral link: A suspension component used to reduce side-to-side movement of a wheel.

lateral runout: A tire that has excessive variations in width; the measured amount of sideways wobble on a rotating tire.

lateral weight transfer: The momentary shift of a vehicle weight from the inside tires to the outside tires, or outside to inside, due to cornering forces.

launch: A good start off the starting line in drag racing.

lay a patch: The same as to *lay rubber.*

lay a scratch: The same as to *lay rubber.*

lay on it: To go fast.

lay on the iron: To cut to the inside of another car on a turn, forcing it away from the apron or apex and out of the groove.

layout: Lines scribed on a piece of material or metal as a guide in bending or forming.

lay rubber: To leave streaks of rubber on the pavement during rapid acceleration. Also *lay a patch* and *lay a scratch.*

lay the crank: A reconditioned crankshaft supplied with the appropriate bearings and an installation kit.

LCD: An abbreviation for *liquid crystal display.*

lead: **1.** An element, Pb, often used as a body filler. **2.** To use lead as a body filler. **3.** Short for *tetraethyl lead,* a compound formerly used to increase the octane rating of gasoline.

lead: To be out in front.

leaded gasoline: Gasoline to which a small amount of *tetraethyl lead* is added to improve engine performance and reduce detonation, a practice no longer allowed due to EPA regulations.

leadfoot: A person who drives faster than necessary.

leading link: A component of the suspension system that is attached to the chassis behind the wheel and positioned to resist fore-and-aft movement of the wheel.

leading shoe: A brake shoe in a nonservo brake that pivots around a fulcrum in the direction of normal drum rotation. Also known as *forward shoe.*

lead sled: **1.** A term used for *kustom* or *kemp.* **2.** A car with excessive amounts of lead or body filler over mediocre, sheet-metal work.

lead sulfate: A hard, insoluble layer that slowly forms on the plates of a discharging battery that may be reduced only by slow charging.

leaf spring: A rear, vehicle-suspension spring featuring one or more flat leaves of spring steel with graduated lengths. It has an "eye" at one end to connect to the vehicle frame and it is connected to the axle with a U-bolt.

leaf-spring bushing: Bushings that are used to dampen noise and vibration from the road to the frame of the car.

leaf-spring center bolt: A bolt passed through a hole in the center of each spring leaf for holding the leaves together. The bolt head is used to locate the spring position on the axle-housing.

leaf-spring hanger: The vehicle frame-attachment bracket for the front eye of the rear leaf spring that allows the spring to pivot.

leaf-spring shackle: The small arm, or swing attachment for the rear of the leaf spring located between the frame and spring eye. It allows the spring to shorten and lengthen during normal driving conditions.

leaf-tip insert: Small, replaceable pads of plastic, rubber, or composition placed between spring leaves near their ends to promote slippage between the leaves during flexing.

leakage current: A small amount of current that flows through insulation when a voltage is present and heats the insulation due to it's resistance, resulting in a slight power loss.

leak detector: **1.** A device using visual or audible signals to indicate a leak. **2.** A dye-type fluid that may be injected into a system that will indicate the presence of a leak.

leakdown test: A test using 100 *psi* (689.5 *kPa*) air pressure injected in each cylinder via the spark-plug hole or injector port to determine the leakage past the rings, gaskets, or valves.

leaker: **1.** A vehicle not well prepared. **2.** A vehicle leaking oil or coolant.

Leak Find: A trade name for a dye solution that can be injected into a system to find difficult leaks.

lean: A term often used for *lean mixture* or *lean out.*

Lean Burn: A Chrysler electronic engine control that appeared in the mid 1970s. It maintains precise control of the spark timing to allow a very lean mixture to burn properly, reducing emissions using an analog computer.

leaner and later: Early calibration strategies for *air/fuel mixture* and ignition timing to reduce HC and CO formation.

lean misfire: A condition caused by a vacuum leak or open *EGR* valve that results in an *air/fuel mixture* too lean to sustain combustion, causing one or more cylinders to pass unburned fuel into the exhaust system, resulting in an increase in hydrocarbon (HC) emissions.

lean mixture: An *air/fuel mixture* with too much air.

lean on it: A term often used for *lay on it* or *lay on the iron.*

lean out: **1.** To increase the portion of air or decrease the portion of fuel in an *air/fuel mixture.* **2.** To decrease the portion of *nitro* in a fuel mixture.

lean roll: To turn carburetor idle-mixture screws in enough to effect a slight *rpm* drop, causing a leaner mixture.

leaver: A drag-race driver that red lights by leaving the line too soon.

left-hand thread: A thread pattern on a bolt or nut that requires it to be turned to the left or counter-clockwise for tightening.

leg it: To go fast.

leg out of bed: A connecting rod that has broken through a cylinder block wall.

Le Mans: A road-racing circuit in France.

Le Mans start: A method of starting a race by having the drivers stand across the track and, at the start signal, race to their vehicle, jump in, belt up, start their engines, and take off.

lemon: A vehicle, generally a new one, with several defects or a defect that can't be resolved.

lemon law: Federal and state laws that assure customer satisfaction for repairs to new vehicles by the dealer or replacement by the manufacturer.

length-to-diameter ratio: The ratio of a coil-spring wire diameter to its overall length, a measure of the spring's effectiveness.

level: **1.** To be on a horizontal plane. **2.** The amount of liquid in a system.

level control: A term used for *automatic level control.*

Lexan: The trade name for the shatterproof, heat-resistant clear plastic used in race-car rear windows.

L-head: A term used for *L-head engine.*

L-head engine: An engine with intake and exhaust valves in the block, parallel to the pistons and cylinders.

LH-Jetronic: An electronic fuel-injection system by Bosch that uses a *mass airflow sensor* with a digital control unit.

LHV: An abbreviation for *lower heating value.*

lid: The cylinder head on a *flathead* or *L-head engine.*

life expectancy: The anticipated longevity of a part, assembly, or component.

lift: **1.** The amount of the opening of a valve. **2.** The amount of rise generated by a lob of the camshaft. **3.** An upward force caused by the airflow around a vehicle.

liftback: A rear luggage compartment that is an extension of the passenger compartment with access gained through an upward-opening, hatch-type door.

lifter: **1.** A part between the camshaft and push rod on an *OHV* engine. **2.** A part between the camshaft and valve stem on an OHC engine. **3.** Also known as *follower.*

lifter bore: The hole in which a valve lifter is located.

liftgate: The rear opening of a *hatchback* or *liftback.*

lift it: Get off the throttle.

lift kit: A suspension package designed to raise the vehicle body above the frame and tires.

lift-throttle oversteer: A loss of grip on the drive wheels of a rear-drive vehicle when the throttle is lifted during fast cornering causing the rear of the vehicle to swing toward the outside of the turn.

light-duty vehicle: Any motor vehicle rated at 8,500 pounds (3,856 kilograms) GVWR or less, having a curb weight of 6,000 pounds (2,722 kilograms) or less, and having a frontal area of 45 square feet (4.2 square meters) or less.

light it off: Start an engine.

light-off, mini-oxidation catalytic converter: A small catalyst mounted just behind the exhaust manifold that gets hot very quickly after the engine is started, so that it begins working in time to neutralize much of the extra pollution that is produced during cold running.

lights: The timing lights at the end of a drag strip.

light the rugs: To smoke the drive tires at the start of the race.

light the weenies: Same as *light the rugs.*

light up: To perform a *burn out.*

A B C D E F G H I J K L M N O P Q R S T U V W X Y Z

A B C D E F G H I J K L M N O P Q R S T U V W X Y Z

lightweight leaf spring: A *fiber-composite spring.*

lightweight-skin spare tire: A bias-ply, compact spare tire with a reduced tread depth to provide about 2,000 miles (3,218 kilometers) of tread life; designed for emergency use only, and driving speed limited to 50 mph.

lime deposits: A condition that exists in a cooling system when the lime, present in water-based coolants, comes out of solution and coats the engine's water passages.

limited-production option *(LPO):* An item of new car equipment available for a limited market, such as a high-performance police package.

limited-slip differential: A differential having special friction mechanisms tending to keep both rear-axle shafts rotating at the same speed, regardless of unequal tire-to-road surface friction.

limited-slip differential gear oil: A specially formulated gear oil required in limited-slip differentials because of the extreme pressures on the clutch cones or clutch plates and discs.

limiting valve: A term used for *front-axle limiting valve.*

limits: The maximum and minimum values designated for a specific element.

limo: Short for *limousine.*

limousine: **1.** A chauffeur-driven formal sedan having a glass partition separating the driver from the passengers. **2.** A bus or van used to carry people to and from an airport or train station.

line-and-hose tape: A type of *insulation tape.*

Linear EGR: An AC Rochester *EGR* system using a linear motor to move the valve's pintle in small steps, which provides precise control of recirculation.

linear-rate coil spring: A coil spring with equal spacing between the coils, one basic shape, and constant wire diameter having a constant deflection rate regardless of load.

line contact: **1.** The contact made between the cylinder and the torsional rings, usually on one side of the ring. **2.** The contact made between the valve and the valve seat.

line job: Beating a drag competitor from the start.

line mechanic: **1.** A mechanic who works on the repair line at a dealership. **2.** A mechanic who is skilled in a particular automotive system.

line pressure: **1.** The base pressure established in a transmission by the pump and pressure regulator valve. **2.** The pressure present in a line or hose.

liner: **1.** The synthetic, gum-rubber material bonded to the inner surface of the tire to seal it. **2.** A sleeve used to repair a worn cylinder. **3.** An insert used to repair a worn valve guide. **4.** Short for *streamliner.*

line ream: To ream bearings or bushings to size after they have been installed.

line-setting card: A card provided by the vehicle manufacturer that lists its specifications and equipment.

line-setting tag: A tag provided by the vehicle manufacturer that lists its specifications and equipment.

lines of force: A term used for *magnetic lines of force.*

line static pressure: A 10 to 15 *psi* (69 to 103 *kPa*) hydraulic pressure maintained in a drum-brake system when the brakes are not applied to keep pressure on wheel-cylinder cup lips in order to seal fluid in, as well as air and dirt out.

lining: The friction material attached to the drum-brake shoe or disc-brake pad that contacts the brake disc or drum when the brakes are applied.

link: **1.** A lever or rod to transmit movement from one part to another. **2.** The portion of a chain's structure that connects two adjacent joint or pitch centers.

linkage: A series of levers or rods used to transmit movement from one part to another part.

linkage power steering: A type of steering system that has a power assist connected directly to the steering rods.

link plate: A side plate of a pin link or a roller link in a roller chain.

liquid-coolant circulation: A term used for *coolant circulation.*

liquid-cooled engine: An engine that is cooled by circulating coolant around the cylinders and through passages in the heads.

liquid cooling: A method of engine cooling that relies on coolant circulation through water jackets inside the cylinder head and block, then on to a radiator to maintain proper operating temperature.

liquid-crystal display *(LCD):* A type of alpha-numerical readout display on some instruments using a liquid-crystal film sandwiched between glass plates that become opaque when an electrical current is applied.

liquid-filled gauge: A mechanical instrument that is filled with a clear liquid to damp out oscillations that could affect accuracy.

liquid line: **1.** The line connecting the receiver-drier outlet with the expansion-valve inlet. **2.** The line from the condenser outlet to the receiver-drier inlet is sometimes called a liquid line.

liquidus: The lowest temperature at which a metal or an alloy is completely liquid.

liquid-vapor separator: An evaporative emissions or fuel-supply-system control component that allows vaporized fuel to return to the fuel tank, where it condenses for reuse. Uncondensed vapors flow through tubing into the *charcoal canister.*

liquified natural gas *(LNG):* Methane gas (CH_4) that has been converted to a liquid by chilling.

liquified petroleum gas *(LPG):* A predominately propane gas (C_3H_8) that contains some butane gas (C_4H_{10}) that is compressed to a liquid state.

liquifier: A term used for *condenser.*

liter: A metric unit of liquid measure.

litmus: An absorbent paper strip impregnated with a purple-colored matter that turns blue if dipped into an alkaline solution and red if dipped into an acid solution.

little end: The smaller end of a connecting rod.

live axle: An axle through which power is applied via a differential and half shaft.

L-Jetronic: A Bosch pulsed, electronic, fuel-injection system that uses input on the volume of intake air to calculate fuel delivery.

LNG: An abbreviation for *liquified natural gas.*

load: **1.** The demand for power placed on an engine. **2.** The cargo of a truck or other vehicle. **3.** The amount of weight placed on a tire. **4.** The heat quantity imposed on an air conditioner. **5.** The fuel for a drag racer, such as nitro.

load-carrying ball joint: The lower ball joint on low-mounted coil springs or torsion bars or the upper ball joint on high-mounted, coil-spring suspensions; compression or tension loaded, supporting the weight of the vehicle.

loading: An enrichment of the *air/fuel mixture* to the point of rough idling, usually resulting in black smoke from the exhaust.

load-leveling shock absorber: Inflatable *shock absorbers,* often used with an electronic height-control system, that are pressurized with air to increase their load-carrying capability.

load-proportioning valve *(LPV):* A device used to distribute hydraulic pressure to the front and rear brakes based on the vehicle load.

load range: **1.** A term used for *load rating.* **2.** The weight a tire will support at a given inflation pressure.

load rating: A term designating the maximum weight that a particular tire is designed to support, usually related to a specific air pressure.

load test: **1.** A starter-motor test in which the current draw is measured under normal cranking load. **2.** A battery test whereby the battery is loaded for testing.

lobe: That concentric part of a camshaft that causes the valves to open and close by actuating a valve lifter.

lobe centers: The number of degrees of a camshaft between intake and exhaust lobes.

local preheating: The preheating of an isolated or specific portion of a structure.

location table: A term used for *component location table.*

locked rear end: A rear axle without a differential where both rear wheels turn at the same speed on curves as well as on a straightaway.

locker: A term used for *locking rear end.*

locking hubs: A *dog clutch* in a wheel that permits it to be disengaged from the axle shaft and made free-wheeling when a driving force is not required, such as when being towed.

locking rear end: A final drive differential that acts as a locked rear end on the straightaway, but allows one wheel to be free-wheeling on curves.

locking torque converter: A hydraulic torque converter in an automatic transmission having a mechanical clutch that locks at cruising speeds.

lock nut: **1.** A nut that is designed to lock when tightened or torqued. **2.** A second nut that is screwed down tightly against the first nut.

lock pin: A pin used in some ball sockets to keep the connecting nuts from working loose, and on some lower ball joints to hold the tapered stud in the steering knuckle.

lockplate: A metal washer-like device with tabs that may be bent around a nut or bolt head to prevent them from turning.

lockstrap: A manual adjustment mechanism that allows adjustment of free travel of a lever or rod.

lock-to-lock: The number of turns required to turn the front wheels from one extreme to the other.

lockup: The point at which braking power overcomes the traction of the vehicle's tires and skidding occurs, causing loss of control, long stopping distances, and flat-spotting of the tires.

lockup clutch: A type of clutch in which the *torque converter turbine* is locked up with the engine, eliminating slippage.

lockup torque converter: A fluid clutch designed with a clutching assembly in an automatic transmission to improve coupling efficiency at a predetermined vehicle speed.

lock washer: A split-type of washer that helps prevent a bolt or nut from working loose.

Loctite: A trade name for a line of sealants and adhesives that are popular in the automotive industry.

long arm: The throw of a *crankshaft* after it has been *stroked.*

longbed: The cargo area of a long-wheelbase pickup truck.

long block: An assembled engine block that contains all of the components from the *intake manifold* to the exhaust ports.

longest leaf: A term used for the *main leaf.*

long gear: A final drive with a low gear ratio providing high gearing.

long, short arm suspension: A front suspension system where the upper control arm is shorter than the lower control arm, allowing each wheel to compensate for changes in the road surface while not greatly affecting the opposite wheel in a manner that allows the wheel and tire assembly to rise and fall vertically as it goes over bumps.

loop: A *spin* in track racing, usually deliberate to avoid an accident.

loops: Formed eyelets with minimal gaps at the ends of extension springs.

loop scavenging: A method used to remove the exhaust gases from the cylinder in a two-stroke engine.

loose: 1. The tendency of a vehicle to oversteer. 2. A slick track or driving surface.

loose roller clutch: A one-way clutch which has the rollers individually placed between the cam and race, not located by a *gauge.*

loper: A big, high-performance engine that idles roughly due to a high profile cam.

lo po: Low performance.

lose fire: To stall an engine.

lost foam casting: A casting method using a model part made of styrofoam.

loud pedal: An accelerator.

lover cover: The *scattershield* between the driver's legs in a front-engined, single-seat drag car.

louvers: 1. Slotted openings in a hood or body panel to admit or emit ambient air. 2. A vertical blind-type shutter to reduce or block airflow through a device, such as a radiator.

love taps: The bumping and shoving that often occurs when cars are running in a closely packed group in circle-track racing.

low beam: A headlamp intensity for use when meeting or following another vehicle.

low end: Low engine speed.

low end power: The engine horsepower output during the first 25–30% of engine *rpm* range.

lower: To reduce the ride height of a vehicle by modifying its suspension.

lower A-arm: The lower member of a double A-arm suspension system.

lower control arm: A front suspension component connected between the pivoting attachment point on the car frame and the lower ball joint, which is fastened to its outer end.

lower end: The *crankshaft, main bearings,* and *connecting rod bearings* assembly in an engine.

lower entry: The bottom of a wet cylinder sleeve in an engine block.

lower entry sleeve: A cylinder sleeve used to repair a *lower entry.*

lower heating value *(LHV)*: The latent *heating value* of water that is exhausted as steam.

lowering block: A device that may be used to reduce the riding height of certain vehicles.

lower mount: A support for the engine or transmission that is below the crankshaft centerline.

lower radiator hose: The radiator hose from the outlet of the radiator to the inlet of the water pump.

lowest achievable level *(LAL)*: A term used to define the lowest amount of emission permitted for any substance considered toxic or otherwise hazardous.

low gear: A speed obtained from a planetary gear set when the internal gear is held and power is applied to the sun gear, producing an increase in torque.

low gearing: A drive ratio that provides maximum output at a low road speed.

low head: A term used for *low-head pressure.*

low-head pressure: The high-side pressure that is lower than expected for a given condition.

low-lead fuel: A gasoline that contains less than 0.018 ounces (0.5 grams) per gallon (3.785 liters) of *tetraethyl lead,* no longer sold in this country due to the environmental impact.

low-loss fitting: A fitting designed to close automatically or manually to prevent fluid or vapor loss when used at connection points between hoses, service valves, vacuum pumps, recovery, or recycle machines.

low maintenance battery: A conventional, vented lead-acid battery that requires periodic maintenance.

low-mounted coil spring suspension: A type of suspension having a coil spring located above the lower control arm, with the top end of the spring contacting the car frame, found primarily on vehicles having a separate or stub frame.

low pedal: 1. A condition where excessive clearance, at some point in the braking system, or a low fluid level, causes almost full pedal movement for the application of the brakes. 2. The clutch-pedal position of a Model T when engaged in low gear.

low-pressure control: An electrical or mechanical device used to control pressure in the low side of a system.

low-pressure cutoff switch: An electrical switch that is activated by a predetermined low pressure to open a circuit during certain low-pressure periods.

low-pressure line: A hose or line used to carry low pressure vapor, liquid, or air.

low-pressure side: 1. Usually refers to the return side of a fluid or air system having a low pressure. 2. Often referred to as *suction side.*

low-pressure switch: A switch that is actuated due to a fall in pressure.

low-pressure vapor line: A term used for *suction line.*

low rider: A vehicle with small wheels so that it has been lowered as much as possible.

Low Risers: Standard-height cylinder heads developed in the early 1960s by Ford for 406, 427, and 428 cid engines.

low side: A term used for *suction side.*

low-side pressure: The pressure in the low side of an air conditioning system, from the evaporator inlet to the compressor inlet, as may be noted on the low-side pressure gauge.

low-side service valve: A device located on the suction side of the system that allows the service technician to check low-side pressures or perform other necessary service operations.

low-speed system: A circuit in the carburetor that provides fuel to the air passing through it during part-throttle, low-speed operation.

low-suction pressure: Pressure that is lower than normal in the suction side of the system due to a malfunction of the unit.

LPG: An abbreviation for *liquified petroleum gas.*

LPO: An abbreviation for *limited-production option.*

LPV: An abbreviation for *load-proportioning valve.*

LSR: An abbreviation for *land speed record.*

lubricant: 1. A substance, usually petroleum based, used to coat moving parts to reduce friction between them. 2. The new synthetic product poly alkaline glycol (PAG) and ESTER used with new refrigerants. 3. A term often used to identify an organic mineral-based grease or oil product.

lubrication guide book: A specially prepared publication, detailing required lubrication services with related information for each make and model of automobile.

lubrication interval: Manufacturer's recommended mileage and/or time limit when periodic lubrication services should be performed as a part of a preventative maintenance program.

lubrication system: The oil pump, filter, hoses, lines and passages in an engine that facilitate the oiling of all moving parts.

lug bolt: A hex-headed, threaded bolt used to hold the wheel on a vehicle.

lugging: Running an engine at less than normal *rpm,* causing it to *balk.*

lug nut: A hex-sided, threaded device used to hold the wheel on a vehicle having *lug studs.*

lug stud: A threaded protrusion used with a *lug nut* to hold the wheel on a vehicle.

LU-Jetronic: A United States version of a Bosch pulsed-electronic, fuel-injection system having a *lambda sensor.*

lumbar support: A seat support for the lumbar portion of the occupant's body.

lump: A large, heavy engine.

lunch: To severely damage or destroy something, such as an engine.

ma: An abbreviation for *milliampere.*

machinability: The relative ease with which materials can be shaped by cutting, drilling, or other chip-forming processes.

machine: **1.** A car, usually a late-model car. **2.** A device capable of doing work.

MacPherson strut: A type of front suspension having a shock absorber mounted directly below the coil spring.

MacPherson strut rear suspension: An independent, rear-suspension system having a shock-absorber strut assembly on each side, two parallel control arms, a tie rod, a strut rod, a forged spindle, a jounce bumper, and a bracket assembly.

MacPherson strut suspension: A front-end, independent suspension system in which the combined strut, steering knuckle, and spindle unit, supported by the coil spring at the top, is connected from the steering knuckle to an upper-strut mount.

MACS: An abbreviation for *Mobile Air Conditioning Society.*

mag: **1.** An abbreviation for *magneto.* **2.** An acronym for *magnesium.* **3.** A term used for a lightweight wheel, such as one made of aluminum or magnesium.

magnaflux: A dry, nondestructive magnetic test to check for cracks or flaws in iron or steel parts.

Magnaglow: A trade term for a wet, magnetic, and ultraviolet nondestructive test to check for cracks or flaws in iron or steel parts.

magnesium: An element, Mg, that is the lightest in weight of all structural metals.

magnetic clutch: A coupling device used to turn the compressor on and off electrically.

magnetic field: The area of influence of a permanent- or electro-magnet that exists between its north and south poles.

magnetic filter: A magnet or magnet assembly located in a fluid system to attract and retain ferrous metal, nickel, and cobalt particles which may be present. Composite particles in which a ferro-magnetic material is entrained may also be present.

magnetic lines of force: The invisible magnetic lines set up between the north and south pole of a magnet.

magnetic particle test: A non-destructive test using a magnet and magnetic particles, such as iron filings, to check iron or steel parts for cracks.

magnetic pick up: The system of an electronic ignition that triggers the amplifier to generate voltage to fire the plugs.

magnetic pickup coil: A small coil of wire wound on an iron core in an electronic ignition system that is magnetically affected as the *reluctor teeth* pass by.

magnetic pole: The point at which the magnetic lines of force enter or leave a magnet.

magnetic switch: A switch energized electrically with a coil of wire serving as an electromagnet, such as in a *relay.*

magnetic timing: A procedure for checking or adjusting an engine's ignition timing by inserting a magnetic probe into a receptacle near the crankshaft harmonic balancer or flywheel.

magnetism: The natural or electrical ability to attract a ferrous metal.

magneto: An electrical device requiring no outside power source, that generates and delivers current to fire the spark plugs.

mags: Wheels.

mag wheels: **1.** A magnesium wheel. **2.** A term used to describe any chromed, aluminum offset, or wide-rim wheel of spoked design.

main: 1. The main bearing. 2. The feature event.

main bearings: The bearings that locate and support the crankshaft in an engine block.

main-body structural components: The assembly made up of the dash panel, underbody, roof, body panels, doors, and deck lid to form the passenger and luggage compartments.

main cap: The structural device that holds the *crankshaft* in place in an engine block.

main hoop: A *rollbar* placed just behind the driver's seat in a race car.

main jets: The nozzle in a *carburetor* that provides fuel during part- or full-throttle operation.

main journals: The journals that fit into the engine block to support the crankshaft.

main leaf: Usually the top leaf on multiple-leaf springs that provides the main vehicle support and contains the spring-mounting "eyes."

mainline pressure: The pump-developed, hydraulic-regulated pressure that operates apply devices, such as bands, and is the source of all other pressures in an *automatic transmission.*

mains: A term used for *main bearings* and *main cap.*

Maintenance Awareness Program *(MAP):* A coalition of automobile repair shops, suppliers, car companies, and associations dedicated to building trust between the aftermarket and its customers.

maintenance-free battery: A sealed battery having no practical provision for the addition of water to the cells, since periodic maintenance is unnecessary.

maintenance manual: A manufacturer's or independent publication containing comprehensive maintenance information for a certain vehicle.

maintenance specifications: A listing of interval oil and lube grades, service points, and capacities for periodic and routine preventative maintenance service.

major diameter: 1. The largest in diameter of a cylinder. 2. The largest in the diameter of a bolt, its threads.

major overhaul: An engine overhaul just short of a *rebuild,* where all worn or damaged parts are rebuilt or replaced.

major tune-up: A conventional, ignition-engine maintenance procedure that may include points, capacitor, cap, rotor, plugs, plug wires, valve adjustment, and a carburetor overhaul.

make: 1. To close a switch. 2. A distinctive name given to a group of cars by a manufacturer, such as Riviera, by Buick.

make and break: 1. The term used when a switch is closed or opened. 2. A pair of contact points, one stationary, and the other operated by a cam that makes the brake, such as in a conventional distributor. 3. To construct something then tear it down.

make the show: To qualify for a race.

male: A term universally applied to any part which fits into another part, such as the shaft on which a pulley or gear hub fits.

male and female: Terms that apply to inner and outer members which fit together, such as a bolt and nut.

malfunction: An improper or incorrect operaton of a device or system.

malleability: The ability of a metal to be formed through hammering or bending.

manganese: An element (Mn), used in alloys of iron, steel, aluminum, and copper, often confused with magnesium (Mg).

manganese bronze: An alloy of copper, zinc, and up to 3.0% manganese used for toothed wheels and gears.

manganese steel: A steel alloy with 12–14% manganese; used for drill bits.

manifold: 1. A device used to hold two or more gauges with fluid passages and flow provisions; used for testing and servicing purposes. 2. A conduit-like device used to channel the *air/fuel mixture* into an engine.

3. A conduit-like device used to duct the exhaust gas out of an engine.

manifold absolute pressure ***(MAP):*** The pressure in an *intake manifold* relative to atmospheric pressure.

manifold absolute-pressure sensor: A variable resistor used as a sensor to signal an engine-control computer relative to the vacuum conditions in the *intake manifold.*

manifold air temperature: The temperature of the intake stream in the *intake manifold,* as increased by a heat riser or an Early Fuel-Evaporation system, and/or converted to an engine-control computer input by a sensor.

manifold gauge set: **1.** A set of gauges used to service an air-conditioning system. **2.** A set of gauges used to measure vacuum or pressure in an engine's intake or exhaust system.

manifold heat control: A flapper-type valve in the *exhaust manifold* that diverts heat to the *intake manifold.*

manifold pressure: A positive pressure in the *intake manifold* measured in pounds per square inch or bars.

manifold vacuum: The negative pressure in an engine's *intake manifold* produced by the engine's pumping action and measured in inches of mercury.

manomometer ***(u-tube):*** A u-shaped tube with a graduated scale used for measuring the pressure of gases or vapors.

manual bleeding: A two-technician technique for bleeding brakes; one pumps the brakes as the other opens and closes the bleeder screws when required.

manual control valve: A spool valve, located in the valve body, that determines fluid flow from the valve body to various hydraulic circuits controlled by the vehicle driver through the selector lever.

manual low: The position of the units in an automatic transmission when the driver selects the Low or first-gear position of the quadrant.

manual operation: **1.** The act of operating the power steering system in the event of an engine system failure; increased effort is required. **2.**The ability to override or cancel any automatic or semi-automatic function.

manual selection: The vehicle operator's ability to: **1.** Move the gear selector lever by hand, such as with a manual transmission. **2.** Select any condition or function using a keypad, switch, or lever.

manual slide release: The release mechanism for a sliding *fifth wheel,* which is operated by hand.

manual transmission: A manually shifted gearing device in the power train that allows variation on the relationship between engine speed and road speed.

manufacturer: A person, firm, or corporation engaged in the mass production or assembly of vehicles or any other product.

manufacturer's code: **1.** An alpha-numerical code to identify a product. **2.** A lettered marking code on tire sidewalls indicating tire manufacturer, plant, tire size, type of construction, and date of manufacture.

Manufacturers of Emission Controls Association ***(MECA):*** A trade association of the manufacturers of emission-control systems and components.

manufacturer's service manual: A book containing service information and technical data that is provided by the vehicle manufacturer.

manufacturing: The science of planning, designing, managing, and scheduling people, goods, material, equipment, and machinery to produce a useable and sellable product.

map: A flat graphic or pictorial topographic representation of a particular geographic area or location.

MAP: **1.** An acronym for *Maintenance Awareness Program.* **2.** An acronym for *manifold absolute pressure.*

MAP sensor: A device used to measure *MAP.*

marbles: Small pieces of dirt, rubber, or other debris on the outer edges of a race track, outside the racing line.

marcel: The spring assembly in the clutch-drive plate that absorbs the energy when the clutch lever is released.

margin: The material between the face and head of an intake or exhaust valve.

Mark: **1.** One of a series of cars usually identified by a Roman numeral. **2.** A term often mistakenly used for *marque.*

markings: The identifying marks on an assembly or component.

marque: A French term meaning "make," as in make of car.

marriage: The installation of a vehicle's power train in its chassis on an assembly line.

MAS: An abbreviation for *mass airflow sensor.*

masking: **1.** To cover up, as in prior to painting. **2.** The modifying of intake ports with dams to redirect the air flow.

masking area: A preparation stage in a paint shop in which the parts that are not to be painted are covered with paper and held in place with masking tape.

mass: The measure of the quantity of matter that is contained in a given quantity of solid, liquid, or gas.

mass airflow sensor *(MAS):* A device found at the inlet of the *intake manifold* of an electronic fuel-injection system which supplies the computer with input as to the volume of air entering the manifold, using the temperature differential between a heated platinum wire or a plastic film and the passing air, to generate a signal of varying voltage.

mass production: The manufacture of a product in large quantities.

master cylinder: A brake-system component having an integral reservoir filled with hydraulic brake fluid where pressure is developed when the driver depresses the brake pedal causing a linkage to move a piston.

master kit *(MK):* An engine kit that contains *crankshaft* and *camshaft bearings, lifters, oil pump, pistons, rings, gasket set,* and *core plugs.*

master leaf: A term used for *main leaf,* as in a spring.

match bash: Two out of three, or three out of five drag racing events.

match race: The same as *match bash.*

material expanders: Fillers that are used in the place of active material in a battery; the primary difference between a 3- and 5-year battery.

Material Safety Data Sheet *(MSDS):* A sheet required of the product manufacturer by *OSHA,* containing specific information about hazardous materials to be displayed in any workplace where such materials are used.

mat fabric: A fiberglass fabric with an irregular strand pattern.

maxi-fuse: A large, cartridge-type fuse having a higher current-capacity use because they are less likely to cause an underhood fire when overloaded.

MCU: An abbreviation for *microprocessor control unit.*

MEA: An abbreviation for *Mechanics' Education Association.*

mean: **1.** A top performing driver. **2.** A car tough to beat in a race.

mean effective pressure *(MEP):* The average pressure developed within an engine cylinder during a four-stroke cycle.

mean motor scooter: A top-performing car in drag racing.

mean rating: A more accurate rating than *absolute rating* or *nominal rating.*

measuring: The act of determining the size, capacity, or quantity of an object.

meat: The structural metal often found in an engine block.

meats: **1.** Big tires. **2.** Drag racing slicks.

MECA: An abbreviation for *Manufacturers of Emission Controls Association.*

A B C D E F G H I J K L M N O P Q R S T U V W X Y Z

mechanical efficiency: The percentage of input power as related to the output power in a mechanical transfer.

mechanical fuel pump: A device which draws fuel from the gas tank and delivers it to the carburetor by use of a mechanically operated diaphragm.

mechanical properties: The characteristics of a material that are displayed when a force is applied to the material; usually relating to the elastic or inelastic response of the material.

mechanical seal: A seal that is formed with direct metal-to-metal contact.

mechanical valve lifters: An adjustable valve lifter that provides direct cam-to-push-rod contact.

Mechanics' Education Association *(MEA):* An association that offers upgrade educational courses to independent shops.

medium riser: Standard-height cylinder heads offered for the Ford 427 cid engine in the 1960s.

meet: **1.** A term often used for competition, such as in "drag meet." **2.** The point where two or more pieces are joined together.

M85: A fuel blend of 85% methanol and 15% gasoline.

MEK: An abbreviation for *methyl ethyl ketone.*

melt: To destroy or damage by overheating.

melt down: A description among drivers of what happens to pistons when they seize in an engine due to preignition or too lean a mixture.

melting capacity: A term used for *ice-melting capacity.*

melting range: The temperature range between solidus and liquidus.

melting point: The temperature at which a solid becomes a liquid.

melting time: **1.** The time required for an overcurrent to sever a fuse element. **2.** The time required for a solid to become a liquid.

MEMA: An abbreviation of *Motor and Equipment Manufacturers Association.*

member: **1.** An essential part of a machine or an assembly. **2.** A person that belongs to a club or a group.

memory chip: A semiconductor device that stores information in the form of electrical charges.

memory seat: A power-seat feature which allows the driver to program different seat positions that can be "recalled" at the push of a button.

memory steer: An occurrence when steering does not return to the straight-ahead position after a turn, and attempts to continue in the original turn direction caused by a binding condition in the steering column or in the steering-shaft universal joints.

meniscus: The top of a column of liquid in a tube.

MEP: An abbreviation for *mean effective pressure.*

mercury: **1.** An element (Hg). **2.** A term used to designate a vacuum on the English scale, such as *inches of mercury* or *in-Hg.*

meshing: The engaging or mating of the teeth of two gears.

metal conditioner: A chemical used to remove light rust from a metal before surfacing or painting.

metal filament: The electrical conductor that glows when heated, as in an incandescent lamp.

metal inert gas *(MIG):* A gas used to shield a weld to prevent oxidation.

metal inert gas welding *(MIG welding):* A welding method using an inert gas shield.

metallic bond: The principal bond that holds metals together.

metallic brake lining: Brake linings made of metal particles that have been bonded or *sintered* together.

metallic friction material: A *sintered* friction material formulated with metallic or metallic-ceramic materials.

metallic paint: A paint that contains very small metallic particles causing the paint to seem to sparkle.

metal matrix composite *(MMC):* A mixture of metal and ceramic that is light weight and resistant to high temperature; used as an insulator.

metal pickling: A condition where metal is exposed to acid or is electroplated, causing it to become brittle.

metalworking area: **1.** The area in an auto-body repair shop where a vehicle body is repaired. **2.** The metal fabrication area of a service facility engaged in such activities as installing truck beds.

meter: **1.** A term used for *dynameter.* **2.** A metric unit for linear measure. **3.** To regulate the flow of a liquid or gas. **4.** An instrument, such as a *voltmeter* or *ammeter.*

metering device: Any device that meters or regulates the flow of a liquid or vapor.

metering rods: The tapered rods that regulate the flow of gasoline in a carburetor.

metering slit: A small, narrow opening used to meter fuel output in a fuel injector.

metering valve: **1.** A valve in a carburetor to deliver a precise amount of fuel or air. **2.** A valve in the brake system that delays brake action to front-disc brakes until after application of the rear-drum brakes.

methane: An odorless, colorless natural gas.

methanol: A form of alcohol used as a fuel for *Indy cars* and some drag-race classes: 85% methanol/15% gasoline.

methyl alcohol: A form of alcohol used as a fuel in *Indy cars* and some drag-race classes.

methyl ethyl ketone *(MEK):* A highly flammable fluid used for cleaning.

methyl tertiary butyl ether *(MTBE):* An oxygenated compound used to raise the octane rating of gasoline.

metric system: A measuring system used by most of the free world, except the United States which uses the English system. The United States is slowly moving to the metric system.

metri-pack connector: A connector similar to the *weather-pack connector,* but without the seal on the cover half.

Mexineering: The improvised back country repairs using only the tools and materials at hand.

mica: An insulating material used to separate the commutator bars of a generator or starter motor.

Mickey Thompson Entertainment Group *(MTEG):* An organization to promote stadium races for off-road vehicles.

microchip: A tiny silicon chip with thousands of electronic components and circuit patterns etched onto its surface.

microetch test: A test in which the specimen is prepared with a polished finish, etched, and examined under high magnification.

microfinish: A machine process to finish a surface of a part to reduce friction.

micron: A metric unit of linear measure equal to one-millionth of a meter.

microprocessor: The basic electronic arithmetic, logic, and control computer elements required for processing. Widely used as the control devices for microcomputers; very useful and extremely important to the automotive industry.

microprocessor control unit *(MCU):* **1.** A device referred to as an *electronic control unit* or *ECM.* **2.** The "black box" that controls various electrical and mechanical functions of a vehicle.

mid-engined: A chassis layout whereby the engine is behind the driver, but in front of the rear wheels.

midget: A small, oval-track racing car.

midnight auto parts: Parts that were stolen. Also known as *midnight auto supply.*

midnight auto supply: Same as *midnight auto parts.*

midpositioned: The position of a stem-type service valve where all fluid passages are interconnected.

midseated: An incorrect term used for *midpositioned.*

MIG: An acronym for *metal inert gas.*

MIG welding: A term used for *metal inert gas welding.*

mil: Short for *millimeter.*

mild steel: A steel alloy with a low carbon content.

miles per gallon *(mpg):* The distance, in miles, a vehicle will travel per gallon of fuel.

miles per hour *(mph):* An indication of distance/time traveled.

military standard *(MS):* A system of grading the quality of fasteners and other items.

mill: An engine.

milliampere *(ma):* A metric unit of electrical measure.

millimeter *(mm):* A metric unit of linear measure.

minimum forward reduction: Simple planetary-gear-set combination with the ring gear as input, the planetary carrier output, and the sun gear held.

minimum thickness: The lower limit of the thickness required after metal has been removed from an item such as a brake disc, to retain its integrity.

minispare tire: A *compact spare tire.*

mini stock: A category for four-cylinder, subcompact vehicles on an oval track.

minor diameter: **1.** The smallest diameter of a cylinder. **2.** The smallest diameter of a bolt thread.

minor overhaul: An engine-repair procedure that includes a valve job, replacing the piston rings, the rod bearings, and the gasket set.

minor tune up: An engine-maintenance procedure performed on distributor ignition vehicles that includes only the replacement of parts found to be defective.

miscible: Capable of being mixed in all proportions.

misfire: An intermittent or continuous failure to ignite the *air/fuel mixture* in one or more cylinders.

miss: A brief engine hesitation such as a *misfire.*

missing teeth: **1.** A condition found in a synchronous belt drive that may be caused by under-tensioning, misalignment, or excessive shock loads. **2.** Teeth broken off a gear.

Mitsubishi Jet Valve: A tiny third valve that admits nothing but air to churn up the air/fuel charge and promote lean running and a complete burn.

mixing: **1.** To blend or combine. **2.** To use parts, such as tires, of two or more sizes or types.

mixing chamber: That part of an apparatus in which a fuel gas, and oxygen or air are mixed.

mixture adjustment: To adjust the portions of a mixture, such as air and fuel.

MK: An abbreviation for *master kit.*

MKT: An engine parts kit that contains all of the parts found in an MK plus the timing gears.

MMC: An abbreviation for *metal matrix composite.*

Mobile Air Conditioning Society *(MACS):* A nonprofit organization founded in 1981 for the dissemination and distribution of comprehensive technical information, training, and communications to its members consisting of automotive air-conditioning shops, installers, distributors, suppliers, and manufacturers in the United States and Canada.

mode: A term generally used to describe a particular set of operating characteristics.

model year: The year of vehicle manufacture, as designated by the manufacturer; not consistent with the calendar year.

modesty panel: The panel below the bumpers that conceals the chassis components. Also known as *modesty skirt.*

modesty skirt: Same as *modesty panel.*

modification: To alter or change from the original.

modified: A vehicle that has been reworked for high-performance operation.

modified MacPherson strut: A type of MacPherson strut suspension that uses shock struts with coil springs mounted between the lower arms and spring pockets in the cross member, to absorb minor road vibrations by the chassis rather than fed back to the driver through the steering system.

modified strut: A strut suspension where the coil spring is not part of the assembly and is independently located between the lower control arm and the frame.

modular wheel: A wheel made of different sections that are bolted or riveted together.

modulated vacuum: A vacuum signal regulated to a particular level.

modulator: **1.** A device that varies the frequency amplitude and phase of electromagnetic waves. **2.** A device that regulates hydraulic line pressure in a transmission to meet varying load conditions.

module: A semi-conductor control for an electronic-ignition system.

modulus of elasticity: The point at which a material has been bent too far to snap back into shape.

moisture: Humidity, dampness, wetness, or small droplets of water.

moisture ejector: A valve mounted to the bottom or side of the supply and service reservoirs that collects water and expels it every time the air pressure fluctuates.

mold: **1.** A hollow form into which molten metal is poured to form a part. **2.** A fungus. **3.** To form or shape an object.

molded: To rejoin body panels using a filler material to conceal the seam.

molded connector: A male or female electrical connector usually having one to four wires that are molded into a one-piece component.

molded curved-radiator hose: A term used for *preformed radiator hose.*

molded hose: A section of hose permanently formed to fit a particular application.

molecular sieve: A drying agent.

molecule: The smallest particle a substance can be divided and still remain that substance.

moly: Short for *molybdenum.*

molybdenum: An element (Mo) used in some steel alloys to add hardness and strength.

molybdenum disulfate: A combination of molybdenum (Mo) and sulfur (S) sometimes added to oil and grease to improve their lubricating qualities.

moly lube: A lubricant popular for coating engine parts during rebuilding to avoid an initial dry start up.

moly ring: A piston ring with a *molybdenum* coating.

moment of inertia: The tendency of a body to resist angular or rotational acceleration, such as a vehicle resisting a cornering effort.

MON: An abbreviation for *motor octane number.*

monel: An alloy of nickel (Ni) and copper (Cu).

A B C D E F G H I J K L M N O P Q R S T U V W X Y Z

money grabber: A driver who enters and starts a stock car race but leaves after a minimum number of laps, collecting appearance money but not racing seriously.

money pit: A vehicle that is expensive to restore and/or maintain.

monitor: An electronic sensor, such as the *lambda* sensor.

monkey motion: The excessive movement of a mechanical device, such as an improperly adjusted carburetor linkage.

monochlorodifluoromethane: The chemical term for refrigerant-22 (HCFC-12).

monocoque: A unitized, autobody-frame structure with stressed-sheet-metal body panels.

monogoggle: Eye protection that may be worn over glasses.

monoleaf spring: A spring that is made up of a single steel leaf.

monolithic: A device or structure made as a single unit.

monolithic catalytic converter: A catalytic converter which has its catalytic materials coating a ceramic honeycomb, as distinguished from the pellet-bed converter.

monolithic timing: The use of an electronic timing device to make precise adjustments in engine timing when the engine is running.

monster truck: A truck, usually a pick-up, having huge, oversized wheels and tires with a heavy duty power train to drive them.

Montreal protocol: An agreement signed by representatives of the participating countries to reduce *CFC* and *HCFC* emissions into the atmosphere by restricting and regulating the manufacture and sale of such products.

moon disc: A smooth hub cap that completely covers the wheel.

motor: A device that converts electrical energy into mechanical energy, such as a fan motor.

Motor and Equipment Manufacturers Association *(MEMA):* An association of manufacturers of automotive aftermarket parts, original equipment parts, tools, chemicals, and other products.

motor home: A self-contained vehicle built on a bus or truck chassis containing a driver's compartment and complete living facilities.

motor mounts: Supports made of hard rubber for the engine and transmission to be secured to the vehicle's frame.

motor mouth: An overly talkative person.

motor octane number *(MON):* A term used for *octane number.*

motor oil: A lubricant expressly formulated for use in an engine.

motor town: Detroit, Michigan.

motor vehicle: A machine, usually on rubber tires, that is propelled by means other than muscle power, and which does not operate on rails.

Motor Vehicle Manufacturers Association *(MVMA):* A trade association of the major automobile and truck manufacturers in the United States.

MoTown: **1.** The name of a Detroit-based record company. **2.** Short for *motor town* or Detroit.

Motronic: A Bosch electronic fuel-injection and ignition-management system.

mountain motor: A *big block V–8,* usually Chevrolet or Ford, enlarged to 500 cid or more; often to as much as 600 to 800 cid.

mount and drive: Pulleys, mounting plates, belts, and fittings necessary to mount a compressor and clutch or any other engine-driven assembly on an engine.

mounting grommet: **1.** The replaceable rubber bushings used at both ends of the shocks for mounting to reduce the transfer of sounds and jarring motions between the shocks and their mounting points on the vehicle. **2.** A reinforced eyelet through which a fastener is attached.

mounting position: The position at which a device is mounted, such as horizontal or vertical.

mouse: A computer input device for moving a cursor or data around on the display screen.

mouse milk: An additive for oil or fuel.

mouse motor: A Chevrolet *small-block V-8.*

moving vehicle resistance: A measure of the air resistance of an auto body design.

mpg: An abbreviation for *miles per gallon.*

mph: An abbreviation for *miles per hour.*

MS: An abbreviation for *military standard.*

MSDS: An abbreviation for *Material Safety Data Sheet.*

MTBE: An abbreviation for *methyl tertiary butyl ether.*

MTEG: An abbreviation for *Mickey Thompson Entertainment Group.*

mud and snow tire: **1.** An *all-season tire.* **2.** A term used for *snow-and-mud tire.*

mud bogging: Racing an off-road vehicle through a 100–200 foot (30–61 meter) mud hole.

mud flap: A flap hanging down behind a tire to prevent the tire from throwing debris against the lower panel of the vehicle or into the path of a following vehicle.

mud plug: A cap installed in the center of a wheel to keep out mud and other debris.

muffler: **1.** A hollow, tubular device used in the lines of some air conditioners to minimize the compressor noise or surges transmitted to the inside of the car. **2.** A device in the exhaust system used to reduce noise.

muffler bearing: A non-existent part.

muffler clamp: A clamp that secures the muffler, or pipe, to the bracket or hanger.

muffler hanger: A hanger that is used to secure the muffler.

mule: **1.** A yard tractor not suitable for over-the-road service. **2.** A road race car used for practice or testing.

multi-fuel vehicle: A vehicle having a modified engine that runs on more than one fuel, such as LPG and gasoline.

multi-leaved spring: A flat suspension spring having more than one leaf.

multiple disc: A clutch with several driving and driven discs.

multiple-disc clutch: A clutch having a large drum-shaped housing that can be a separate casting or a part of the transmission housing.

multiple-leaf spring: Leaf springs having a series of flat, steel leaves of varying lengths clamped together with a center bolt extending through them to maintain their position.

multiple pass: A term that applies to a recovery/recycle unit that removes refrigerant from an air-conditioning system and circulates it through the recovery/recycle unit to remove contaminants before it is pumped into the recovery cylinder.

multiple-strand chain: An assembly made up of two or more rows of roller links joined into a single structure by pins extending transversely through all rows.

multiple-viscosity oil: An engine oil that has a low viscosity when cold and a high viscosity when hot.

multiple-wire hard-shell connector: A connector, usually having a removable, hard-plastic shell, that holds the connecting terminals of separate wires.

multi-point fuel injection: A type of fuel-injection system that has a separate fuel injector for each of the engine's cylinders to deliver better performance and lower emissions than throttle-body injection (TBI) systems.

multi-point injection: A term used for *multi-port fuel injection.*

multi-purpose vehicle: A term used for *sports vehicle.*

multi-valve head: A cylinder-head design having more than one exhaust and/or intake valve per cylinder.

multi-viscosity oil: An oil having a low viscosity when cold and a high viscosity when hot.

Muncie: A popular four-speed manual transmission built by General Motors in Muncie, Indiana.

Muroc: An early, dry-lake site in the Mojave Desert, now known as Rogers dry lake; it is a space shuttle landing site at Edwards Air Force Base.

muscle car: A high-performance car with a big-block engine in a light-weight chassis on heavy-duty suspension with a two-door body.

mushroomed valve stem: A valve stem that is worn so much that its end has spread and metal is hanging over the valve guide.

mushroom lifter: A valve lifter having a foot diameter larger than the body diameter.

mutual induction: A condition whereby a voltage is generated in a secondary coil due to the application of voltage in an adjacent primary coil, such as the coil used in a conventional ignition system.

MVMA: An abbreviation for *Motor Vehicle Manufacturers Association.*

N: **1.** The neutral position of an automatic transmission. **2.** The chemical symbol for *nitrogen.*

NAAA: An abbreviation for *National Auto Auction Association.*

NACA duct: A bottle-shaped, low-drag air intake design.

NACAT: An abbreviation for *National American Council of Automotive Teachers, Inc.*

NADA: An abbreviation for *National Automobile Dealers Association.*

nail it: To apply full throttle.

NARSA: An abbreviation for *National Automotive Radiator Service Association.*

NASA duct: A frequent but incorrect reference for *NACA duct.*

NASCAR: An acronym for *National Association for Stock Car Auto Racing.*

nasty car: A *NASCAR* stocker.

NATEF: An abbreviation for *National Automotive Technicians Education Foundation.*

National American Council of Automotive Teachers, Inc. *(NACAT):* An association of secondary and post secondary automotive technology instructors.

National Association for Stock Car Auto Racing *(NASCAR):* The major sanctioning authority for stock car racing.

National Auto Auction Association *(NAAA):* A trade association of auto auction owners.

National Automobile Dealers Association *(NADA):* A trade association representing franchised new car and truck dealers.

National Automotive Radiator Service Association *(NARSA):* A trade association for those engaged in the marketing, service, or repair of cooling system components, tools, and equipment.

National Automotive Technicians Education Foundation *(NATEF):* A non-profit organization administered by *ASE* that sets standards for training programs in Vocational-Technical schools.

National Highway Traffic Safety Administration *(NHTSA):* An agency of the Department of Transportation responsible for establishing and enforcing highway safety regulations.

National Hot Rod Association *(NHRA):* The major authority for drag racing events.

National Institute for Automotive Service Excellence: A non-profit corporation founded in 1972 to promote and encourage high standards of automotive service and repair through voluntary testing.

National Institute for Occupational Safety and Health: A branch of the Federal Government, through the *Occupational Safety and Health Administration,* that is responsible for overseeing and enforcing occupational safety and health issues.

National Institute for Standards and Technology *(NIST):* A national organization that helps to set standards and technology for mobile air-conditioning systems.

National Lubricating Grease Institute: A trade association of manufacturers and distributors of grease.

National Street Rod Association *(NSRA):* A sanctioning body for street rods.

National Tire Dealers and Retreaders Association, Inc. *(NTRDA):* A trade association of tire dealers and tire retreaders.

National Tractor Pullers Association *(NTPA):* A sanctioning body for tractor-pull events.

National Vehicle Leasing Association *(NVLA):* A trade association of vehicle leasers, dealers, and funders.

National Wheel and Rim Association *(NWRA):* A trade association for the distributors of vehicular wheels, rims, and brake drums.

natural gas: A gas occurring naturally in the earth, such as methane, ethane, butane, and propane.

naturally aspirated: An engine that uses atmospheric pressure to force the air into the cylinders.

NC: An abbreviation for *normally closed,* usually expressed nc. Sometimes used to indicate no connection.

NDRA: An abbreviation for the *Nostalgia Drag Racing Association.*

NDT: An abbreviation for *non-destructive test.*

near side: The side of a vehicle closest to the curb.

necking: A narrowing area of the exhaust pipe due to a buildup of exhaust by-products.

necking knob: A knob attached to the steering wheel that permits rapid one-hand steering.

needle and seat: A metering assembly to admit fuel into the *carburetor.*

needle bearing: A bearing that contains needle-like rollers.

needle valve: The small, tapered male part of a *needle and seat.*

negative: **1.** One of the two poles of a magnet. **2.** One of the two poles of a battery, representing ground.

negative back-pressure EGR valve: An exhaust-gas-recirculation valve having a normally closed bleed hole which opens when back pressure drops, reducing vacuum above the diaphragm and cutting the recirculation flow.

negative camber: The inward vertical tilt of the wheels on a vehicle.

negative offset: A wheel rim that has been placed inward from the center of its mounting flange.

negative pole: The negative terminal on a battery. Also known as *negative post.*

negative post: Same as *negative pole.*

negative terminal: The side of a battery or electrical device nearest ground potential, often identified by a minus (–) sign.

neoclassic: The modern design of a car to resemble a classic of the late 1920s and early 1930s.

neoprene: A type of synthetic rubber that is resistant to heat, light, oil, and oxidation.

nerf: To bump or shove another car in an oval-track race.

nerf bars: Small, tubular bumpers at the front and rear of an oval-track race car.

net horsepower: The maximum engine horsepower as measured on a dynamometer.

net torque: The maximum engine torque as measured on a *dyno.*

net valve lift: The lift, less the running clearance, of a valve.

neutral: **1.** The position of an automatic transmission when the engine is disengaged from the drive train. **2.** A condition of a planetary gear set when all members, gears, and planet carriers, are free to rotate.

neutral flame: An oxyfuel gas flame that has neither oxidizing nor reducing characteristics.

neutral safety switch: An electrical switch used on vehicles equipped with automatic transmissions to open the starter control circuit when the transmission shift selector is in any position except PARK or NEUTRAL. Also known as *neutral start switch.*

neutral start switch: A term used for *neutral safety switch.*

neutral wire: A grounded conductor in a three-wire electrical system that does not carry current unless the system is unbalanced.

neutron: A particle in the nucleus of an atom.

new old stock *(NOS)*: Old shelf-worn parts that have not been used.

Newton's laws of gravity: **1.** Every mass in the universe exerts an attractive influence on every other mass. **2.** For two objects, the force of gravitation acting between them is directly proportional to the product of their masses, and inversely proportional to the square of the distance between their centers of mass.

Newton's laws of motion: There are three laws: **1.** A body in motion remains in motion, and a body at rest remains at rest, unless some outside force acts on it. **2.** A body's acceleration is directly proportional to the force applied to it, and the body moves in a straight line away from the force. **3.** For every action, there is an equal and opposite reaction.

NG: An abbreviation for *no good.*

N-heptane: A hydrocarbon with an octane number of 0.

NHRA: An abbreviation for the *National Hot Rod Association.*

NHTSA: An abbreviation for the *National Highway Traffic Safety Administration.*

nickel: An element, Ni, used in alloys and electroplating.

Nikasil: A trade name of a popular coating applied to the walls of cylinder sleeves to reduce friction and promote sealing at high temperatures.

nitride: A compound containing electronegative nitrogen and an electropositive element or metal.

nitrile seal: Seal made by mixing butadiene and acrylonitrile.

nitro: A short term used for *nitromethane.*

nitrogen: A colorless, odorless, tasteless gas that makes up 78% of the atmosphere and is part of all living tissues.

nitrogen oxide: Any chemical compound of nitrogen and oxygen (NO_X) as a by-product of combustion that forms smog in the presence of sunlight.

nitromethane: A highly combustible liquid that is the main ingredient in drag-racing fuels.

nitrous: A term used for *nitrous oxide.*

nitrous oxide: A non-flammable, non-explosive gas (N_2O) used as an oxidizing agent with gasoline or methanol to increase the rate and efficiency of combustion thereby increasing the horsepower.

Nitrous Oxide Systems *(NOS)*: A popular manufacturer of nitrous oxide injection systems.

NO: An acronym for *normally open,* usually given as no.

noble metals: A rare and expensive group of metals that resist corrosion such as gold (Au), iridium (Ir), platinum (Pt), palladium (Pd), ruthenium (Ru), and rhodium (Rh), some of which may be found in catalytic converters.

nodular cast iron: Cast iron used to make engine blocks that is treated, while molten, with an alloy that causes the formation of graphite in very small lumps.

no good *(NG)*: Not fit for service.

no-load test: A starting-motor test to determine the *rpm* and current draw of a starter motor. A mechanical tachometer is placed on the drive end of the armature shaft and the motor is then connected to a proper power source. An ammeter, attached to the power means, determines current required.

Nomex: A popular trade name for a fabric that is used in race drivers' apparel.

nominal rating: The lowest rating of a system or device.

nominal size: A designation used for general identification, such as the inside diameter of a tube.

nonconductor: **1.** A material that does not conduct electricity. **2.** A material that is resistant to the conduction of heat.

A B C D E F G H I J K L M N O P Q R S T U V W X Y Z

noncorrosive flux: A soldering paste, usually composed of resin-based materials, that does not chemically attack the base metal.

non-destructive test *(NDT)*: A method of testing without causing damage.

nonfriction modified: An automatic transmission fluid without friction modification agents.

non-leaded gasoline: A term used for *unleaded gasoline.*

non-live axle: A non-driven truck axle that is held off the road when the vehicle is empty and put on the road when a load is being carried, used extensively in eastern states with high per-axle weight laws.

non-load bearing ball joint: A term used for *non-load-carrying ball joint.*

non-load-carrying ball joint: A ball joint designed with a preload which provides damping action that maintains suspension component location but does not support the chassis weight. Also known as *stabilizing ball joint.*

nonmetallic leaf spring: A term used for *fiber-composite spring* or *fiber-composite leaf spring.*

non-miscible: Not capable of being mixed in any proportion.

nonparallel drive shaft: A drive shaft with working angles of equal length; the companion flanges and/or yokes are not parallel.

nonpolarized gladhand: A *gladhand* that can be connected to either the service- or the emergency-gladhand.

non-servo: A type of drum brake in which the brake shoes work independently of each other.

non-slip differential: A differential that is designed so that when one wheel slips, the major portion of the driving power is transferred to the non-slipping wheel.

nonsynchronized mode: A mode of operation on a throttle body, fuel-injection system in which the injector is pulsed every 12.6 milliseconds, independent of a distributor reference pulse.

normal evacuation: An evacuation to remove air and moisture from a system with only one *pump down.*

normally aspirated: An engine that is not equipped with a forced means of inducing air.

normally closed *(nc)*: A term referring to a device, such as an electrical switch, that is closed in its normal position.

normally open: A term referring to a device, such as an electrical switch, that is open in its normal position.

normal wear pattern: The wear pattern of a part or assembly that is normal or expected.

north-south: Placing an engine longitudinally in a vehicle, as in most front-engined, rear-drive vehicles.

NOS: **1.** An abbreviation for *Nitrous Oxide Systems.* **2.** An abbreviation for *new old stock.*

nose: **1.** Front end of a vehicle. **2.** The uppermost part of a camshaft lobe. **3.** The front of a semi-trailer.

nose over: An engine or other component that is starting to self-destruct.

nose plug: A threaded, tapered plug used to repair cracks in the diesel engine cylinder head around the injector holes.

Nostalgia Drag Racing Association *(NDRA)*: An organization staging drag races for older cars.

no-tilt convertible fifth: A *fifth wheel* with fore/aft articulation that can be locked out to produce a rigid top plate for applications that have either rigid and/or articulating upper couplers.

NO_X: *Oxides of nitrogen,* not to be confused with *nitrous oxide* (N_2O).

NO_X control system: A device or system used to reduce the oxides of nitrogen *(NO_X)* produced by an engine.

nozzle: The orifice or jet through which fuel passes when it is discharged into a *carburetor.*

NSRA: An abbreviation for the *National Street Rod Association.*

NTP: An abbreviation for normal temperature and pressure.

NTPA: An abbreviation for *National Tractor Pullers Association.*

NTRDA: An abbreviation for *National Tire Dealers and Retreaders Association, Inc.*

NVLA: An abbreviation for *National Vehicle Leasing Association.*

NWRA: An abbreviation for *National Wheel and Rim Association.*

number failure code: A term used for failure code.

number of cylinders: The total number of cylinders (1, 2, 4, 6, or 8) contained in an engine.

numeric tire rating: An alpha-numeric group molded into the tire sidewall indicating tire application, aspect ratio, and rim size.

nut: A fastener having internal threads used on a *bolt* to secure two or more pieces together.

O: Orange.

OAH: An abbreviation for overall height.

OAL: An abbreviation for overall length.

OASIS: An acronym for *On-Line Automotive Service Information System.*

OAW: An abbreviation for overall width.

OBD: An abbreviation for *on-board diagnostics.*

observed horsepower: The brake output of an engine as observed on a *dynamometer.*

Occupational Safety and Health Administration *(OSHA):* A branch of the Federal Government that regulates and oversees the occupational environment as related to health and safety.

occupational safety glasses: Protective eye wear designed with special high-impact lenses, frames, and side protection.

octane: A gasoline's ability to resist detonation. The higher the octane number, the greater the fuel's resistance to detonation.

octane number: The number representing the average blend of iso-octane with an index of 100, and other hydrocarbons with an index of 0, usually between 87 and 92.

octane rating: A term used for *octane number.*

octane requirement: The minimum octane rating of a fuel required to operate an engine without a spark knock.

o.d.: An abbreviation for *outside diameter.*

odd fire: A *V–6* engine with a 60- or 90-degree block having a conventional *crankshaft* with two connecting rods attached to each of three journals that produces uneven firing.

odometer: A mechanical or electronic counter in the *speedometer* that indicates trip or total miles accumulated on the vehicle.

OEM: An abbreviation for original equipment manufacturer.

off highway: A term used for *off-road.*

off-highway vehicle *(OHV):* A vehicle intended for off-road use.

office: The driver's compartment.

office manager: An employee or company official whose duties vary from general office procedures to serving as a parts manager and bookkeeper.

Offie: **1.** An Offenhauser racing engine. **2.** Any product, such as a manifold or valve cover, manufactured by the Offenhauser Equipment Company.

off pavement: A term used for *off-road.*

off-road: Any unpaved, rough, or ungraded terrain on which off-road activities take place.

off-road vehicle *(ORV):* Any vehicle designed for use in off-road activities.

offset: **1.** A term used for *wheel offset.* **2.** A condition when two parts are not directly in line with each other.

offset link: A type of chain link that utilizes bent, or offset link plates and is assembled with a pin at one end and a bushing and roller at the other, so as to act as a combination link.

offset rod: A rod on which the bran section is not directly centered over the bearing housing.

offset section: A factory-assembled section of roller chain, made up of a roller link and an offset link, and used to connect strands of chain having an odd number of pitches.

offside: The side of a vehicle away from the curb.

off the line: A good start from the starting line in a drag race.

off-the-road vehicle *(ORV)*: A term used for off-road vehicle.

ogee: An S-shaped curve.

OHC: An abbreviation for *overhead cam* or *overhead camshaft.*

ohm: A metric unit of electrical resistance.

ohmmeter: An analog or digital instrument used to measure electrical resistance in ohms.

Ohm's law: The law that states that the direct current through an electric circuit is proportional to the voltage applied; given by the equation $I = E/R$, where I is current, E is electromotive force, and R is resistance.

OHV: **1.** An abbreviation for *off-highway vehicle.* **2.** An abbreviation for *overhead valve.*

oil: **1.** A diesel fuel. **2.** A liquid lubricant used to reduce friction between moving parts.

oil/air separator: A device used to prevent the aeration of oil.

oil bath filter: An engine air filter that channels the air through an oil bath that traps dust and debris.

oil bleed line: An external line that usually bypasses a metering device to ensure positive oil return to the compressor at all times.

oil bleed passage: Internal orifice that bypasses a metering device to ensure a positive oil return to the compressor.

oil burner: **1.** A diesel. **2.** An older car or truck that uses excessive oil, generally evident by a smoking exhaust.

oil canning: A sheet-metal panel popping from convex to concave and/or vice versa.

oil clearance: The small space between the main bearing and crankshaft journal, usually 0.001 to 0.003 inch (0.025 to 0.076 mm), for lubricating oil to circulate.

oil control ring: The bottom piston ring that scrapes the oil from the cylinder wall.

oil cooled piston: A piston that is cooled by a jet of oil sprayed under the dome of some diesel- and endurance-race engines.

oil cooler: **1.** A device used to cool oil or automatic transmission fluid. **2.** A device used to cool race-car manual transmission and final-drive lubricants.

oil dilution: The thinning of oil in the *crankcase,* usually caused by gasoline seepage past the *piston rings* from the *combustion chamber.*

oil dipper: A small scoop located at the bottom of the connecting rod in early, low-performance engines to dip into the oil to lubricate the *rod bearings* and *crankshaft.*

oil filter: A component, located near the oil pump, that removes abrasive particles from the motor oil by a straining process as the oil circulates through the lubrication system.

oil-fouled plug: A wet, oily deposit on a spark plug that may be caused by oil leaking past worn piston rings.

oil gallery: Passages drilled or cast into the *cylinder heads, engine block,* and *crankshaft* to receive pressurized oil from the *oil pump* for distribution throughout the engine.

oil-injection cylinder: A special calibrated cylinder that may be used to inject a measured amount of refrigeration oil into the system.

oil injector: A term used for *oil-injection cylinder.*

oil-intake screen: A strainer located at the lower end of the oil pickup tube in sump to remove large contaminant particles.

oil level indicator: A term used for *dipstick.*

oil pan: A removable part of the engine that contains the oil supply.

oil-pan rail: The flat-flanged portion of the oil pan with many bolt holes used to secure it to the engine.

oil pressure: The pressure, 15 to 75 *psi* (103 to 517 *kPa*), developed by the oil pump to force oil through the lubrication system.

oil-pressure gauge: An instrument used to display the oil pressure of the engine lubrication system.

A B C D E F G H I J K L M N O P Q R S T U V W X Y Z

oil-pressure indicator: A term used for *oil-pressure gauge.*

oil pump: A pump, driven directly or indirectly by the camshaft, that draws oil from the oil pan and forces it, under pressure, through the engine lubrication system.

oil pumping: The leakage of oil past the rings and into the *combustion chamber.*

oil reservoir: **1.** A shock-absorber section containing an extra fluid supply to meet operational requirements. **2.** The engine-oil pan. **3.** A container for an oil reserve supply.

oil ring: *Piston ring* that scrapes oil from the cylinder wall to control cylinder wall lubrication and prevent excessive oil loss past the *piston* and into the *combustion chamber.*

oil-ring expander: A thin metal strip used to maintain a constant pressure on the oil-ring rail against the cylinder wall.

oil-ring rails: Two thin metal rings used to scrape oil from the cylinder wall.

oil-ring separator: A term used for *oil-ring expander.*

oil scraper ring: A ring used to scrape oil off the cylinder wall that may have been left by the *oil control ring.*

oil seal: A seal around a rotating shaft or other moving part to prevent oil leakage.

oil seal and shield: A pair of devices that are used to prevent or control oil leakage past the valve stem and guide into the combustion chamber ports.

oil separator: A device used to separate oil from air, or oil from another liquid.

oil slinger: A metal disc that is located between the engine pulley and timing gear to force oil away from the timing-gear cover seal.

oil sludge: An accumulation of thickened oil, water, carbon, and dust particles inside an engine.

oil squirt hole: A small hole located near the lower end of the connecting rod, providing a lubrication path to the *cylinders* and *camshaft.*

oil strainer: A wire-mesh screen placed at the inlet of the oil-pump-pick-up tube to prevent dirt and debris from entering the oil pump.

oil sump: A term used for *oil pan.*

omnibus: The origin of "bus," from the French term *voiture omnibus,* or "carriage for all."

on a pass: Making a good, fast run in drag racing.

onboard computer: A resident computer in a vehicle, such as an automobile.

onboard diagnostics *(OBD):* A special, standardized diagnostic software and hardware system used to detect performance problems that adversely affect emissions and engine performance.

onboard fire-extinguisher system: A fire-extinguishing system permanently installed in the driver's compartment of a vehicle; required by most race-sanctioning bodies.

one off: A custom-built vehicle with no plans for mass production.

one-piece oil ring: An oil ring having the expander and rails combined in a single part.

one-way clutch: A friction or ratcheting device that permits motion in only one direction.

one-way roller clutch: A term used for *one-way clutch* or *overrunning clutch.*

one-wire system: An electrical system using body and/or chassis metal as a ground circuit.

On-Line Automotive Service Information System *(OASIS):* A computerized information system for service technicians.

on-road: Refers to paved or smooth-graded surface terrain on which a tractor/trailer will operate; generally considered to be part of the public highway system.

on the bubble: The last position on the grid subject to be bumped if another driver achieves a better qualifying speed.

on the cam: The operation of an engine at its most efficient *rpm.*

on the grid: A starting position for a race. Also known as *on the line.*

on the line: Same as on the *grid.*

on the piano: A term used when something is misplaced.

on the wood: An accelerator pedal pressed to the floor.

oodle: To operate an engine at idle speed.

open-chamber head: A cylinder head in which the cylinders have no *quench area.*

open circuit: A circuit in which there is a break in continuity.

open course: That part of a race track that extends beyond the finish line into a shut-off area.

open-end spring: A coil spring having its end loops apart from the spring coils.

open loop: In engines with a computer and oxygen-sensor control system, a mode of operation during which the computer ignores the signal from the oxygen sensor, typically before the engine reaches normal operating temperature.

open structural member: A flat body panel having an open access from the rear.

open system: A *crankcase,* emission-control system having no tube from the *crankcase* to the air cleaner; drawing air through the oil filter cap only.

open the tap: Increase the speed.

operational control valve: A device used to control the flow of compressed air through the *brake system.*

operational specifications: Specifications used to show how the vehicle operates, such as acceleration, tire inflation, and other general information.

operational test: A term used for *performance test.*

opposed engine: An engine with cylinder banks at 180 degrees, such as the Volkswagen flat four.

opposite lock: Turning the steering wheel in the opposite direction of a turn to control or correct oversteer.

optical horn: A Chrysler term for a *flash to pass* dimmer switch feature.

or: Orange.

organic brake lining: Brake linings that are made of a carbon-based compound combined with non-organic magnesium silicate and/or glass and synthetic fibers, replacing asbestos which has been determined to be hazardous to health and the environment.

organic friction material: A friction material having organic binders substantially formulated with nonmetallic fibers.

orifice: A small hole or opening.

Orifice Spark Advance Control *(OSAC):* A Chrysler emissions control system which slows vacuum advance of ignition timing by means of an orifice in a component mounted on the air cleaner.

Orifice Spark Advance Control Valve *(OSAC Valve):* A device used on some older Chrysler engines to limit *oxides of nitrogen* formation by delaying the vacuum signal to the distributor advance during idle and part-throttle operation.

orifice tube: A term used for *expansion tube* or *fixed-orfice tube.*

O-ring: A round ring having a square or round cross section used as a seal, such as at the end of a hydraulic line.

ORV: An abbreviation for *off-road vehicle.*

OS: An abbreviation for *oversize.*

OSAC: An abbreviation for Orifice Spark Advance Control.

OSAC Valve: An abbreviation for *Orifice Spark Advance Control Valve.*

oscillating: A device moving back and forth or to and fro, like a clock pendulum.

oscillating fifth wheel: A term used for *fully oscillating fifth wheel.*

oscillation: The rotational movement, either in a fore and aft or side-to-side direction around a pivot point, such as in a *fifth wheel* design in which such articulation is permitted.

oscillation damper: A *shock absorber* may be considered a damper that controls energy stored in the springs under load.

oscilloscope: An instrument that produces a visible image of one or more rapidly varying electrical quantities with respect to time or with another electrical quality.

OSHA: An acronym for *Occupational Safety and Health Administration.*

Otto cycle: The basic principle of operation of the four-stroke piston engine: intake, compression, power, and exhaust.

Otto-cycle engine: A four-stroke cycle engine.

O_2 sensor: A term used for *oxygen sensor.*

outboard brakes: A brake assembly that is mounted at the outer or wheel end of an axle half shaft.

outer cam clutch: A *one-way roller clutch* which has the cam profile surfaces on the outer element.

outer race: The race nearest to the outside of the hub of a roller bearing.

outgas: To release gas from a solid, such as plastic, as the result of heat generated during machining or cutting.

out of phase: A term that relates to a drive shaft having two universal joints that are not in phase with each other, producing a jouncing or jerking motion as the drive shaft turns.

out of round: The condition of a circular part when it is not circular, as in slightly oval shaped.

out of square: The condition of a square or rectangular part when its vertical sides are not at right angles to its horizontal sides.

out of the box: A vehicle, device, or component that is absolutely stock without modification.

out of the chute: Off the starting line in a drag race. Also known as *out of the gate* and *out of the hole.*

out of the gate: Same as *out of the chute.*

out of the hole: Same as *out of the chute.*

output driver: An electronic switch in a computer used to turn ON and OFF an actuator in a vehicle.

output shaft: The main shaft of a transmission.

outside diameter *(o.d.)*: The major diameter of the outside of a tube or tube-like device.

outside snap ring: A *snap ring* used on the outside of a shaft or part to hold the assembly in place.

outside-vehicle lubrication services: Work performed outside the vehicle during a chassis lubrication such as checking tire pressure, lubricating door and trunk hinges and locks, cleaning the windshield, and inspecting the safety-related items.

out to lunch: **1.** Anything of no value. **2.** A worthless car.

oval: **1.** Out of round or egg-shaped. **2.** An elliptical race track.

oval piston: A term used for *cam-ground piston.*

oval port: An exhaust or intake port that is oval in shape.

over: A term used for *bore* or *overbore.*

overbore: To enlarge the cylinders of an engine to a size larger than stock diameters.

overcenter preload: The adjustment of a steering-gear-sector shaft's resistance to turning.

overcenter spring: A spring used on some clutch linkages to reduce the effort required to depress the pedal.

overcharge: **1.** A term used when too much refrigerant or oil is added to the refrigeration system. **2.** A condition where the charging system is supplying too much voltage/current to the battery.

overcharging: The continual charging of a battery after it has reached its normally charged condition.

overcord: The covering bonded to the tensile member of a synchronous belt that protects it from frictional wear if a backside idler is used or if power is transmitted from the backside of the belt.

overdrive: A transmission having a ratio of less than 1:1 where the output shaft turns at a greater *rpm* than does the input shaft.

overdrive band: A transmission brake device that is engaged in *overdrive*.

overdrive ratio: A ratio identified by the use of a decimal point, such as 0.80, indicating less than one driving input revolution compared to one 1.0 output revolution of a shaft, or 0.85:1.0.

overflow: **1.** The spilling of the excess of a substance. **2.** To run or spill over the sides of a container.

overflow tank: **1.** A tank at the top of a radiator to receive heated coolant and vent trapped air due to expansion. **2.** A device found in some fuel tanks to prevent gas escaping due to expansion. **3.** A term used for *expansion tank*.

overhaul: To rebuild a device or an assembly, such as an engine or a transmission.

overhaul and maintenance specifications: Specifications used to service vehicle components such as an engine, differential, or transmission.

overhead cam *(OHC)*: A term used for *overhead camshaft* or *overhead camshaft engines*.

overhead camshaft: A *camshaft* mounted in the cylinder head.

overhead camshaft engine: An engine in which the *camshaft* is mounted over the cylinder head.

overhead position: The position in which welding is performed from the underside of the joint.

overhead valve *(OHV)*: An I-head arrangement where the valves are located over the piston in the cylinder head.

overhead-valve engine: An engine in which the valves are mounted in the cylinder head over the combustion chamber.

overheat: To become excessively hot.

overinflation: The condition of a tire that is inflated to more than the recommended pressure, decreasing the contact area, increasing the rolling diameter, and stiffening the tire resulting in excessive wear at the center of the tread.

overlap: The interval of valve timing when the *intake valve* starts to open before the exhaust valve is fully closed.

overlay cam: A *camshaft* having a hard face material welded to the nose and flank of the lobes to help decrease wear and increase lift.

overlubrication: Term referring to the application of lubricant amounts in excess of factory recommendations that may overload or damage grease seals.

overpull: The pulling of a body member beyond its specifications with the expectation that it will snap back to its intended shape when the pulling effort is discontinued.

overrev: To run an engine at excessive *rpm*.

overrunning clutch: A device used when two members are to run freely relative to each other in one direction, but are to lock in the other direction.

overrunning-clutch drive: An *overrunning clutch* used for the engagement and disengagement of a starting motor.

oversize *(OS)*: A part that is larger than the original to make up for wear and machining.

oversize valve stem: A valve having a stem diameter that is larger than the stem diameter of the original valve. It is used to fit a worn valve guide that is reamed oversize.

overspray: **1.** A paint spray mist that drifts onto a surface where it is not wanted. **2.** The overlap of new paint over old paint.

oversquare: A cylinder with a bore greater than its stroke.

overstaging: Staging a drag racer ahead of the usual staging position.

oversteer: A condition in cornering when the rear wheels of a race car tend to break loose and slide outward.

oxidation: The combination of a substance with oxygen forming an oxide, such as rust.

oxidation catalyst: A two-way catalytic converter which promotes the oxidation of HC and CO in an engine's exhaust stream, as distinguished from a three-way or reduction catalyst.

oxidation inhibitor: An additive to reduce chemicals in gasoline that react to oxygen.

oxide: A compound formed when a substance combines with oxygen.

oxides of nitrogen (NO_x): Harmful, gaseous emissions of an engine composed of compounds of nitrogen and varying amounts of oxygen which are formed at the highest temperatures of combustion.

oxidize: To form an oxide.

oxidizer: **1.** A material that causes oxidation. **2.** An additive that increases the oxygen content of an *air/fuel mixture.*

oxidizing agent: The same as *oxidizer.*

oxidizing flame: An oxyfuel gas flame in which there is an excess of oxygen, resulting in metal vaporization.

oxyacetylene welding: An oxyfuel gas welding process that uses acetylene as the fuel gas.

oxyfuel gas welding: A group of welding processes, with or without a filler metal, that produces a merger of work pieces by heating them with an oxyfuel gas flame.

oxygen: A colorless, gaseous, tasteless, element (O) that makes up 21% of the atmosphere.

oxygen cutting: The process of cutting metal at a high temperature with the chemical reaction of *oxygen.*

oxygenerator: A term used for *oxidizer.*

oxygen sensor: A device found in the *exhaust manifold,* which generates a small voltage dependent on the amount of oxygen present in the exhaust stream, used as a signal to the engine-control computer to determine the amount of fuel necessary to maintain a proper *air/fuel ratio.*

ozone: A molecule of oxygen, an unstable pale-blue gas (O_3), which is formed by exposure of O_2 to an electrical discharge. It has a penetrating or pungent odor and a strong oxidizing effect.

p: Pink.

P: The designation for the park position of an automatic transmission.

pace car: A vehicle used in closed-course racing, usually a convertible passenger car, to lead the field up to speed for a *rolling start.* Also used to lead the field when a caution flag is displayed.

pace lap: The last lap just before the start of a closed-course race, as the *pace car* leads the field up to speed.

pacer: A driver that runs at a constant or steady speed in oval-track racing.

package: A combination of optional equipment or accessories offered for a new car buyer usually discounted when compared to individual component pricing.

package tray: The shelf behind the rear seat in a sedan. Early trunk-mounted air conditioners used ducts through the package tray as the intake and outlets of the unit.

pad: A *brake lining* and metal back riveted or molded together. Two pads provide stopping friction by rubbing against both sides of a rotor or inside a drum when the brakes are applied.

paddock area: A designated area behind or near the pit where race cars are tuned and prepared for qualifying and racing.

pad-wear indicator: A mechanical or electrical warning device on *disc brakes* that indicates need for pad replacement.

PAFT: An abbreviation for *Program for Alternative Fluorocarbon Toxicity Testing.*

PAG: An abbreviation for *poly alkaline glycol.*

paint: A material, generally containing binders, solvents, and pigments, that is applied as a liquid to a surface, and forms a solid film for decoration and/or protection.

Paint, Body, and Equipment Association *(PBEA):* A trade association of manufacturers, wholesalers, and distributors of vehicle paint and body-shop equipment.

paint shop: The area in a repair facility where the vehicle is refinished by refinishing technicians or painters.

pallet: A special platform, usually wooden, used to support and transport, usually by fork lift, components in the shop, or for storage between manufacture and assembly.

pal nut: A thin pressed-steel nut used to lock a regular nut in place.

pan: A term often used for *oil pan.*

pancake engine: An engine having opposed pistons, such as a *flat four* or a *flat six.*

P&G check: The measurement of an engine's specifications, such as displacement, to determine the legality of competition using instruments provided by the P&G Manufacturing Company which allow such checks without the disassembly of the engine.

panel beater: A body shop worker.

panel cutter: Special "snips" used to cut through body sheet metal, designed to leave a clean, straight edge that can be easily welded.

panel nut: A thin nut used to hold parts or fittings to a firewall or bulkhead.

panel truck: An enclosed light truck or van having no windows in the cargo area.

panhard rod: A rod that is attached to the vehicle frame at one end and to the axle at the other end to prevent the chassis from moving side to side relative to the axle; used on a *beam axle* or *de Dion axle* type rear suspension.

pan rails: The sides of an *oil pan* that bolt to the *engine block.*

A B C D E F G H I J K L M N O **P** Q R S T U V W X Y Z

paper: Sheets of fibrous absorbent materials made from organic fibers that will "swallow up" liquids and may well disintegrate under such action if the liquid is a solvent for the binder.

parade lap: A lap or laps in closed-course racing before the *pace lap* to give the drivers an opportunity to warm up their tires and engines and to give the fans a chance for a good look at the cars at a slow speed.

parallel circuit: An electric circuit having two or more paths that the current flows through at the same time. The current is divided with more of it flowing through the path of least resistance.

parallel-joint type: A drive shaft installation whereby all companion flanges and/or yokes in the drive line are parallel with each other with the working angles of the joints of a given shaft being equal and opposite.

parallel linkage: A steering linkage having equal-length tie rods.

parallelogram linkage: A steering linkage system having a short idler arm mounted on the right side in such a manner that it is parallel to the pitman arm.

parasitic drag: Any interference of the aerodynamic efficiency of a vehicle body such as may be caused by outside mirrors, door handles, antenna, and windshield wiper arms and blades.

parasitic load: An electrical load that is present when the ignition switch is in the OFF position.

parent metal: The original metal in a body panel to which another metal panel has been added.

park contacts: Electrical contacts, inside a windshield-wiper motor assembly, to provide current to the motor after the control switch has been turned to the OFF or PARK position, allowing continued motor operation until reaching the park position.

parking brake: A mechanically applied brake system, usually to the rear wheels or drive shaft, to prevent a parked vehicle's movement.

parking-brake cable: A cable or cables that transmit brake-actuating force in the parking-brake system.

parking-brake equalizer: A device used to equalize the pull between the parking brake actuator and two rear wheels.

parking-brake system: A brake system, intended to hold a vehicle stationary, in which one or more brakes may be held in the applied position without continued application of force to the control.

particle: A very small piece of dirt, metal, or other debris which can be contained in oil, air, or fuel that can be removed with a *filter.*

particulate: **1.** A solid matter, mainly soot from burned carbon, in an internal combustion engine's exhaust. **2.** A form of solid air pollution such as microscopic solid or liquid matter that floats in the air.

particulate emissions: Solid particles, such as carbon and lead, found in vehicle exhaust; soot. A problem, especially in diesels.

particulate trap: An emissions control device in the exhaust system of a diesel engine which is used to capture particulate before they can enter the atmosphere.

partition coefficient: The ratio of the solubility of a chemical in water as compared to its solubility in oil.

parting edge: The *excess flash* material that is found around the edge of a part that has been cast in a two-piece mold.

part number *(PN)*: The alphanumeric designating of a part as listed in a catalog or parts list.

part out: To sell an assembly, such as an engine, piece by piece.

parts changer: One who, without proper diagnostic skills, randomly changes parts until the remedy to a problem is found.

parts chaser: **1.** A vehicle used for errands. **2.** A person who runs errands, especially for buying parts.

parts counterman: One very knowledgeable in part numbers, and/or determining requirements who generally works in a parts-supply wholesale or retail outlet.

parts distribution: A wholesale, discount, or retail establishment that sells parts to the trade or do-it-yourselfer.

parts manager: One responsible for ensuring that the in-house stock of parts is adequate so the customer's parts are readily available to the service technician when needed.

parts per million *(ppm):* **1.** A unit used to measure the amount of contamination in a substance, such as moisture in refrigerant. **2.** A measurement of the emissions of a motor vehicle given as the number of parts of a particular chemical within one million parts of exhaust gas.

parts requisition: A form that is used to order parts.

parts specialist: One who sells automotive engine and vehicle parts.

parts washer: A machine that is used to clean parts.

pass: To run down a *drag strip.*

passages: A term often used for *coolant passages* or *water jacket.*

passenger car: A four-wheeled, motor-driven vehicle that carries ten passengers or less, intended for use on streets and highways.

passive: **1.** Receiving or subjected to an action without responding or initiating a corresponding action. **2.** Accepting without resistance or objection.

passive restraint: Occupant restraints, such as air bags or seat belts, that operate automatically with no action required on the part of the driver or occupant, as mandated by Federal regulations to be used in all vehicles sold in the United States after 1990.

passive seal: A seal that has no extra springs or tension devices to help make the seal, such as the O-ring seals on valves.

patch: **1.** The area of contact of a tire with the road surface when the tire is supporting the weight of the vehicle. **2.** A small, rubber-like piece of material for repairing a puncture in an inner tube. **3.** A term often used for a temporary repair.

patch area: A term used for *tire contact area.*

PAW: An acronym for *plasma arc welding.*

pawl: A ratchet tooth that is used to lock a device.

payload: The weight of the cargo that may be carried by a truck, determined by subtracting the curb weight of the vehicle and 150 pounds (68 kg) for each passenger from the *gross vehicle-weight rating.*

PBEA: An abbreviation for *Paint, Body, and Equipment Association.*

PC seals: Valve stem seals made by Perfect Circle that have Teflon inserts to wipe the stems clean.

PCV: An abbreviation for *positive crankcase ventilation.*

PCV hose: A Neoprene- or synthetic-rubber hose that is connected to one or both ends of a *PCV valve.*

PCV valve: A vacuum-controlled metering device that regulates the flow of crankcase fumes in the positive crankcase ventilation system by allowing more flow at high speed than at low speed, and acts as a system shutoff in case of engine backfire to prevent an explosion in the crankcase.

PDI: An abbreviation for *predelivery inspection.*

peak: Maximum, as in peak horsepower, meaning maximum horsepower.

peaked: A body panel having raised beading for a styling effect.

peak out: The engine speed at which maximum horsepower is developed.

pearlescent paint: A color paint with fine mica particles blended into the pigment.

pedal clearance: The amount of downward brake-pedal movement, 1/4 inch (6.35 mm) or so, before the pushrod contacts the *piston.*

pedal to the metal: The accelerator pedal pressed all the way to the floor.

A B C D E F G H I J K L M N O **P** Q R S T U V W X Y Z

peel: To leave streaks of rubber on the pavement during hard acceleration.

peel rubber: A term used for *lay rubber;* the same as to *peel.*

peel test: A destructive method of testing that mechanically separates a lap joint by peeling it.

peen: **1.** To shape metal by pounding it. **2.** The ball-shaped end of a hammer head.

peg: **1.** A pin at the top end of a speedometer or tachometer to prevent instrument damage. **2.** The highest possible reading on a speedometer.

PEL: An abbreviation for *permissible exposure limit.*

pellet-bed catalytic converter: A General Motors catalytic converter design having a stainless steel shell and a bed of catalyst-coated ceramic pellets that can be replaced using special vibrator/aspirator equipment.

pellet thermostat: An engine-coolant thermostat having a wax pellet as a power element, which grows when heated and shrinks when cooled, connected through a *piston* to a *valve.*

pen: To draw or design.

PERA: An abbreviation for *Production Engine Remanufactures Association.*

percent: The ratio of one material to another in a mixture as in a fuel mixture.

percent of grade: A value that is determined by dividing the height of a hill by its length, often used to determine the power requirements of trucks or for determining maximum pay load.

percolation: A condition in which the fuel-bowl vent fails to open when the engine is turned off and internal pressure forces raw fuel through the main jets into the *manifold.*

performance chart: A chart that has been produced from a *dynamometer* showing the *horsepower, torque,* and *fuel consumption* of an engine at various speeds.

performance test: **1.** Readings of the temperature and pressure, under controlled conditions, to determine if an air-conditioning system is operating at full efficiency. **2.** A test to determine if a system or sub-system is performing at maximum efficiency.

perimeter frame: A conventional chassis frame design that is similar in construction to a *ladder frame,* having full-length side rails that support the body at its greatest width, providing optimum protection to passengers in the event of a side impact.

permanent magnet: A piece of ferrous material, such as steel, that retains its magnetic properties without the use of an electric current.

permanent strainer: A device generally of a Y-configuration having an accessible cylindrical strainer element used in horizontal and vertical lines. The element, retained by a plug end which may be plain or fitted with a valve, can be opened for "blow-through" cleaning.

Permatex: A tradename for a brand of engine and transmission sealants.

permissible exposure limit *(PEL):* The maximum length of time a person should be exposed to a hazard or hazardous material established by *OSHA* and expressed as a time-weighted average limit or ceiling exposure limit.

petcock: A small faucet-like valve used for draining liquids, such as that found at the bottom of some radiators.

petroleum: The crude oil from which gasoline, lubricant, and other such products are manufactured.

phaeton: A four or five passenger, two- or four-door, open-body style that was most popular in the 1920s and 1930s.

pH level: A measure of the acidity or alkalinity of a solution on a scale where pH0 is most acid, pH14 is most alkaline, and pH7 is neutral.

phosgene gas: A highly toxic gas, carbonyl chloride ($CCOCl_2$). Until recently, it was believed that phosgene gas was produced when CFC refrigerants, such as *R-12,* came into contact with heated metal or an open flame. It is now known that little or none of this gas is produced in this manner.

phosphor bronze: A hard, tough alloy of copper (Cu), lead (Pb), tin (Sn), and phosphorus (P) low in friction and resistant to wear.

phosphoric acid: A colorless and odorless acid (H_3PO_4) used to remove rust from steel and cast iron.

photochemical smog: A noxious, unhealthy gaseous compound in the atmosphere, formed by the interaction of various chemicals such as the pollutants hydrocarbons (HC) and oxides of nitrogen (NO_X) in the presence of sunlight.

photovoltaic diode: A device having a junction of two dissimilar metals that produces an electrical signal proportional to the amount of light that strikes it.

physical hazards: Any personal hazards, such as excessive noise, temperature, pressure, and rotating equipment.

physical properties: Properties that pertain to the physics of a material, such as melting point, density, electrical and thermal conductivity, specific heat, and coefficient of thermal expansion.

pickup: **1.** Vehicle acceleration. **2.** A term used for *pickup truck.*

pickup coil: An engine-speed sensor in an electronic ignition system.

pickup truck: A type of vehicle having an open cargo bed behind an enclosed cab.

pickup tube: **1.** A tube extending from the outlet of the receiver almost to the bottom of the tank to ensure that 100% liquid is supplied to the liquid line or metering device. **2.** A tube used to transfer fuel or oil from a storage tank.

piece work: A method of payment whereby a worker receives an agreed fixed amount per piece produced.

pig: **1.** An unattractive vehicle. **2.** A vehicle that is not performing well or as expected. **3.** A bar of cast metal.

pigment: Small particles added to the paint to influence properties such as color, corrosion resistance, and mechanical strength.

Pikes Peak: A 14,110 foot (4,300 meters) high mountain peak in central Colorado where an annual auto-racing hill climb is held around the Fourth of July.

pilot: **1.** A driver. **2.** A device that guides the operation of a machine or part.

pilot bearing: The bearing at the output end of the engine's crankshaft that supports the transmissions input shaft.

pilot hole: A small hole drilled into a part to serve as a guide for a larger size drill bit.

pilot model: The first vehicle of a new design to be constructed on the assembly line to test the assembly procedures and fixtures before full-scale production begins.

pilot operated: A small valve used to energize or regulate a large valve.

pilot-operated absolute *(POA):* A term used for *pilot-operated evaporator pressure regulator.*

pilot-operated evaporator pressure regulator *(POEPR):* An evaporator pressure-regulated valve that is regulated by an internal pilot-valve pressure.

pilot-operated absolute suction throttling valve *(POASTV):* A term for *pilot-operated evaporator pressure regulator.*

pilot shaft: A shaft that is used to align parts during assembly.

pin: A threaded, tapered metal part used to repair cracks in castings such as an engine block, using a process called *pinning.*

PIN: An acronym for *product identification number.*

ping: A mild knock noise from the engine.

pink slip: A vehicle ownership certificate.

pinion: **1.** A small gear such as at the end of the steering shaft of a *rack-and-pinion steering system.* **2.** A small ring gear such as the planet gears in a planetary gear set.

pinion bearing: A bearing that is used to support the pinion gear in the differential housing.

pinion gears: A term used for *pinion.*

pinion seal: An oil seal for the pinion gear in a ring and pinion.

pinion shaft: A shaft used to support a *pinion bearing* and/or *pinion.*

pinning: Using a series of *pins* to repair a crack in a casting by screwing them into a series of predrilled holes in a manner so that they overlap each other from one end of the crack to the other.

pintle: A valve-like part of a fuel injector that controls the fuel-spray pattern.

pintle hook: A pin installed in the rear of some vehicles for attaching a tow bar.

pintle valve: The ball and seat found inside a thermostatic expansion valve, attached to the diaphragm, which causes it to open and close in response to pressure changes.

pipe: An exhaust system.

pipe brace: A brace that extends between opposite hangers on a spring or air-type suspension.

pipes: A dual-exhaust system.

pipe dream: An illusionary fantasy plan, or way of life.

piston: **1.** An engine part that reciprocates in the cylinder and transfers the force of the expanding gases via the *piston pin* and *connecting rod* to the *crankshaft.* **2.** A round, caliper component in a disc brake that is moved outward by fluid pressure to press the pads against each rotor face. **3.** An aluminum or sintered-iron component of a drum brake inside a wheel cylinder that supports the cylinder cup. **4.** That part of a compressor that is driven by a *crankshaft* to compress vapor.

piston boss: That part of the piston which supports the *piston pin.*

piston crown: The top of the piston.

piston displacement: The cylinder volume displacement by the piston as it travels from the bottom to the top of the cylinder during one complete stroke.

piston markings: Marks on a *piston* which are used to identify piston oversize and the front of the *piston.*

piston material: Metal used in *piston* construction, usually cast- or forged-alloy aluminum.

piston oiler: A device that injects oil into the *piston* as an aid to carry away heat from the *piston crown.*

piston oil ring: A term used for *oil ring.*

piston oil squirt hole: A term used for *oil squirt hole.*

piston pin: A round precision-ground part, usually hollow, used to attach the piston to the connecting rod by press fit or held in place with internal snap rings.

piston-pin bushing: A bushing that is pressed into the upper end of the connecting rod when free-floating pins are used.

piston-ring land: The surface of a piston between ring grooves.

piston rings: Rings that fit into grooves on the outer wall of a piston, just below the *crown,* that seal the combustion chambers and scrape oil from the cylinder walls.

piston rod: A plated rod attached to the shock-absorber piston, usually extending from the top of the shock to provide attachment to the vehicle.

piston-rod seal: A non-replaceable oil seal around a movable *piston rod,* located at the upper end of the hydraulic cylinder.

piston skirt: The lower part of the piston that makes contact with the cylinder wall.

piston-skirt clearance: The allowable space between the piston skirt and cylinder wall, usually 0.001 to 0.002 inch (about 0.025 to 0.051 mm) at room temperature.

piston-skirt expander: A device that may be inserted behind the *piston skirt* to force the skirt out toward the cylinder wall to reduce skirt-to-wall clearance.

piston slap: The noise that is made by an undersize or loose *piston skirt* as it makes contact with the cylinder wall.

piston speed: The velocity of a *piston,* given in feet per minute, as it reciprocates in a cylinder.

piston stops: The tabs, or protrusions, on a backing plate so positioned as to prevent the wheel cylinder pistons from leaving the wheel cylinder.

piston temperature: The temperature of the *piston* at different areas, varying from over 450°F (232°C) at the crown, to 200°F (93°C) near the bottom of the *skirt.*

piston-to-valve clearance: The distance between the intake and exhaust valves and the *piston crown* when the valves are at overlap period and the piston is at top dead center.

piston valve: A shock-absorber component used to control the flow of oil and produce the pressure and vacuum in the chamber above the *piston* during compression and rebound.

pitch: **1.** The rotating motion of a vehicle's spring mass about its lateral axis, compressing the springs at one end of the vehicle and extending the springs at the other end. **2.** The uniform spacing of adjacent elements of a series of points, lines, joints, planes, blades, or teeth, as in a broach or gear. **3.** The center-to-center dimension between chain joints. **4.** For a sprocket, the dimension between the centers of rollers bedded against the bottoms of adjacent tooth spaces. **5.** The distance between corresponding points on adjacent threads. **6.** The angle of setting of some tools. **7.** The distance from center-to-center of wire in adjacent coils in an open-wound spring.

pit crew: **1.** Those responsible for servicing, timing, and the communications of a race car during a race on the track side of the pit wall. **2.** Those who support the team efforts on the infield side of the pit wall.

pitman arm: Arm attached to the pitman shaft that moves the relay rod as the steering wheel is turned.

pits: The trackside service facilities for servicing vehicles competing in a race.

pit stop: A visit to the pits during a race for routine or emergency service or to ask directions.

pitting: Metal-surface irregularities as a result of corrosion.

pit wall: A wall that divides the pit service facilities on the trackside and infield side.

pivot: A pin or short shaft upon which another part rests or turns, or about which another part rotates or oscillates.

pivot bushing: A rubber bushing in the front suspension located at the inner end of the control arms where pivoting occurs.

pizza cutters: The extremely narrow front wheels of a drag-race car used to reduce rolling resistance.

pk: Pink.

placard: **1.** A notice for display in a public (obvious) place. **2.** A label or poster.

plain ends: End coils of a helical spring having a constant pitch and with the ends not squared.

plain vanilla: **1.** Ordinary and without frills. **2.** A low-priced economy car.

planetary carrier: The framework holding the planetary pinions within the sun gear.

planetary gear set: A group of gears named after the solar system because of their arrangement and action. This unit consists of a center (sun) gear around which pinion (planet) gears revolve. The assembly is placed inside a ring gear having internal teeth. All gears mesh constantly. Planetary gear sets may be used to increase or decrease *torque* and/or obtain neutral, low, intermediate, high, or reverse.

planetary gear system: A gear set consisting of a central sun gear surrounded by two or more planet gears which are, in turn, meshed with a ring gear, used in overdrives and automatic transmissions.

planetary gear train: Gears of the transmission in either the Simpson or Ravigneaux design.

planetary pinion gears: Small gears fitted into a framework called the planetary carrier.

planet carrier: The cradle that holds the planet gears in a *planetary gear set.*

planet pinions: Small gears that orbit around the sun gear, meshing with and rotating between the sun and internal gears.

planing: A condition that exists when a tire is unable to stay in contact with the ground or pavement due to water. Also known as *hydroplaning* and *aquaplaning.*

plant it: To apply full throttle.

plasma: An inert gas used in welding, such as *argon* or *nitrogen.*

plasma arc welding *(PAW):* An arc welding process that uses a constricted arc between a non-consumable electrode and the weld pool (transferred arc) or between the electrode and the constricting nozzle (non-transferred arc) with shielding obtained from the ionized gas issuing from the torch, which may be supplemented by an auxiliary source of shielding gas.

plastic deformation: A micro-structural change as a result of exceeding the yield point of the material and generally including elongation of the grain structure and increased hardness.

plastic filler: A two-part, putty-like material that may be used to fill dents in body panels.

plastic gasket compound: A plastic paste that is squeezed out of a tube to make a *gasket* in any shape or configuration.

plastic spring: A term used for *fiber-composite spring.*

Plastigauge: The tradename for a small, thin, plastic strip that is used to help determine the clearance between main and/or connection *rod bearings* and the crankshaft journals.

plate: **1.** A rectangular sheet of spongy lead in a battery. **2.** A platform on which an *assembly* may be mounted.

plate group: All of the positive plates and all of the negative plates that make up one battery cell.

platform: The basic understructure of a vehicle including all running gear.

platinum *(Pt):* A rare, valuable metallic element which is highly resistant to corrosion, and is used as a catalytic agent in automotive catalytic converters of the oxidizing type.

plenum: **1.** A chamber containing air under pressure. **2.** A chamber containing a gas such as air or an *air/fuel mixture* under higher than atmospheric pressure.

plenum-blower assembly: Located on the engine side of the firewall, this *assembly* contains air ducts, air valves, and a blower that permits the selection or air from the outside or inside of the car and directs it to the evaporator or to the heater core if desired.

plenum chamber: An area, filled with air at a pressure that is slightly higher than the surrounding air pressure, such as the chamber just before the blower motor.

Plexiglas: A brand name of a transparent, lightweight plastic material often used for windows in an enclosed race car.

plies: The cord plies that surround both beads and extend around the inner surface of the tire to enable the tire to carry its load.

plow: A condition in cornering when the slip angles of the front tires are greater than the slip angles of the rear tires and the front end tends to break loose and slide.

plug: **1.** A spark plug. **2.** A threaded, tapered metal pin used to repair cracks in castings, such as an engine block.

plug weld: A weld made in a circular hole in one member of a *joint,* fusing that member to another member.

plunger: That part of a *hydraulic valve lifter* that moves up or down depending on *oil pressure.*

ply rating *(PR):* A method used to indicate relative tire strength, formerly used to designate the load-carrying capacity of a tire.

The ply rating does not necessarily indicate the actual number of plies; a four-ply rated tire, for example, may have two plies but is considered as strong as a standard four-ply tire.

PM: An abbreviation for *preventative maintenance.*

PMA: An abbreviation for *Professional Mechanics Association.*

PN: An abbreviation for *part number.*

POA: An abbreviation for *pilot-operated absolute,* as in a *suction throttling valve.*

POA valve: A term used for *suction throttling valve.*

POASTV: An abbreviation for *pilot-operated absolute suction throttling valve.*

POEPR: An abbreviation for *pilot-operated evaporator pressure regulator.*

pogo stick: The air- and electrical-line support rod mounted behind the cab to keep the lines from dragging between the tractor and trailer.

point of intersection: Apex of the extended centerline of the pivot axis and the extended centerline of a front wheel.

points: A term used for *breaker points.*

polarity: The quality of an electrical circuit or component that determines the direction of current flow.

polarizing a generator: The correction of the generator-field polarity so it will build up polarity in the proper direction to charge the battery.

polar moment of inertia: **1.** The tendency of a body to resist angular or rotational acceleration. **2.** The tendency of a vehicle to resist cornering.

pole: The starting position on the inside of the front row in oval-track or road racing.

police options: A high-performance equipment package available on new cars for law enforcement authorities.

police package: A term used for *police options.*

polished: Intake and exhaust passages that have been finished to the smoothest possible surfaces to remove the rough surfaces left by factory casting.

pollutant: **1.** Any substance that adds to the *pollution* of the atmosphere. **2.** Any substance in the *exhaust gas* from an engine or evaporating from the *fuel tank* or *carburetor.*

pollution: **1.** Any gas or substance in the air that makes the air less fit to breathe. **2.** Excessive noise from machinery or vehicles.

poly alkaline glycol *(PAG)*: The synthetic lubricant of choice in many automotive air conditioners used with *HFC-134a refrigerant.*

poly ester: A synthetic lubricant that is used in some *HFC-134a* air-conditioning systems.

poly lock: A type of rocker-arm lock nut that is secured with a set screw.

polymerization: A chemical reaction whereby many small molecules combine to form larger, more complex molecules having a higher molecular weight with different chemical properties.

polyurethane: A synthetic material often used as a filter media.

poncho: A Pontiac.

ponies: Horsepower.

pony car: Any compact, two-door sports car similar to a Mustang.

Pooch: A Porsche.

pop: A racing fuel, generally with nitromethane.

pop back: A condition when the *air/fuel mixture* is ignited in the *intake manifold* and may cause a "pop back" through the *carburetor.*

pop-off valve: A spring-loaded pressure relief valve.

poppet valve: An intake or exhaust valve consisting of a disc at the end of a shaft used in a four-stroke cycle engine.

pop rivet: A fastener used to hold two pieces together.

A B C D E F G H I J K L M N O **P** Q R S T U V W X Y Z

pop the clutch: To engage the clutch suddenly.

porcelain: A hard, ceramic material as used in spark plugs to insulate the electrode from the shell.

porcupine head: The cylinder head on a *rat motor.*

port: A term used for *intake port* or *exhaust port.*

portable cable: An electrical cable used to transmit electrical power to mobile equipment.

port bowl: The area of the port next to the *valve head.*

ported vacuum: Engine vacuum, available above the throttle plates of a carburetor, used to advance ignition timing when the throttle is opened above its idle position.

ported-vacuum switch *(PVS):* A valve which permits or stops the passage of a *ported vacuum* to a vacuum-operated component, such as a distributor advance mechanism that may be thermostatically operated, or controlled by electric current or the movement of a mechanical component.

porting: To enlarge the intake and exhaust passages.

port injection: A fuel-injection system that injects fuel directly into the individual intake ports.

port runner: A partition in the *intake manifold* that directs the *air/fuel mixture* to the individual cylinders.

Pos-A-Traction: A brand of tires, tubes, and wheels.

posi: A generic term for a *limited-slip differential.*

positioned weld: A weld made in a joint that has been placed in a position to facilitate making the weld.

position sensor: A term used for *EGR valve-position sensor.*

positive: **1.** One of the two poles of a magnet. **2.** One of the two terminals of an electrical device, such as a battery.

positive absolute-suction throttling valve: A suction throttling valve having bronze bellows under a nearly perfect vacuum, which is not affected by atmospheric pressure.

positive back-pressure EGR valve: A common type of exhaust gas recirculation valve which uses exhaust system backpressure to sense engine load, thus more accurately metering the amount of exhaust recycled.

positive camber: The outward tilt of a wheel on a vehicle.

positive crankcase ventilation *(PCV):* An engine emissions-control system, operating on engine vacuum, that picks up crankcase gases and meters them into the intake stream to be burned.

positive displacement pump: An engine-driven air or liquid pump that displaces the same amount of air or liquid, per revolution, regardless of engine speed.

positive offset: A wheel rim placed outward of the center of the mounting flange.

positive pole: **1.** The *positive terminal* of an electrical device. **2.** The north pole of a magnet.

positive post: The *positive terminal* of a battery.

positive terminal: The terminal to which electrons flow in a complete electrical circuit, often identified by the symbol +.

Positive Traction: A limited-slip differential by Buick.

positive wiping seal: A seal that maintains constant contact with a stem or shaft to wipe off excess oil.

Posi-Traction: A limited-slip differential by Chevrolet.

post: A battery terminal.

post ignition: Ignition that occurs after the engine-ignition system is shut off, due to carbon buildup in the *combustion chamber.*

post start: The time from a cold start to the warm up of an engine.

pot: 1. A *carburetor.* 2. An abbreviation for *potentiometer.*

potential energy: Energy of a body or system with respect to the position of the body or the arrangement of the particles of the system.

potentiometer *(pot):* A three-wire variable resistor that acts as a voltage divider to produce a continuously variable output signal proportional to a mechanical position.

pounds-feet: An English measure of *torque.* One pound (lb) raised one foot (ft) is equal to one pounds-feet (lb-ft).

pounds-inch: An English measure of *torque;* the energy required to raise one pound, one inch.

pounds per horsepower: The weight of the vehicle divided by its horsepower, a measure of the vehicle's performance.

pounds per square inch *(psi):* An English measure for pressure or stress.

pounds per square inch, absolute *(psia):* An English measure for pressure or stress taken from absolute pressure.

pounds per square inch, gauge *(psig):* An English measure for pressure or stress taken from atmospheric pressure.

poured bearings: Bearings that are formed by first pouring molten babbitt material into the bearing cavity, allowing it to cool, then boring it to a specific size.

pour on the coal: To accelerate rapidly.

pour point: The lowest temperature at which a fluid will flow freely.

powder puff race: An auto-racing event strictly for women drivers.

power: 1. A measure of work being done. 2. The rate at which work is being done.

power brakes: A system utilizing energy from the vehicle's engine to reduce the amount of brake-pedal pressure that must be exerted by the driver to stop the car.

power break: In an automatic transmission-equipped drag racer, the act of holding the brake pedal firmly while revving the engine, then simultaneously releasing the brake while flooring the accelerator to get the quickest possible start.

power cable: Cable used to supply electrical power (current).

power control module: A module or computer used in an electronic transmission to aid in control of the shift solenoids.

power cylinder: 1. A shell containing the power-brake operating parts. 2. Linkage-type steering component attached between the frame and the steering relay rod.

power flow: The flow of power from the input shaft through one or more sets of gears, or through an *automatic transmission* to the output shaft.

power hop: The tendency of an axle housing to rotate slightly with the wheels then snap back during hard acceleration.

power mirror: Outside mirrors, having reversible permanent-magnet motors, that are electrically positioned from the inside of the driver's compartment.

power on/off watchdog circuit: Supplies a reset voltage to the microprocessor in the event that pulsating output signals from the microprocessor are interrupted.

power oversteer: The loss of traction of the rear wheels while cornering and accelerating, causing the rear of the vehicle to swing toward the outside of the turn.

power piston: A component acted on by pressurized fluid to assist wheel turning on integral or linkage-type power-steering systems.

power plant: An engine.

power section: The section of a gas-turbine engine containing the power turbine rotors which, through reduction gears, turn the wheels of the vehicle.

power servo: A *servo* unit used in automatic temperature control which is operated by a vacuum or an electric signal.

power shift: The rapid forced shifting of a manual transmission without releasing the clutch or accelerator.

A B C D E F G H I J K L M N O P Q R S T U V W X Y Z

power slide: A controlled four-wheel skid in dirt-track racing to maintain speed through a turn.

power steering: A power assisted steering system that uses hydraulic pressure to increase the *torque* (turning effort) applied to the steering wheel by the driver.

power steering fluid: Special hydraulic fluid formulated to withstand extremely high system pressures and temperatures.

power steering, integral-type: A power-steering system having the power assist mechanism and related controls inside the steering gearbox.

power steering, linkage-type: A power-steering system having the control valve and power cylinder attached to the steering linkage under the car.

power steering pump: A hydraulic pump driven by a belt from a crankshaft pulley to provide up to 1,300 *psi* (8,964 *kPa*) "boost" pressure necessary to operate the power-steering system.

power stroke: The *piston* stroke, with both valves closed, in which combustion takes place, forcing the piston from *TDC* to *BDC.*

power synchronizer: A device in the transmission that speeds up the rotation of the main-section gearing for smoother automatic downshifts and to slow down the rotation of the main-section gearing for smoother automatic upshifts.

power take off *(PTO):* An output shaft found on some four-wheel drive vehicles for driving an external accessory, such as a winch.

power team: The engine, transmission, rear end, and axle combination.

power tools: Tools that use compressed air (pneumatics), electricity, or hydraulic pressure to generate and multiply the force required to accomplish the work.

power-to-weight ratio: The relationship of a vehicle's horsepower to its weight as given in horsepower per pound.

power train: A combination of the *engine, transmission,* and *final drive.*

power valve: A valve in the *carburetor* that opens during acceleration to increase fuel flow.

ppm: An abbreviation for *parts per million.*

PR: An abbreviation for *ply rating.*

prang: A collision of vehicles.

pre-cat converter: A term used for *light-off, mini-oxidation catalytic converter.*

precision insert bearing: Bearings that may be installed in an engine without having to be bored, reamed, honed, or ground.

precombustion chamber: A second *combustion chamber* placed directly off the main *combustion chamber,* used to ignite a rich mixture of air and fuel which then ignites a lean mixture in the main *combustion chamber.*

predelivery inspection *(PDI):* The process of inspecting, adjusting, and fine tuning a new car prior to delivery to the customer.

preformed radiator hose: A large-diameter hose connecting the radiator to the engine cooling system that is molded in the proper shape to fit on a certain engine.

preheater: A glow plug that is used to heat the precombustion chamber of a diesel engine before it is started.

preheating: Heating the weld area of a metal before welding, to avoid thermal shock and stress.

pre-ignition: The ignition of the *air/fuel mixture* in the *combustion chamber* by means other than the spark; usually caused by hot spots in the *combustion chamber* due to sharp edges, carbon accumulation, or *spark plugs* with a heat range that is too hot.

preload: **1.** The transfer of weight from one side of a vehicle to the other to compensate for the lateral weight transfer that occurs when cornering. **2.** The pressure applied to a part during assembly or installation.

prelube: To apply lubricant to the parts of a rebuilt engine before starting it up.

prerun: **1.** To examine and appraise an off-road race route. **2.** To test a rebuilt engine before installing it in the vehicle.

prerunner: **1.** A vehicle that is built to be used to examine and evaluate the off-road route of an upcoming race. **2.** A vehicle often used as an off-road race-chase vehicle for parts, fuel, and other supplies.

press fit: The fitting of two parts with very close tolerances, one inside the other, generally by using a hydraulic press. Also known as *force fit.*

pressure: Force per unit area or force divided by area, usually expressed in *psi, psig,* or *psia (kPa* or *kPa absolute).*

pressure balancing ring: A compression ring having a ceramic coating to ensure sealing against the cylinder wall.

pressure bleeder: A device used to facilitate the removal of air from the *brake system.*

pressure bleeding: Bleeding the *brake system* using a *pressure bleeder* to charge the system.

pressure cap: A cap placed on the radiator to increase the pressure of the cooling system, reduce cavitation, protect the radiator hoses, and prevent or reduce surging.

pressure connector: A connector applied using pressure to form a *cold weld* between the two parts.

pressure control valve: A device used to control and regulate the pressure of a system or sub-system.

pressure drop: The loss of pressure due to a restriction, a long line, or a small diameter line.

pressure differential valve: A valve used in a dual-brake system to turn on a dash warning light if the pressure drops in either part of the system.

pressure-fed oil system: A type of lubrication system that uses an oil pump to force oil to the various parts of an engine.

pressure hold: One condition in an anti-lock brake control module in which no more braking pressure can be produced in the *master cylinder.*

pressure hose: A special, reinforced high-pressure hose through which pressurized fluid flows.

pressure line: **1.** A term that generally refers to the *discharge line* of an air-conditioning system. **2.** A term that may apply to brake or power-steering hoses.

pressure plate: That part of a manual clutch that is mounted on and rotates with the flywheel and exerts pressure against the friction disk.

pressure plate assembly: An assembly of various parts that operate to "press" the clutch disc against the flywheel, engaging power flow to the clutch shaft, or "release" the disc, disengaging power flow.

pressure reduction: A drop of pressure due to a restriction or metering device.

pressure regulator: **1.** A device that operates to prevent excess pressure buildup in a system. **2.** A valve in some *automatic transmissions* that opens to release oil from a line when the oil pressure reaches a specific maximum limit.

pressure release grill: An air vent that prevents pressure from building up inside the car while the air-conditioning system is operating.

pressure relief groove: A small groove cut in the *piston-ring land* between the grooves for the compression and oil rings to equalize combustion pressures between them.

pressure relief valve: **1.** A valve held closed by a spring or other means, which automatically relieves pressure in excess of its setting. **2.** A carburetor fuel bowl vent operated by vapor pressure from evaporated fuel. **3.** An air-injection system valve that functions to relieve pump-output pressure that exceeds a predetermined amount, usually at speeds over 45 mph (72.4 km/hr). **4.** A special valve in an engine lubricating system having a calibrated spring working against a moveable plunger or ball to regulate maximum oil pressure.

pressure test: A procedure for testing the entire cooling system for leaks.

pressure tester: A device used to pressure test the cooling system and *pressure cap.*

pressurize: To apply more than atmospheric pressure to a device or system.

pressurized carburetor: A *carburetor* on a turbocharged engine that mixes fuel with air under pressure from the turbo.

prevailing torque nut: A nut designed to develop an interference fit between the nut and bolt threads.

preventative maintenance *(PM):* A systematic and scheduled inspection and service of a vehicle to ensure proper maintenance to provide a standard for satisfactory and sound operational conditions.

price guides: **1.** A reference that is used to determine a proper fee for repair labor based on time. **2.** A reference providing the value of used vehicles based on geographic location, condition, and accessories.

primary: The low-voltage circuit of an *ignition system.*

primary brake shoe: The brake shoe in a set which initiates self-energizing action.

primary circuit: **1.** The electrical circuit of the low-voltage winding of an *ignition coil.* **2.** The main fuel passages in a *carburetor.*

primary damage: The damage caused by the initial impact of a collision.

primary resistance: A calibrated resistance wire or ballast resistor in a conventional ignition system inserted between the *ignition switch* and *coil* to reduce the primary voltage at low engine speeds.

primary seal: A seal between the compressor shaft seal and the shaft to prevent the leakage of refrigerant and oil.

primary shoe: The brake shoe toward the front of the vehicle in a dual-servo drum brake having a lining that may be shorter, thinner, and have different frictional characteristics from the *secondary shoe.*

primary winding: The outer, low-voltage winding of an *ignition coil.*

primer: A type of paint applied to a surface to increase its compatibility with the topcoat or to improve the corrosion resistance of the substrate.

primer sealer: A type of paint that provides adhesion for the topcoat and seals old, painted surfaces.

primer surfacer: A type of paint having a high-solids content which fills small imperfections in body work and which usually must be sanded.

priming spring: A coil spring that holds the movable slide in the maximum position in a vane-type pump.

prindle: The standardized shift pattern for an automatic transmission, *PRNDL.*

printed circuit: An electrical circuit made by applying a conductive material to a *printed circuit board.*

printed circuit board: A thin, insulating board used to mount and connect various electronic components, such as resistors, diodes, switches, capacitors, and microchips in a pattern of conductive lines.

PRNDL: The shift pattern for many automatic transmissions.

product identification number *(PIN):* A four-digit code that is used to identify hazardous material during transportation.

production: **1.** A term used for *stock.* **2.** A standard factory version.

Production Engine Remanufactures Association *(PERA):* A trade association of those who rebuild gasoline and diesel engines or manufacture and/or distribute engine parts and engine rebuilding equipment.

Professional Mechanics Association *(PMA):* An association of professional mechanics that offers members savings on tools, travel, insurance, training, equipment, and shoes.

PROCO: An abbreviation for *programmed combustion.*

profile: A term used for *aspect ratio* and *tire profile.*

Program for Alternative Fluorocarbon Toxicity *(PAFT)* Testing: A testing agency founded by refrigerant manufacturers for the purpose of study and testing of new refrigerants to determine toxicity levels.

Program for European Traffic with Highest Efficiency and Unprecedented Safety *(Prometheus):* A research and development effort by major European auto makers for the purpose of studying if telecommunications technology can improve the efficiency and safety of European roads.

programmable read-only memory *(PROM):* A read-only memory chip that can be programmed by electrical pulses.

programmed combustion *(PROCO):* A research type of stratified charge engine developed by Ford.

programmed protection system: A bypass valve system to protect the catalyst and their containers from being destroyed due to overheating conditions that may be caused by certain modes of operation or from an engine malfunction.

progressive linkage: A carburetor linkage system used on multiple *carburetors* to progressively open the secondary circuits.

progressive-rate spring: A spring used in a vehicle that stiffens under load.

PROM: An acronym for *programmable read-only memory.*

Prometheus: An abbreviation for *Program for European Traffic with Highest Efficiency and Unprecedented Safety.*

prony brake: A machine for testing the power of an engine while running against a friction brake.

proof loading: Subjecting chain to a tensile loading of some predetermined percentage of the chain's rated strength.

propane: A combustible, liquefied petroleum gas (C_3H_8) that becomes a liquid when compressed.

propane enrichment: A service procedure used to set idle mixture where a metered amount of *propane* gas is added to the intake stream and a resulting *rpm* increase is observed.

propeller shaft: **1.** A term used for *drive shaft.* **2.** A hollow, steel shaft that connects the transmission output shaft to the differential drive pinion yoke through *universal joints* at each shaft end.

property class: A number stamped on the end of a metric bolt to indicate the hardness of the bolt.

proportioning valve: A valve in the brake hydraulic system that reduces pressure to the rear wheels to achieve better brake balance.

pro stock: A category for compact, passenger drag-racing cars with a big block V-8 engine having a maximum of 500 cubic inches (8.2 liters).

pro street: A compact passenger car with a modified *V-8 engine* and *rear drive* built for street use.

protective atmosphere: A gas or vacuum envelope surrounding the work pieces, used to prevent or reduce the formation of oxides and other detrimental surface substances, and to facilitate their removal.

protective tube: A *shock absorber* component used to keep dirt and road dust away from the *seals* and *piston rod.*

protest: **1.** To object to a rule or conditions prior to the competition. **2.** To accuse a competitor of a rules violation after a race competition.

proto: An abbreviation for *prototype.*

proton: A positively-charged particle in the nucleus of an atom.

prototype *(proto):* **1.** An individually built test version of a new design of a car or system. **2.** An individually built vehicle in sports-car racing when an insufficient number of vehicles have been manufactured to be placed in a category.

prove-out circuit: A function of the *ignition switch* that completes the warning light circuit to ground when it is in the START position, so the warning lights will be on during engine cranking to indicate to the driver that the bulbs are working properly.

A B C D E F G H I J K L M N O **P** Q R S T U V W X Y Z

proximity effect: The distortion of current density due to magnetic fields increased by conductor diameter, close spacing, frequency, and magnetic materials such as steel conduit or beams.

Prussian blue: A blue pigment in solution applied to the surface area used to determine the contact area between two mating surfaces.

psi: An abbreviation for *pounds per square inch.*

psia: An abbreviation for *pounds per square inch, absolute.*

psig: An abbreviation for *pounds per square inch, gauge.*

psych: To intimidate a competitor in competition.

PTO: An abbreviation for *power take off.*

pu: Purple.

puck: **1.** A brake pad. **2.** A valve-adjusting shim.

puffer: A supercharger.

pull: A condition whereby the vehicle veers to one side or the other when the brakes are applied.

pull circuit: A circuit that brings a tractor cab from a fully tilted position up and over the center.

pulldown: A term used for *pumpdown.*

pulley: A wheel-shaped, belt-driven device used to drive engine accessories.

pull rod: A body-working tool in which the curved end of the rod is inserted in the drilled hole to pull out dents.

pull-type clutch: A type of clutch whereby the release bearing is pulled toward the *transmission.*

pulsation: A surge that may be felt in the brake pedal during low-pressure braking.

pulsed air-injection system: A device found in the exhaust emission system that injects pulses of *ambient air* into the exhaust as an aid in the burning of exhaust gases.

pulsed air system: A shortened term for *pulsed air-injection system.*

pulse-delay variable resistor: A resistor in the electrical circuit of the wiper system used to delay the wiper motion from 0 to 25 seconds.

pulse period: A term used for *pulse width.*

pulse ring: A ring on the *crankshaft* to trigger a sensor used with an onboard computer to control ignition timing.

pulse time: A term used for *pulse width.*

pulse width: The time, in milliseconds, the injectors are energized and held open, which determines the amount of fuel injected. Also known as *pulse period* and *pulse time.*

pump: A device that transfers gas or liquid from one place to another.

pump air bleed: In a fuel-supply system, small vent holes in the mechanical fuel-pump body that allow air to enter and leave the pump as the *diaphragm* moves back and forth.

pumpdown: A term used for *evacuate,* as in an air-conditioning system. Also known as *pulldown.*

pump gas: Standard gasoline without any additives.

pump impeller: **1.** That part of a fluid pump that causes the fluid to move. **2.** The output drive member that receives its power from the engine.

pumping brakes: A term that refers to the practice of depressing the brake pedal several times in quick succession.

pump link: In a fuel-supply system, a short, connecting part between the pump rocker arm and the diaphragm on certain models.

pump reservoir: A container located on or adjacent to the *pump,* providing fluid and reserve for proper system operation.

pump rotors: Parts of a positive displacement pump that uses rotating rotors to develop flow.

pump screen: A filter to remove debris from a fluid before it enters the pump.

pump up: A condition where a *hydraulic-valve lifter* has developed more pressure than required and prevents the valve from seating properly.

punch: A racing fuel, generally with nitromethane.

punch it: **1.** To apply full throttle. **2.** A term used for *plant it* or *punch off.*

punch off: A term used for *punch it* or *plant it.*

puncture-sealing tire: A specialty tire made to permanently seal tread punctures up to 3/16 inch (4.76 mm) in diameter with a resultant rubber-compound sealer applied to the inside of the tire by the manufacturer.

pup converter: A term used for *light-off, mini-oxidation catalytic converter.*

pure stock: A vehicle that conforms to all original factory specifications.

purge: **1.** To remove moisture and/or air from a system or a component by flushing with a dry gas, such as nitrogen (N). **2.** To remove all refrigerant from an air-conditioning system.

purge valve: A valve in some evaporative emission-control system charcoal canisters to limit the flow of vapor and air to the *carburetor* during *idle.*

purity test: A static test that may be performed to compare a suspect refrigerant pressure to an appropriate temperature chart to determine its purity.

pur sang: A french term for "pure blood" used to describe a particularly fine car.

push: A condition in cornering when the slip angle of the front tires is greater than the slip angle of the rear tires and the front end of the vehicle tends to break loose and slide toward the outside of the turn.

push angle: **1.** The travel angle when the electrode is pointing in the direction of weld progression. **2.** Also used to partially define the position of guns, torches, rods, and beams.

push bar: A bar in the front of a *push car* to prevent damage to either vehicle when push-starting a race car.

push car: Usually a truck; a vehicle used to *push start* a race car.

push circuit: A circuit that raises a tractor cab from the lowered position to the desired tilt position.

pusher: A rear-engine vehicle.

pushrod: **1.** A solid or hollow rod which serves as a link between the *valve lifter* and *rocker arm.* **2.** The rod that transmits the movement and force of the driver from the brake pedal lever to the *master cylinder.* **3.** The rod that transmits the movement and force of the wheel cylinder piston to the *brake shoe.*

pushrod tube: The passage from the engine block to the cylinder head that contains a *pushrod.*

push start: To start a vehicle by pushing it with another.

push-type clutch: A type of clutch in which the release bearing is not attached to the clutch cover.

pussycat: A smooth, quiet running vehicle.

put on the trailer: To decisively defeat a racing competitor.

putter: A custom car that is not intended for high performance.

put to the wood: A term for applying full throttle.

putty: Body filler.

PVC valve: A valve used in *positive crankcase ventilation systems* to meter blowby into the *intake stream.*

PVS: An abbreviation for *ported-vacuum switch.*

pyroconductivity: Electric conductivity that develops with changing temperature, and notably upon fusion in solids that are practically nonconductive at atmospheric temperatures.

pyrolysis: The decomposition of a compound due to its exposure to high heat.

pyrolytic oven: An oven used to remove grease, oil, and rust from the surfaces of engine parts.

pyrometer: An instrument used to measure high temperature.

pyrophoric: A substance having the quality of spontaneous ignition when exposed to air.

Q-jet: A Rochester Quadrajet four-barrel carburetor.

Q-ship: An innocent looking vehicle with outstanding performance.

quad: **1.** A four-barrel carburetor. **2.** Four headlamps, two on either side of the vehicle. **3.** A four-wheel drive vehicle.

quad carburetor: A four-barrel carburetor.

quad-4: A high-performance, four-cylinder engine developed by General Motors.

quadrant: **1.** A device used to indicate the position of the gear selector of an *automatic transmission.* **2.** One quarter of a circle.

Quadra-Trac: A full-time, four-wheel drive system.

quad ring: A rubber or plastic sealing ring with square sides.

quad valve head: A cylinder head with four valves, two exhaust and two intake, per cylinder.

qualify: To earn a starting position in a race.

quarter: A term used for a *quarter-mile drag strip.*

quarter elliptic spring: One half of a semi-elliptic spring having one end attached to the chassis frame and the other attached to the axle.

quarter mile: **1.** A distance of 1,320 feet (402 meters). **2.** The standard distance for a drag strip.

quarter-mile drag strip: The most popular standard distance for a competitive drag-racing strip.

quarter panel: A term used for the metal work for either the front or rear corner of a vehicle body.

quarter race cam: A *camshaft* that has been reground to provide better than standard performance.

quarter-speed cam: A *camshaft* that rotates once for each four revolutions of the *crankshaft.*

quartz-halogen: A type of high output lamp containing a gaseous halogen; used for headlights, driving lights, or fog lights.

quartz-iodine: A lamp containing gaseous *iodine* (I).

Quattro: A full-time, four-wheel drive system by Audi.

quench: To suddenly cool a heated metal or alloy by dousing it with water or oil.

quench area: Any internal portion of a *combustion chamber* which causes combustion to cease because of the temperature drop in the air/fuel charge where it meets this area.

quick change: Any system or sub system that is designed for ease and speed for changing or replacing.

quick charger: A battery charger that is designed to charge or boost a *battery* in a short time.

quick coupler: A coupler that allows hoses to be quickly connected and/or disconnected. Most shop air hoses, for example, are equipped with quick couplers.

quick release valve: A device used to exhaust air as close as possible to the service chambers or spring brakes.

A B C D E F G H I J K L M N O P **Q** R S T U V W X Y Z

R

r: Red.

race: **1.** That element of a one-way roller clutch providing the cylindrical surface through which the rollers and *cam* transmit *torque.* **2.** A groove, edge, or track on which a rolling or sliding part moves.

race for the pink: To race for actual ownership of the competitor's vehicle.

racer: **1.** A competition driver. **2.** A competition vehicle.

racer's tape: *Duct tape.*

races: The metal rings on which ball or roller bearings rotate.

rack and pinion: A type of steering assembly that has a gear or pinion at one end of the steering shaft engaging a horizontal-toothed bar or rack having tie rods at either end that are attached to the steering arms.

rack-and-pinion steering: A term used for *rack and pinion.*

racy bopper: An auto-racing groupie.

rad: A term often used to describe a radically modified engine or vehicle.

radial compressor: A space-saving compressor used on small cars.

radial motion: A motion extended to either, or both, extremes of a *radius.*

radial ply: A term used for *radial ply belted tire.*

radial ply belted tire: A tire having the ply cords placed at right angles to the beads, plus belts under the tread section providing the least tread distortion while moving, thereby minimizing tread wear and rolling friction.

radial runout: Variations in tire diameter; the measured amount of out-of-roundness on rotating tires.

radial tire waddle: A term used for *tire waddle.*

radiation: A natural process by which energy is transmitted.

radiation dose: The amount of energy per unit of mass of material deposited at each point of an object undergoing radiation.

radiator: A heat exchanger used to remove heat from the coolant in the cooling system containing a vertical- or horizontal-finned tubing section connected between two tanks.

radiator cap: A term used for *radiator pressure cap.*

radiator core: The center of the radiator, made of tubes and fins, used to transfer heat from the coolant to the air.

radiator fan: A term used for *fan.*

radiator hose: An oil- and ozone-resistant synthetic-rubber hose that connects the *radiator* to the thermostat outlet housing and water pump inlet housing.

radiator hose clamp: A term used for *hose clamp.*

radiator pressure cap: A cap that seals in pressure from hot expanding coolant until a predetermined limit is reached, then the valve opens, allowing excess pressure to escape, generally to a coolant-recovery tank.

radiator shutter system: An engine temperature-control system that controls the amount of air flowing through the radiator by use of a shutter system.

radius: A line extending from the center of a circle to its boundary.

radius arms: Longitudinal suspension arms used to position a *beam axle.*

radiused: **1.** A procedure that is used to reduce the radius diameter at the area the valve stem meets the valve head. **2.** A valve type that is ground with a radial grinder to aid in air flow around the valve. **3.** Wheelwells that have been cut to a circular shape.

radius ride: A condition where the *crankshaft* rides on the edge of the *bearing.*

radius rods: A term used for *radius arms.*

radix: **1.** A base. **2.** The base number in a number system.

ragged edge: The absolute limit of a vehicle's potential.

rag top: A convertible.

rail: An early dragster with exposed frame rails.

rail job: A term used for *rail.*

rake: A suspension or structural design to lower one end of the vehicle in relation to the other end.

rally: A sports-car driving contest for driver performance as opposed to vehicle performance.

RAM: An acronym for *random access memory.*

ram air: Air forced through the radiator and condenser coils by the movement of the vehicle or the action of the fan.

ram air cleaner: An air cleaner for high-performance cars that opens an air scoop on the hood to provide a ram effect when the *throttle* is wide open.

ram induction: An *intake manifold* designed to cause a resonant effect at a specific predetermined engine speed.

ram tubes: Short, tuned, tubular stacks on the top of *carburetors.*

ramp: The sloping section of a camshaft lobe which raises the lifter.

ramp angle: An angle formed by lines adjacent to the static-loaded radius of the front and rear wheels intersecting at the point of the lowest ground clearance under the middle of the vehicle.

ram tuning: The tuning of an intake manifold to ensure that the passages are of sufficient length to cause a resonant effect at a specific predetermined engine speed.

Ranco control: A tradename often used for a *thermostat.*

R&D: An abbreviation for *research and development.*

random access memory *(RAM)*: A computer memory into which the user can enter information and instructions (write), and from which the user can call up data (read).

random intermittent welds: Welds on one or both sides of a joint in which the weld increments are made without regard to spacing.

R&R: An abbreviation for *remove and replace.*

range shift cylinder: Located in the auxiliary section of the transmission, this component, when directed by air pressure via low and high ports, shifts between high and low range of gears.

range shift lever: Located on the shift knob, this lever allows the driver to select low- or high-gear range.

raster pattern: Also known as *stacked pattern.*

rat: A Chevrolet big block V-8 engine.

ratio: **1.** The relative amounts of two or more substances in a mixture. **2.** The comparison of two numbers as in teeth on gears.

ratio valve: A device used on the front or steering axle of a heavy-duty vehicle to limit the brake application pressure to the actuators during normal service braking.

rat motor: A Chevrolet big block V–8.

Ravigneaux gear train: A planetary gear train with two sun gears, three long and three short planetary pinions, planetary carrier, and ring gear.

RC engine: A shortened term for *rotary combustion engine.*

RCRA: An abbreviation for *Resource Conservation and Recovery Act.*

reaction time: **1.** The amount of time required to physically apply the brakes of a vehicle after mentally being aware of the necessity to do so. **2.** The amount of time required from the moment the Christmas tree light turns green until the vehicle trips the starting-line timing light.

reactivity: The characteristics of a material that cause it to react violently with another material, such as air, heat, or water.

read-only memory: A solid-state storage chip that is programmed at the time of its manufacture, cannot be reprogrammed by the computer user, and retains its program when the computer's power is turned off.

rear-axle assembly: A group of parts that operate to transfer driving torque from the *drive shaft* to the rear wheels, and includes mounting pads for the rear springs and shocks.

rear-axle housing: The basic framework of the *rear-axle assembly* inside which the individual parts, including the differential and axles are mounted and/or operate.

rear-camshaft plug: A plug driven in the rear of the block behind the camshaft to contain the oil supplied to the rear camshaft bearing.

rear clip: The rear end of a vehicle's bodywork extending from the door pillar back.

rear control arm: Horizontal arms that connect the rear axle housing to the frame when coil springs are used in the rear suspension system to maintain axle alignment and handle the driving and torque loads.

rear drive: A term used for *rear-wheel drive.*

rear end: The differential and final drive assembly on a rear-wheel drive vehicle.

rear-end torque: The twisting reaction of a rear-axle assembly in a direction opposite to that of wheel rotation when power is applied, controlled by the control arms in coil-spring models and by the leaf springs in leaf spring models.

rear engine: An engine arrangement where the engine is placed behind the rear wheels, driving them through a transaxle assembly.

rear leaf spring: A term used for *leaf spring.*

rear roll center: The center as determined by the rear suspension geometry around which the rearward part of a vehicle has a tendency to roll.

rear shock: A type of *shock absorber* that is compatible with a particular type of driving and its load requirements.

rear-spring shackle: A device bolted to a rubber bushing in the rear main leaf eye, and having an upper shackle bolt that extends through a similar rubber bushing in the rear-spring hanger, providing fore and aft movement with variations in spring length.

rear steer: A steering gear that is positioned behind the front-wheel centerline.

rear suspension: An integral part of the total suspension system designed to keep the rear axle and wheels in their proper position under the car body, typically of two types; solid-axle type and independent rear-suspension type.

rear-suspension coil spring: A *coil spring* that is used in the rear suspension system.

rear-suspension rebound stop: A function of the rear shocks as they limit the distance the rear-axle assembly can move downward.

rear-wheel drive: A *drivetrain* layout that provides power to the rear wheels only. Often referred to an *rear drive.*

reassembly: Putting the disassembled parts of a device back together, replacing those parts as necessary.

rebore: To increase the diameter of a cylinder.

rebound: The outward extension of the springs and shocks in a vehicle suspension system.

rebound clip: Metal clamps placed at three or four intervals around multi-leaved springs to prevent the leaves from becoming separated on rebound.

rebound travel: The downward movement of a wheel from its normal position when the spring and shock are expanding, as the sudden drop of a wheel into a depression and a weight transfer away from the wheel.

rebound valve: A calibrated piston valve mounted on the shock piston that provides variable resistance to fluid flow during rebound.

recall: A notice issued by the vehicle manufacturer that a certain make and model vehicle should be returned to the dealer for a correction of a suspected or known problem.

recall bulletins: A dealer notice pertaining to vehicle service or the replacement of parts related to a *recall.*

recapping: To recover a tire casing with a new tread.

receiver: **1.** A term sometimes used for *drier, receiver drier,* or *receiver dehydrator.* **2.** A container for the temporary storage of liquid refrigerant.

receiver dehydrator: A combination container for the storage of liquid refrigerant and a *desiccant.*

receiver drier: A term used for *receiver dehydrator.*

receiver tank: The second tank in a radiator is often referred to as a receiver because it receives coolant after passing through the many tubes of the radiator core. The receiver, on automatic transmission equipped cars, contains a *transmission oil cooler.*

receptacle: A contact device, installed at an electrical outlet, extension cord, or drop light for the connection of an attachment plug and flexible cord to supply power to portable equipment.

recharging: **1.** The action of forcing electrical current into a battery, reversing the chemical reaction between the plates and electrolyte. **2.** Filling an air-conditioning system with *refrigerant* after repairs are made.

reciprocating: A back-and-forth, up-and-down, or to-and-fro motion, such as that of a piston in a cylinder.

reciprocating engine: A powerplant in which the pistons move in a *reciprocating motion.*

reciprocating motion: The motion of an object between two limited positions in a straight line back and forth or up and down.

recirc door: An abbreviated form of *recirculate door.*

recirculate door: A door in the *plenum* that regulates the amount of recirculated air flow.

recirculating ball: A worm and gear steering assembly with balls between the *worm and sector* to reduce friction.

recirculating-ball-and-nut steering gear: A type of steering gear having a nut meshing with a worm gear sector, with balls that circulate between the nut and worm threads.

recirculating-ball steering gear: A low-friction steering gear box having a worm-gear meshing with a cross shaft sector through about 40 recirculating balls that travel in matching grooves inside the ball nut and outside the worm shaft which acts as a rolling thread.

reclaim: To process used *refrigerant* to new product specifications by means that may include distillation. This process requires that a chemical analysis of the *refrigerant* be performed to determine that appropriate product specifications are met.

recombination battery: Sometimes called dry-cell batteries because they do not use a liquid electrolyte solution.

recon: A short term for *reconditioned.*

reconditioned: A used part that has been repaired, tested, and determined to be in good order. Also known as *recon.*

recovery: The recovery of *refrigerant* is to remove it, in any condition, from a system and to store it in an external container without necessarily testing or processing it in any way.

recovery cylinder: A cylinder for the storage of used *R-12* and/or *R-134a* must meet *DOT* specification 4BA-300, characterized by a combined liquid/vapor valve located at the top. A *recovery cylinder* should be painted grey with a yellow shoulder.

recovery/recycle systems: A term often used to refer to the circuit inside the recovery unit used to recycle and/or transfer *refrigerant* from the air-conditioning system to the *recovery cylinder.*

recovery/recycle unit: A term used to identify the complete unit used to recover and/or recycle *refrigerant* from the *air-conditioning system.*

recreational vehicle *(RV):* A motor-driven, van-like vehicle having sleeping, food preparation, and bathroom facilities.

rectifier: An electrical device used to convert alternating current to direct current.

rector: The *stator* in a *torque converter.*

recycle: To recycle is to clean the *refrigerant* for reuse, by oil separation and passes through other devices such as filter-driers, to reduce moisture, acidity, and particulate matter.

Red Book: A used-car price guide.

red dye trace solution: The dye that reveals the exact location of a leak in a fluid system by depositing a colored film around the leak area.

red flag: A signal for all drivers to come to a stop.

red light: **1.** In normal driving situations, a warning to stop. **2.** In drag racing, to jump from the line before the start signal. **3.** In competition, to foul and become disqualified for an event.

red line: **1.** The absolute minimum an auto dealer will accept for a vehicle. **2.** The absolute maximum recommended engine speed.

reducer: **1.** A tool that reduces the size of a device, such as a pipe. **2.** A paint thinner. **3.** An adapter for using a smaller tool.

reducing atmosphere: A chemically active protective atmosphere, which at elevated temperature will reduce metal oxides to their metallic state.

reducing flame: An oxyfuel gas flame with an excess of fuel gas.

reduction catalyst: The section of a three-way catalytic converter that breaks NO_X (oxides of nitrogen) down into harmless nitrogen and oxygen through a reduction reaction.

reed valve: A thin leaf of steel located in the valve plate of automotive compressors, serving as suction and discharge valves.

reference voltage: In computerized engine-management systems, a five-volt signal sent out from the electronic control unit to a variable-resistance sensor such as a throttle-position sensor so the computer can "read" the voltage value of the return signal.

refrigerant: The chemical compound used in a refrigeration *system* to produce the desired cooling effect.

refrigerant-containment device: A device introduced on some car lines to guard against high pressure resulting in *refrigerant* loss by controlling the compressor and/or condenser fan motor.

refrigerant lines: Specially designed vapor-barrier hoses reinforced with woven nylon mesh and fabric capable of withstanding the high temperatures and pressures of the system.

refrigerant recovery: The act of recovering *refrigerant.*

refrigerant recycle: The act of recycling *refrigerant.*

Refrigerant CFC-12 *(R-12):* The refrigerant, dichlorodifluoromethane (CCl_2F_2), used in automotive air conditioners, as well as other air-conditioning and refrigeration systems.

Refrigerant HCFC-22 *(R-22):* A refrigerant, monochlorodifluoromethane (CHC_1F_2), used in some early automotive applications but not to be used for today's automotive air conditioners because of high pressures.

Refrigerant HFC-134a *(R-134a):* The "refrigerant of choice," tetrafluoroethane (CH_2FCF_3), to be used in automotive air conditioners as *R-12* is being phased out.

refrigeration: To use an apparatus to cool; keep cool; or keep chilled under controlled conditions by natural or mechanical means, as an aid to ensure personal safety and comfort.

refrigeration cycle: The complete cycle of the refrigerant back to the starting point, evidenced by temperature and pressure changes.

refrigeration lubricant: A mineral oil or synthetic oil-like lubricant, such as PAG and ESTER, that is formulated for specific use and application in designated refrigeration systems.

refrigeration oil: A highly refined (specially formulated), non-foaming organic mineral oil free from all contaminants, such as sulfur, moisture, and tars, used to lubricate the air-conditioner compressor.

refrigeration tape: A term used for *insulation tape* or *duct tape.*

regeneration system: A system in a gas turbine that converts some of the heat, usually wasted, into usable power.

regenerator: A device placed on a gas turbine to take the heat of exhaust and put it into the intake of the engine.

regional offices and distributorships: Offices owned and operated by the automobile company, and considered to be the link between the manufacturer and the dealerships.

register: A high-speed device used in a central processing unit for temporary storage of small amounts of data or intermittent results during processing.

regular-duty coil spring: A coil spring supplied to handle average loads to which the vehicle is subjected, having a small wire diameter as compared to a heavy-duty spring.

regular gasoline: Gasoline that has an octane number near 85–90.

regular production option *(RPO):* Items of new car equipment available to any buyer for a price.

regulator: A term often used for *voltage regulator.*

reheat principle: A principle that is used in automotive air-conditioning systems to control in-car relative humidity by first cooling, then reheating the air.

Reid vapor pressure *(RVP):* The measure of the volatility of liquid fuels.

relative humidity: The moisture content of the surrounding air.

relative pressure: The difference between intake manifold pressure and the output pressure in a *fuel-injection system.*

relative wheel weights: The weight on each of the vehicle's wheels as measured by individual matched scales under each wheel.

relay: An electro-mechanical switch having a coil and one or more sets of points.

relay/quick-release valve: A valve, similar to a remote control foot valve, used on trucks with a wheel base of 254 inches (6.45 meters) or longer attached to an air tank to main supply line to speed the application and release of air to the service chambers.

relay rod: A term used for *center link.*

release bearing: A term used for *throwout bearing.*

release fingers: A term used for *release levers.*

release levers: A lever-type, pivoting pressure-plate device in the clutch assembly that is moved by throwout bearing movement, causing the clutch-spring pressure to be relieved so the *clutch* is released from the *flywheel.*

reliability run: A *rally* for *hot rod* drivers.

relief valve: A term used for *compression valve* or *pressure relief valve.*

relieved: An engine in which the intake and exhaust passages have been cleaned of all obstructions and ridges.

reluctor: A gear-like part of an electronic ignition system, having the same number of teeth as there are cylinders, near the top of the distributor shaft so that as a tooth passes a pickup coil, resistance to magnetic flux is reduced and the magnetic field is strengthened, which triggers the electronic control unit.

remote back-pressure transducer: An exhaust back-pressure sensing device mounted in the vacuum line leading to the *EGR valve* rather than on the valve itself, to bleed off the vacuum signal to prevent recirculation.

remote bulb: A sensing device connected to an expansion valve or thermostat by a capillary tube, sensing temperature and transmitting pressure to the control for its proper operation.

remote sensing bulb: A term used for *remote bulb.*

remove and repair: To remove a component from a vehicle and then to repair it; it is assumed that it will also be replaced on the vehicle.

remove and replace *(R&R):* To remove a component and replace it with a reconditioned, rebuilt, good used, or new component on a vehicle.

repair order: A written summary report of repairs that are requested or that are made.

replacement springs: New springs having the same characteristics, shape, and type of ends as the springs in the vehicle, all four of which should be replaced at the same time.

replacement tire: A tire purchased to replace an original tire that was supplied by the vehicle manufacturer, having the same size and temperature range as the original tire.

required sidewall information: Information that is required by the *Department of Transportation* to be included on each tire such as size, load rating, maximum inflation pressure, generic name of materials used in construction, actual number of plys, and so on.

required voltage: **1.** The low voltage required by a device or system. **2.** The highest voltage required to fire a *spark plug.*

research and development *(R&D):* As the term implies, to research and develop a concept, system, or product.

research octane number *(RON):* A term used for *octane number.*

reserve capacity: A term used for *reserve capacity rating.*

reserve capacity rating: The number of minutes a fully charged battery at 80°F (26.7°C) can supply power at 25 amperes to run the ignition, lights, and accessories after a charging system failure or when the engine is not running, before dropping to 1.75 volts per cell.

residual magnetism: The magnetism that remains in an electromagnet material after the current is interrupted.

residual pressure: The pressure that is retained in a specific area.

residual stress: Stress present in a joint member or material that is free of external forces or thermal gradients.

resin: The molten plastic part of fiberglass-reinforced plastic.

resistance: **1.** An opposition to the flow of electricity. **2.** An opposition to movement, such as wind against a vehicle.

resistance welding: **1.** A term used for *spot welding.* **2.** A welding process that unites the surfaces with heat, obtained from the resistance of the work pieces, and the application of pressure.

resistor: An electrical device used to reduce voltage.

resistor spark plug: A *spark plug* having a resistance of 5,000 to 10,000 ohms inside the upper part of the insulator, increasing the electrode life and suppressing radio interference.

resonator: A small auxiliary muffler, after the main muffler, used to help reduce engine noise without restricting the flow of exhaust gas.

Resource Conservation and Recovery Act *(RCRA):* A Federal law that requires users of hazardous materials to store used material in a manner consistent with established guidelines until they can be properly disposed.

resto: The restoration of an old car.

restriction: A blockage in an air or fluid line caused by a pinched or crimped line, foreign matter, or moisture freeze-up.

restrictor: An insert fitting or device used to control the flow of air or fluid.

restrictor plate: **1.** A plate with holes drilled in it to restrict the flow of air or liquid. **2.** A plate with holes of a calibrated diameter that is placed between the *carburetor* and *intake manifold* of a *stock car* to restrict the air/fuel flow, thereby reducing engine speed and *horsepower.*

restyle: 1. To customize a vehicle. 2. To change the appearance of a vehicle.

retard: To cause ignition spark to occur later in an engine's cycle.

retarder: 1. A mechanical, electrical, or hydraulic device used to slow a diesel engine when the *throttle* is lifted. 2. A chemical compound that reduces the speed of paint drying.

retard sensor: A term used for *knock sensor.*

retention seal: A seal that has the ability to retain fluid pressure.

retracting spring: A spring that is used to pull the *brake shoes* away from the drum when the brake pedal is released, pushing the wheel-cylinder piston back into its bore and thus returns the *brake fluid* to the *master cylinder.*

retread: A used tire having had its old tread removed and a new tread applied.

retrocar: A modern vehicle that has been restyled to look like an old vehicle.

retrofit: To modify equipment that is already in service using parts and/or materials available or made available after the time of original manufacture.

returnable cylinder: 1. A cylinder of gas, such as oxygen, nitrogen, propane, or acetylene that can be reused. 2. A *DOT* approved refrigerant-recovery cylinder.

return hose: 1. The low-pressure hose in a power-steering system through which fluid returns to the pump reservoir after leaving the *steering-gear assembly* or *control valve.* 2. A non-pressurized return hose in a pressurized fluid or *air system.*

return spring: 1. Springs used on drum brakes to pull the brake shoes away from the drums when the brakes are released. 2. A spring that is used to return a mechanism to its original position.

reusable cylinder: A cylinder, such as a refrigerant-recovery cylinder, that may be reused.

rev: 1. A short term for *revolution.* 2. To noticeably increase engine speed.

rev counter: A tachometer.

reveal file: A small file that is "pulled, not pushed," available in numerous shapes, generally curved to fit tightly crowned areas such as around windshields, wheel openings, and other panel edges for final fitting of *reveal molding.*

reveal molding: The metal trim that outlines an opening, depression, or raised area of an auto body.

reverse: 1. The transmission position enabling the vehicle to back up. 2. To increase wheel track or tread by installing the rims backwards.

reverse bleeding: A method of purging air from a hydraulic system by forcing fluid into the system through a bleeder valve and allowing the air to escape at the *master cylinder* or reservoir.

reverse clutch: A multiple-disc clutch that is engaged in reverse gear.

reversed bias: A term used when a positive voltage is applied to the N-type material and a negative voltage is applied to the P-type material of a diode.

reversed polarity: A condition that exists when battery cables are connected to the wrong terminals of a battery.

reverse Elliot axle: A solid-beam front axle on which the steering knuckles span the axle ends.

reverse flush: A method of cleaning an engine and/or radiator by flushing in a direction opposite of normal coolant flow, under pressure.

reverse flushing: A method used to clean an engine or cooling system by flushing it in the opposite direction of normal coolant flow.

reverse gear: A transmission gear arrangement obtained from a planetary gear set when the planet-gear carrier is held and power is applied to the sun gear.

reverse idler gear: **1.** A gear in most standard transmissions that must be meshed to obtain reverse, and that idles in all forward gears. **2.** A gear that transfers power in reverse only. Rotating on a separate shaft, it reverses the direction of the counter-gear motion.

reverse shoe: A term used for *trailing shoe.*

reversion: The interval during valve timing overlap when both intake and exhaust valves are open and a small mist of air/fuel is forced out back through the *carburetor.*

rev limit: The maximum recommended engine speed.

rev limiter: **1.** A governor. **2.** A mechanical or electrical device that limits engine speed.

revolution: The movement around a center or axis such as the circling of the Earth around the Sun.

revolutions per minute *(rpm):* **1.** The number of times a member, such as a shaft, makes a complete revolution in one minute. **2.** The rate of speed of a rotating part, such as a *crankshaft.*

revs: A term used for *rpm.*

rheostat: A two- or three-terminal variable resistor used to regulate the voltage of an electrical current.

rib: The tread-section element around a tire; the raised section of the tread.

ribbon-cellular radiator core: A type of radiator core that consists of ribbons of metal, usually copper (Cu), soldered together along their edges.

ribbon spring: A spring designed to hold the sprags of a sprag-overrunning clutch at the correct angle for instantaneous engagement.

rich mixture: An *air/fuel mixture* having insufficient air or excessive fuel.

ride: **1.** A personal car. **2.** An opportunity to drive in an event or series of events.

ride and handling: **1.** An indication of the degree of comfort a tire delivers to the passenger. **2.** A measure of the vehicle responsiveness in relation to the driver's steering actions.

ride height: The distance between the road and the bottom of a vehicle.

ride shotgun: To ride in the right front seat of a vehicle.

ridge: A condition in the bores of an engine caused by piston-ring wear of the cylinder walls.

right-hand thread: A common thread for a bolt or nut that is tightened by turning it clockwise, or to the right.

Right-To-Know-Law: A federal law that requires all manufacturers of hazardous materials to notify and make information available to all employees, vendors, customers, and end users relative to the nature of the hazard, and to provide remedies if accidently misused.

rigid axle: A term used for *beam axle* or *solid axle.*

rigid axle suspension: A term used for *solid-axle suspension.*

rigid discs: Steel plates to which friction linings or facings are bonded or riveted.

rigid fifth wheel: A *fifth wheel* that is fixed rigidly to a frame having no articulation or oscillation, generally used in applications where the articulation is provided by other means, such as an articulating upper coupler of a frameless dump.

rigid motor mount: A solid mount for attaching the motor to the frame without any provisions for vibration dampening.

rigid rear-driving axle: A common rear suspension system design constructed of a central differential and tubular axle shaft housing.

rigid rear suspension: A *suspension system* which has both rear wheels attached to the ends of a solid rear axle housing.

rigid transmission mount: A solid mount for attaching the transmission to the frame without any provisions for vibration dampening.

rim: The outer edge of a bare wheel.

rim offset: The distance between the wheel-rim centerline and the mounting face of the disc affecting front suspension loading and operation.

rim safety ridge: The small, metal ridge just inside the tire-bead section to retain the tire position on the rim in case of a flat or blow-out that tends to keep the bead of an uninflated tire out of the rim drop-center area.

ring: **1.** A term that generally applies to *piston ring, compression ring,* or *oil ring.* **2.** A gasket, such as an *O-ring.* **3.** A hollow noise, such as when striking a sound brake drum.

ring and pinion: The combination of a ring gear attached to the differential and the pinion at the end of the *drive shaft.*

ring expander: A flexible spring-steel part placed behind certain *piston rings* to increase ring pressure on the cylinder wall.

ring gap: The gap between the ends of a *piston ring* installed in the cylinder generally 0.004 inch (about 0.1 mm) for each inch (25.4 mm) of cylinder diameter.

ring gear: **1.** The gear around the edge of a *flywheel.* **2.** A large, circular gear such as that in the final drive assembly.

ring grooves: Grooves machined around the piston wall to support the rings.

ring-groove spacer: Strips of thin metal used to fill the gap left between the piston and ring groove after the groove has been machined.

ring joint: A shock-absorber component used to attach the *shock absorber* to the *axle* and *frame.*

ring land: The surface of a *piston* between *ring grooves.*

ring ridge: The ridge left at the top of a cylinder wall as the wall below is worn away by piston-ring movement.

ring spacing: The width of the *ring land.*

riser valve: A term used for *heat-riser valve.*

rivet: A fastener used to hold two pieces together.

riveted lining: Linings on drum-brake shoes or disc-brake pads that are attached to their steel backing with *rivets.*

RMA: An abbreviation for the *Rubber Manufacturers Association.*

RO: An abbreviation for *repair order.*

roach coach: A catering vehicle.

roadability: The steering and handling qualities of a vehicle as it is being driven on the road.

road-draft tube: A pre-emission, control-era device for ventilating the *crankcase* to the atmosphere; a pipe routed under the chassis at an angle that produces a small vacuum as the vehicle travels forward. Fresh air is drawn in through a mesh filter in the oil filler cap, circulated around inside the *crankcase* and exhausted through the *road-draft tube* carrying *blowby* with it.

road feel: A term that relates to the driver's ability to sense the vehicle directional control from the movement transmitted through the steering wheel by the front wheels; not so sensitive on power-steering-equipped vehicles.

road horsepower: *Horsepower* available at the drive wheels of the vehicle.

road load: A constant vehicle speed on level terrain.

road racing: A competition race on an irregular course simulating the variety of terrain that is found on an average public road system.

roadster: **1.** An single-seat *Indycar* of the late 1960s and early 1970s that resembled a two-passenger car. **2.** A two-passenger, open car.

rocker arm: **1.** A pivoting part in a mechanical fuel pump, one end of which rides on a camshaft eccentric or *pushrod,* and the other end of which pulls on the *diaphragm* via a mechanical linkage, causing the *diaphragm spring* to be compressed. **2.** A pivoting lever in the valve train of an overhead valve system which applies motion, directly or indirectly from the *camshaft* to open an intake or exhaust valve.

rocker-arm ratio: The difference between the upper movement of the *pushrod,* acting on one end of the *rocker arm,* and the downward movement of the *rocker arm* acting on the *valve stem.*

rocker-arm shaft: A shaft in each *cylinder head* on which *rocker arms* are arranged.

rocker panel: The sheet metal along the bottom of an auto body, beneath the doors, and between the front- and rear-wheel openings.

rod: **1.** An abbreviation for *hot rod.* **2.** A steering arm. **3.** A suspension arm. **4.** An engine *connecting rod.*

rod bearing: The bearing at the *crankshaft* end of a *connecting rod.*

rod big end: The end of the connecting rod that attaches around the *crankshaft.*

rod bolts: The bolts used to secure the cap to the rod at the *big end.*

rod-end bearing: The spherical bearing found at the end of a *suspension arm* or *rod.*

rod journal: A term used for *rod bearing.*

rod-length ratio: The center-to-center length of a connecting rod divided by the piston stroke.

rod out: To clean a *radiator.*

rod ratio: A term used for *rod-length ratio* or *rod-to-stroke ratio.*

rod seal: A component used to keep oil from leaking past the *piston rod* and into the atmosphere during pressure conditions.

rod small end: The end of the *connecting rod* through which the piston pin passes to connect the *piston* to the *connecting rod.*

rod-to-stroke ratio: A term used for *rod-length ratio.*

roll: **1.** To turn over and over. **2.** To tip or sway side-to-side. **3.** The motion of a vehicle body toward the outside of a turn when cornering or changing directions suddenly.

roll axis: The longitudinal axis of a vehicle defined by an imaginary line running through the front and rear *roll centers.*

roll bar: A tubular bar behind the cockpit of a race car to protect the driver in case of a *rollover.*

roll cage: A tubular, cage-like structure surrounding the cockpit of a race car to protect the driver in the case of a *rollover* and to add strength to the overall structure of the vehicle.

roll centers: The points at the front and rear about which the vehicle's sprung mass will *roll.*

roller: **1.** A race car offered for sale without an engine. **2.** A cylindrical locking element of a one-way roller clutch.

Roller: A Rolls-Royce.

roller bearing: A bearing using rollers within an outer *race* or ring.

roller cam: A *camshaft* having a small roller at the base of each lifter.

roller chain: A timing chain having rollers that engage the gear sprockets.

roller clutch: A one-way clutch containing a number of rollers that operate by wedging on a ramp between an inner and outer race to lock up (drive) when the outer race is turned in one direction, and to slip (overrun) when it is turned in the opposite direction.

rolling diameter: A term used for *tire rolling diameter.*

roller lifter: A valve lifter, used in some high-performance engines, featuring a roller which contacts the camshaft to reduce friction.

rolling radius: The distance between the center of the wheel and the contact point of the tire on the ground under the weight of the vehicle.

rolling resistance: The pounds of force required to overcome the resistance of a tire to rotate.

rolling start: When cars are lined up running *parade laps* and a *pace lap,* behind a *pace car,* they are permitted to start with the display of a *green flag.*

roll out: The distance a race car travels in a drag-racing event at the beginning of a run before the front tires clear the stage beam and start the clock.

rollover: An accident in which the vehicle rolls over and over or turns upside down.

roll steer: The direction and amount that the rear axle may cause the vehicle to steer as it moves through its travel when the body rolls during cornering requiring the driver to *oversteer* or *understeer* to compensate for the problem.

roll stiffness: The resistance, measured in pounds per inch of spring travel, of a suspension system to the rolling of the vehicle's mass.

rolly: A term used for *roller cam.*

RON: An acronym for *research octane number.*

R-134a: A trade term for *refrigerant HFC-134a.*

rookie: A race-car driver in competition for the first season in an event or series of events.

room temperature vulcanizing *(RTV):* The trade name for a rubber-like sealing compound.

root bead: A *weld bead* that extends into, or includes part or all of the joint root.

roots supercharger: A mechanically driven, positive-displacement blower with hourglass-shaped rotors.

rope seal: A type of seal used on *crankshafts* shaped much like a small, thin rope.

rosette weld: A term used for *plug weld.*

rosin: A sticky substance applied to the rear tires of drag-race cars for better traction *off the line.*

rotary: The turning motion around an axis.

rotary combustion engine *(RC engine):* A four-cycle engine having a revolving triangular-shaped rotor to transfer power thrust via eccentric(s) to the *output shaft.*

rotary diesel: A rotary combustion engine operating on diesel principles in which the fuel injected into the *combustion chamber* at the end of the compression phase is ignited by the heat produced during compression, rather than by the spark from a plug.

rotary engine: **1.** A form of radial engine used in early aircraft, outmoded by the end of World War I. **2.** An engine with a three-sided rotor in a slightly hourglass-shaped oval chamber.

rotary flow: Torque converter oil flow associated with the coupling stage of operation.

rotary oil flow: A condition caused by the centrifugal force applied to the oil as the converter rotates around its axis.

rotary vacuum valve: That part of a vacuum control used to divert a vacuum signal for operation of doors, switches, and/or valves.

rotary valve: A semi-circular disc in a two-stroke engine that opens and closes the *intake port.*

rotary valve steering gear: A type of power-steering gear.

rotation: **1.** A term used to indicate that a gear, shaft, or other device is turning. **2.** A term used for *tire rotation.*

rotator: A device in the *cylinder head* that causes the valve to rotate to reduce carbon build-up.

rotor: **1.** Disc-shaped component that revolves with hub and wheel. The lining pads are forced against the rotor to provide a friction surface for the *brake system,* so as to slow or stop a vehicle. **2.** In a conventional ignition system, the part that transfers the secondary voltage from the central terminal to the outer terminals of the *distributor cap* as it rotates inside the cap. **3.** The rotating part of an *alternator* that provides the necessary magnetic field to create a current flow.

rotor face: The flat, parallel surface on each side of the rotor contacted by the pads (linings) during stopping.

rotor oil pump: A type of oil pump in which a pair of *rotors* produce the pressure required to circulate oil to the engine parts.

rotor-type pump: An oil pump that uses two rotors, one having internal and one having external teeth to draw in and dispense oil in much the same manner as the gear-type pump.

roundy-round racing: Any kind of oval-track racing.

rpm: An abbreviation for *revolutions per minute.*

RPO: An abbreviation for *regular production option.*

RTV: An abbreviation for *room temperature vulcanizing.*

RTV/Silicone: A silicone-based *RTV.*

R-12: A trade term for *refrigerant CFC-12.*

R-22: A trade term for *refrigerant HCFC-22.*

rubber bumper: **1.** Rubber stop on the car frame that prevents metal-to-metal contact when the rear-axle housing moves to its maximum upward position. **2.** Rubber stops positioned on or near a *control arm* to limit the maximum upward or downward travel of the arm.

rubberized cork: A mixture of cork and rubber sometimes used as a *gasket.*

Rubber Manufacturers Association *(RMA):* A trade association for the manufacturers of rubber products for motor vehicles.

rubbing block: The insulated section of the movable breaker-point arm that contacts the breaker cam of a conventional ignition distributor.

rules of motion: A term used for *Newton's laws of motion.*

rumble seat: A folding seat in the rear deck of some older two-door coupes, convertibles, and roadsters.

rumper: A big, noisy, rough-idling, high-performance engine.

running board: A flat step between the front and rear fenders to help passengers get in and out of the vehicle.

running gear: The undercarriage of a vehicle and all of the mechanical components attached to it.

running on: The condition that exists when a spark-ignition engine continues to run after the *ignition switch* has been turned off.

run-on: A term used for *dieseling* or *running on.*

run out: The amount a rotating object may wobble out of its plane of rotation.

run whacha' brung: To run whatever one brings to a racing event, especially a drag-racing event.

rust: A metal *oxidation* caused by moisture and oxygen.

rust converter: A liquid that can be painted over that is sprayed on bare metal to eliminate light rust and prevent other rust from forming.

rust inhibitor: A chemical added to the coolant in a *radiator* to reduce the build-up of rust.

RV: An abbreviation for *recreational vehicle.*

RVP: An abbreviation for *Reid vapor pressure.*

RWD: An abbreviation for *rear-wheel drive.*

Rzeppa joint: A constant-velocity universal joint.

S

S: The designation for a *stock car* with a standard transmission.

SA: **1.** A designation for an engine lubricating oil that may be used under the mildest of conditions. **2.** The designation for a *stock car* with an *automatic transmission.*

saddlebag: Air chambers or openings in the left- and right-front corners of the car body between the kickpads and the exterior of the car. The evaporator is sometimes located in the right *saddlebag.*

saddle-clamp access valve: A two-part accessory valve that may be clamped around the metal part of a system hose to provide access to the system for diagnostics and service.

saddle valve: A term used for *saddle-clamp access valve.*

SAE: An abbreviation for *Society of Automotive Engineers.*

SAE horsepower: The corrected brake horsepower of an engine when tested in accordance with *SAE standards.*

SAE standards: A standard for automotive- and aircraft-engine testing and measurement established by the *Society of Automotive Engineers.*

Safe-T-Trak: A *limited-slip differential* by Pontiac.

safety: Prevention of injury or danger.

safety factor: The amount of load that can be absorbed by and through the chassis-frame members of a vehicle.

safety glass: Laminated glass used for vehicle windshields and windows; designed to resist shattering on impact.

safety glasses: Eye glasses with or without side shields to protect the eyes when working in a hazardous environment and to be worn at all times when in the automotive shop.

safety goggles: For those who wear prescription glasses, eye protection from all sides fitting over the glasses and against the face and forehead to seal and protect the eyes from outside hazards.

safety lap: An extra lap taken by the winner of a race to ensure that an error was not made by the official lap scorer.

safety ridge: A small, metal ridge just inside the tire-bead section to retain the tire position on the rim in case of a flat or blowout.

safety rim: A groove around the outer edge of a rim to provide a lock for the *tire bead.*

safety valve: A valve that opens to release excess pressure or heat.

safety wire: A strong wire, usually stainless steel, used to hold pre-drilled nuts or bolts in place, preventing them from turning.

sag: A momentary decrease in throttle rate after the vehicle has gained some speed.

sail panel: The roof rear-quarter panel on a notchback body style that extends from the rearmost side window to the rear window.

sales representative: One who sells new and used merchandise, such as automobiles, to the general public.

saloon: The British term for *sedan.*

salt flats: The dry lake beds in the desert, such as Bonneville, used by hot roders for top-speed runs.

salvage yard: A modern term for *junkyard,* an establishment that sells used auto parts.

Sam sled: A consistently and unnecessarily slow drag-race driver.

sandbag: **1.** To legally secure equipment without having to pay for it. **2.** To hold back in the staging area of a drag race in an attempt to select a specific opponent during eliminations. **3.** To hold speed down in drag racing in an attempt to fall into a favorable bracket during final eliminations. **4.** To deliberately run slower in road- or oval-track racing in an effort to conserve fuel and/or to *psych* the competitors. **5.** A bag filled with sand that is used to help shape metal panels.

sandblast: To clean metal part surfaces by forcing abrasives, usually fine sand, against it with compressed air.

sand-cast: To form a part by pouring molten metal into an impression cavity or mold made by a pattern in sand.

sand hole: An unwanted hole in a sand-cast part.

sanitary: A very clean or well prepared show or race car.

sano: A term short for *sanitary.*

saturated desiccant: A drying agent that contains all of the moisture it can hold at a given temperature.

saturated drier: A drier, accumulator-drier, or receiver-drier having a *saturated desiccant.*

saturated point: The point at which matter must change states at any given temperature and pressure.

saturated temperature: The boiling point of a substance at a particular pressure.

saturated vapor: A term that indicates that the space holds as much vapor as possible and no further vaporization is possible at that particular temperature.

sauce: Racing fuel.

SAW: An acronym for *submerged arc welding.*

SB: **1.** A designation for lubricating oil that is acceptable for medium-duty engines operated under mild conditions. **2.** An abbreviation for *small block.*

SBEC: An abbreviation for *single-board engine controller.*

SC: The designation for lubricating oil that meets the requirements for 1964–1967 gasoline engines in cars and trucks.

scattered: An engine that has literally blown apart.

scattershield: A reinforced housing around the clutch and flywheel to protect the driver from flying parts.

scavenge pump: An oil pump that returns oil to the sump in a dry-sump system.

scavenger: A powerful car that is difficult to beat in a race.

scavenger deposits: White or yellow carbon deposits that normally occur when certain fuels are burned.

scavenging: The forced removal of exhaust gases from a cylinder during the overlap period.

SCCA: An abbreviation for *Sports Car Club of America.*

schematic: A map-like drawing of the electrical system that gives the colors and shows the terminal points; used to trace the circuit for troubleshooting.

Schrader valve: A spring-loaded valve, similar to a tire valve, located inside the service-valve fitting and some control devices to hold vapor or fluid in the system. It requires special adapters for access to the system.

scoop: An opening in the hood or body panel used to take in *ambient air* for cooling or ventilation.

'scope: A term used for *oscilloscope.*

score: A scratch or small dent in the finished surface of a vehicle.

SCORE: An acronym for *Short Course Off-Road Enterprises.*

SCORE International: The actual name for *SCORE.*

scoring: The grooves worn into the friction surface of a brake drum or rotor which may be machined away; if the depth exceeds specified limits, the drum or rotor must be replaced.

SCR: An abbreviation for *silicone controlled rectifier.*

scraper ring: The second ring from the top of a piston used to scrape oil from the cylinder wall.

scrap yard: **1.** A firm selling used parts. **2.** A firm that accepts scrap metal for recycling.

scratch: **1.** To spin the drive wheels, usually enough to leave a mark on the road surface. **2.** A mark on the finish of a body surface made by a scribe or other sharp object. **3.** To make a mark on a finished surface using a sharp object, such as a key.

scratch built: A vehicle built from the ground up, generally of an original design.

screamer: A vehicle with spectacular performance.

screw: The final drive gear.

screw it on: To rapidly accelerate.

screw it on the meter: To install an engine on a *dynamometer.*

screw-thread pitch gauge: A thin material with V-shaped notches that, when matched with a thread of a bolt or nut, indicates the number of threads per inch or millimeter, as well as the thread pitch.

scribe: A sharp, pointed steel tool with a hardened end for marking lines on metal in laying out work.

SCRS: An abbreviation for the *Society of Collision Repair Specialists.*

scrub: A term used for *tire scrub.*

scrub radius: The distance between the centerline of the ball joints and the centerline of the tire at the point when the tire contacts the road surface.

scrub radius area: A term used for *scrub radius.*

scuff: A surface that has been roughened by scraping.

scuff in: To run a new set of racing tires long enough to bring them up to temperature, and wear the manufacturer's protective coating off the tread area. Also known as *scuff off.*

scuffing: A type of wear between two parts where there is a transfer of material from one part to the other.

scuff off: A term used for *scuff in.*

scuffs: New racing tires that have been *scuffed in.*

SD: The designation for a lubricating oil developed for use in 1968–1971 cars and some trucks.

SDV: An abbreviation for *spark delay valve.*

SE: The designation for a lubricating oil that meets the requirements for use in gasoline engines in 1972 and later cars, and in selected 1971 cars and trucks.

sea gull: One at a drag race that complains or squawks all the time.

seal: **1.** A device used around a rotating shaft to prevent fluid or vapor leaks. **2.** A gasket-like material used between two or more parts to prevent fluid or vapor leaks.

sealed battery: A term used for *maintenance-free battery.*

sealed beam headlight: A self-contained glass unit made up of a filament, an inner reflector, and an outer glass lens.

sealed bearing: A replaceable bearing, such as those found on many rear axle shafts or at the front of alternator rotor shafts, that is lubricated and permanently sealed by the manufacturer to contain the grease while keeping out contaminants.

sealed fuel pump: A permanently sealed, non-serviceable pump that has its body and cover crimped together.

sealer: A thick, tacky compound used as a gasket or to seal small openings or irregularities between two mating parts.

seal weld: A weld designed primarily to provide a specific degree of tightness against leakage.

A B C D E F G H I J K L M N O P Q R S T U V W X Y Z

A B C D E F G H I J K L M N O P Q R **S** T U V W X Y Z

seasoned part: A part that has been *stress relieved.*

seat: **1.** The surface on which another part rests. **2.** To wear to a good fit. **3.** The bench-like unit one sits on in a vehicle.

seat adjuster: A device that permits fore and aft and/or up and down movement of a vehicle's front seat.

seat back: The upright portion of a vehicle seat.

seat belt warning system: A warning device to alert the driver and occupants to fasten their seat belts and/or shoulder harness.

seat of the pants *(SOP):* Driving by sheer instinct where instruments cannot be used to determine speed or distance.

secondary air: Air that is pumped to the pollution-control devices to promote chemical reactions that reduce exhaust gas pollutants.

secondary available voltage: High voltage that is available to fire the *spark plug.*

secondary brake shoe: The rear brake shoe in a drum-brake set that is energized by the primary shoe and increases self-energizing action of the brakes.

secondary circuit: **1.** The electrical circuit on the output side of an ignition coil. **2.** That portion of a welding machine that conducts the secondary current between the welding transformer terminals and the electrodes, or electrode and workpiece. **3.** The secondary passage in a *carburetor.*

secondary lock: A component of a fifth-wheel locking mechanism that can be included as a backup system for the primary locks. The secondary lock is not required for the *fifth wheel* to function and can be either manually or automatically applied. On some designs, the engagement of the secondary lock can only be accomplished if the primary lock is properly engaged.

secondary shoe: The brake shoe located toward the rear of the vehicle, in a dual-servo drum brake, having a longer and thicker lining, and providing most of the braking force during forward stops.

second gear: The intermediate or kickdown passing-gear range of a transmission.

second law of motion: A body's acceleration is directly proportional to the force applied to it, and the body moves in a straight line away from the force.

sectioned: A vehicle body that has been lowered by removing a horizontal section of metal all the way around the vehicle.

section height: The straight-distance measurement of a tire from the rim to the tread.

section modulus: A measure of the strength of the vehicle-frame side rails.

section repair: A body repair that is accomplished using a section of another identical vehicle.

section width: The straight distance from one sidewall of a tire to the other sidewall.

sector: A component that is not a complete circle, such as the gear on the pitman shaft of many steering systems.

sector gear: A gear that converts the rotary motion of the worm in a *recirculating ball* or *worm and sector* steering gear to the straight line motion of the pitman arm.

sedan: A two- or four-door vehicle with front and back seats that can accommodate four to six persons.

sedan de ville: A sedan with an open cockpit and enclosed passenger compartment, intended to be chauffeur driven.

segments: **1.** A portion of a larger figure cut off by one or more lines of a plane. **2.** The copper (Cu) bars of a *commutator.*

seize: Two moving parts that suddenly bind together, usually due to lack of lubricant. Also known as *freeze-up.*

seizing: The stiffening or "freezing" of a chain joint or shaft as a result of roughness and high friction caused by galling or lack of lubrication.

self-adjuster: A mechanism on a drum brake that compensates for shoe-lining wear and keeps the shoe adjusted close to the drum.

self-adjusting clutch: A mechanism that automatically takes up the slack between the pressure plate and clutch disc.

self-diagnostics: In automotive computers, especially those for engine control, a program which assesses the condition of the system, including the sensors and the computer itself, and communicates its findings to the technician by means of trouble codes.

self-discharge: The discharging of a battery due to chemical action, although there is no electrical demand.

self-energization: The placing of brake shoes so that the drum tends to drag the lining along with it, resulting in a wedging action between anchor and drum.

self-induction: The induction of a voltage in a current-carrying coil of wire, such as an ignition coil.

self-locking screw: A screw configured so that it locks itself in place when tightened.

self-sealing tire: A tire having a special compound on the inner surface that seals punctures when the puncturing object is removed.

self-tapping screw: **1.** A screw that cuts its own threads without pre-drilling in sheet metal. **2.** A screw that cuts its own threads in a pre-drilled hole in heavy-gauge metal.

SEMA: An abbreviation for *Specialty Equipment Market Association.*

semi: Short for *semitractor* or *semitrailer.*

semiautomatic welding: Manual welding with equipment that automatically controls one or more of the welding conditions.

semicentrifugal clutch: A clutch having weighted components in the pressure plate, such as rollers or release levers, that apply additional force against the pressure plate to hold the disc tighter during high engine rpm.

semiconductor: A solid material, usually germanium (Ge) or silicon (Si), that integrated circuits are made of, with an electrical conductivity between the high conductivity of metals and the low conductivity of insulators used to control the flow of electricity.

semiconductor ignition system: A term used for electronic-ignition system.

semielliptical: Refers to the amount that the ends are higher than the center arch of a leaf-spring configuration.

semielliptical spring: **1.** A leaf spring. **2.** A set of leaf springs of regressive lengths stacked with the longest at the top to the shortest at the bottom.

semifloating axle: A popular automotive system in which the axle shaft provides three functions: transfers *torque* to drive the vehicle, supports the car weight, and retains the wheel.

semifloating rear axle: An axle that supports the weight of the vehicle on the axle shaft and transmits the driving force to the rear wheels.

semi-independent rear suspension: A rear-suspension system in which one rear wheel has a limited amount of movement without affecting the opposite rear wheel.

semiknocked down *(SKD)*: A vehicle that is sold in a partially assembled condition.

semimetallic brake lining: A brake lining that is made of an organic resin to bond steel fibers.

semioscillating: A fifth-wheel-type vehicle that oscillates or articulates about an axis perpendicular to the vehicle centerline.

semitractor: A large truck having a *fifth wheel* used to tow a *semitrailer.*

semitrailer: A load-carrying vehicle equipped with one or more axles and constructed so that its front end is attached to, and supported by the *fifth wheel* of a *semitractor* that pulls it.

sending unit: An electrical or mechanical sensing device that sends information relative to some physical property such as temperature or pressure to a gauge or light. Also known as *sensor unit.*

sensible heat: Heat that causes a change in the temperature of a substance, but does not change the state of the substance that can be felt or measured with a thermometer.

sensing voltage: A condition that provides a means for the voltage regulator to monitor and control the battery voltage charge rate.

sensor plate: A plate used in the air intake of a continuous-flow, fuel-injection system to measure airflow.

sensor unit: A term used for *sending unit.*

separator: **1.** A tank-like device used to remove liquids from a vapor, such as oil from refrigerant. **2.** A nonconductive divider placed between the positive and negative plates of a battery.

separator disc: The metal plates in a multi-plate clutch that separates the friction disc.

sequential fuel injection: A term used for *sequential-port fuel injection.*

sequential-port fuel injection *(SPFI)*: A type of multi-port injection system where individual fuel injectors are pulsed sequentially, one after another in the same firing order as the spark plugs, rather than being pulsed simultaneously. This allows more precise fuel control for lower emissions and better performance. Also known as *sequential fuel injection.*

series circuit: A circuit having only one path through which the current or fluid can flow, having to pass through one component before going on to another.

series number: A term used for *tire aspect ratio.*

series parallel circuit: A circuit in which some components are in series while others are in parallel.

series parallel system: An arrangement where two 12-volt batteries are connected in such a manner as to provide 24-volts for starting and 12-volts for accessories and charging.

serpentine belt: A wide, flat belt having multi-V-grooves to provide frictional contact with the pulleys that winds through all of the engine accessories to drive them off the *crankshaft pulley.*

serrated nut: A self-locking type of nut having serrations on its contact side to prevent it from loosening when tightened.

serrated rod cap: A connecting rod cap that has serrated parting edges to help maintain alignment with the rod.

service access gauge-port adapter: An adapter that is used to connect a test gauge to the service port of a system for nonstandard service-port applications.

service access gauge-port valve: Fittings found on some service valves and some control devices used to access the system for testing and service.

Service Bay Diagnosis System: A computerized information network system that is connected, often by satellite, to the vehicle manufacturer and used to answer service and diagnostic questions.

service bulletin: Technical service information provided by the manufacturer, used as updates for the service manuals and to provide the latest service tips, field repairs, product information, and other related information for the service technician.

service hose: A hose that attaches a test gauge set to the service fitting of the system.

service manager: The person who is generally responsible for the entire service operation of a dealership.

service manual: A manual provided by the manufacturer or other publisher that describes service procedures, troubleshooting and diagnosis, and specifications for a particular car line.

service port: An access fitting found on the service valve and some control devices that the gauge set hoses are connected to for service and testing.

service procedures: A recommended step-by-step procedure to follow to troubleshoot, disassemble, assemble, or repair an automotive system or component.

service rating: **1.** A designation that indicates the type engine an oil is suited for. **2.** The temperature and/or load rating of a tire.

service representative: A manufacturer's agent who works in the local area to provide direct and immediate service to the dealership relative to customer service.

Service Station Dealers of America *(SSDA)*: A national federation of service station owners.

service technician: One actively involved in troubleshooting, maintenance, and repair of the vehicle.

service valves: Special manually operated or Schrader-type valves that allow connecting gauge hoses to a pressurized system during servicing procedures.

service writer: One who writes the work order.

servo: A device that converts hydraulic pressure to mechanical movement, such as a brake-wheel cylinder.

servo action: A method of brake construction in which a primary shoe pushes a secondary shoe to generate self-energization.

servo brake: A drum brake in which brake shoes are linked, so that the braking force of one shoe amplifies the input of the other shoe.

set screw: A type of screw having a point that fits into the matching recess of a shaft to secure a pulley or gear.

setup: The engine transmission/drivetrain and chassis combination that offers improved performance.

severe ring: A piston ring that exerts a high amount of pressure against the cylinder wall, often used in an engine with severe cylinder-wall wear.

sewing machine: A small, foreign car.

shackle: A swinging support for the rear end of a spring that permits it to vary in length as it deflects.

shackle assembly: This assembly is attached to the front spring eye and bushing and is then mounted through a *shackle bushing* to the frame allowing the *leaf spring* to pivot up and down.

shackle bushing: Insulated bushings to help prevent the transfer of noise and road shock from the suspension to the chassis and vehicle interior.

shade-tree mechanic: **1.** An amateur mechanic. **2.** A mechanic with little or no knowledge of the trade. **3.** A rural mechanic that literally works under a shade tree.

shadow graph: A scale using reflected light to indicate a difference in the weight of two parts.

shaft-mounted rocker arms: Rocker arms that are mounted in a straight row on a shaft.

shaft seal: An assembly that prevents vapor or fluid from escaping around a rotating shaft.

shaker: A hood scoop on some muscle cars that channels directly into the air cleaner.

shaky: A Chevrolet among Ford enthusiasts.

shallow staging: A term often used for *back staging.*

shaved: **1.** A vehicle body with the factory chrome trim removed and the holes filled in and painted. **2.** A cylinder head that has been resurfaced.

shear pin: A pin passing through two or more parts, as in securing a gear on a shaft, designed to break, preventing damage if an overload occurs.

shell: **1.** Outer spark-plug casing having a threaded end and hexagonal flats for a

A B C D E F G H I J K L M N O P Q R S T U V W X Y Z

wrench attachment. **2.** Outer front and/or rear metal container of the power-brake unit. **3.** The sheet metal body structure of a vehicle.

shielding gas: Protective gas used to prevent or reduce atmospheric contamination while welding.

shift-bar housing: A component that houses shift rails, shift yokes, detent balls and springs, interlock balls, and pin and neutral shaft; available in standard- and forward-position configurations.

shifter: A floor- or steering column-mounted lever on a motor vehicle used to select and/or shift the transmission gears.

shift forks: A Y-shaped component located between the low/reverse, first/second, and third/fourth gears on the main shaft of a transmission that causes the gears to engage or disengage via the sliding clutches.

shifting forks: A term used for *shift forks.* Also known as *shift yoke.*

shifting rods: The linkage of a manual transmission that connects the *shifter* to the *shift forks.*

shift kit: Parts required to provide high performance of an automatic transmission shifter.

shift lever: **1.** A lever used to change gears in a transmission. **2.** A lever that moves the starter drive pinion in and out of mesh with the flywheel in some applications.

shift rail: A series of grooves in the forks to guide the shift forks, tension balls, and springs to hold the shift forks in gear, allowing them to interlock the rails to prevent the transmission from being shifted into two gears at the same time.

shift tower: The main interface between the drive and transmission, a gearshift lever, pivot pin, spring, boot, and housing.

shift valve: A valve body component acted on by oil pressure, allowing fluid flow to the involved band and/or clutch at the appropriate time, causing the transmission to upshift or downshift.

shift yoke: A term used for *shift fork.*

Shillelagh: A hot Chevrolet V-8 engine.

shim: **1.** A thin metal spacer used to align the clearance of a part. **2.** A slotted strip of metal used to adjust the front-end alignment on many vehicles.

shimmy: A harsh, side-to-side vibration of the steering wheel usually due to front wheel imbalance.

shim stock: Thin metal, usually in a roll, that can be easily cut to be used as a *shim.*

shock: A term used for *shock absorber.*

shock absorber: A hydraulic device used at each wheel of the suspension system to help control the up, down, and rolling motion of a car body by dampening the oscillations or jounce of the springs when the car goes over bumps, thereby contributing to vehicle safety and passenger comfort. Also referred to as *shock.*

shock-absorber function: A typical *shock absorber* has three functions: to dampen the effect of spring oscillation in order to control the ride stabilization of a vehicle, to control body sway, and to reduce the tendency of a tire tread to lift off the road surface (a problem often caused by static unbalance).

shock-absorber lag: The incorrect operation of a *shock absorber* because of aeration due to the mixing of air with oils, causing the *shock absorber* to produce a poor ride.

shock-absorber ratio: A rating of shock-absorber extension control compared to the amount of compression control, varying from 50/50 to 80/20.

shock-absorber strut assembly: In a *MacPherson strut,* the independent rear-suspension system that includes a rubber isolated top mount, upper and lower spring seat, coil spring insulator, and coil spring.

shock compression: The *shock absorber* in its shortened position, which occurs when the wheel moves upward.

shock fluid: Specially formulated hydraulic fluid used inside of *shock absorbers.*

shock foaming: The mixing of air and shock fluid, due to rapid movement of fluid between the chambers, causing the *shock absorber* to develop a lag because the *piston* is moving through an air pocket that offers up resistance. A gas-filled *shock absorber* is designed to reduce oil foaming.

shock hydraulic principles: Fluid is forced through orifices and/or valves at a controlled rate to provide the desired dampening effect.

shock mounting position: The direction and/or angle at which a *shock absorber* is mounted: vertical, horizontal, or slanted inward at the top.

shock mounts: The rubber isolating bushing or grommets attached to the upper shock-mounting piston rod and the lower mounting cylinder tube in which the piston operates.

shock operational check: A method used to check shock efficiency that includes bouncing the vehicle bumper vigorously and observing shock dampening action, or pumping brakes slowly at low speed to see if a vehicle "rocking" motion is set up.

shock piston: The component attached to the bottom of the piston rod containing the rebound valve that moves back and forth inside the inner cylinder.

shock rebound: The rebound travel when the *shock absorber* is in its lengthened position, which occurs when the suspension or spring moves downward.

shoe: The lining and its steel backing on a drum brake that press against the inside of the brake drum to provide stopping power.

shoe box: Any car manufactured in the U.S. having a non-aerodynamic, box-like shape built from the late 1940s through the 1950s.

shoe hold-down spring: A coiled compression spring that applies pressure to hold the brake shoes against the backing plate.

shoe retracting spring: A coiled tension spring that pulls the shoes away from the brake drum after the pedal is released, forcing the brake fluid back into the master cylinder.

shoot: To spray paint, as in painting a vehicle.

shop housekeeping: A term used for *housekeeping.*

short and long A-arms: A double A-arm suspension where one arm is smaller than the other.

short arm: The throw of a crankshaft that has not been stroked.

short arm, long arm suspension: A conventional front-suspension system that uses a short upper-control arm and a long lower-control arm.

short bed: The cargo area of a short-wheelbase pickup truck, usually about six feet (1.83 meters) long.

short block: A new or rebuilt engine block with all internal parts.

short circuit: The intentional or unintentional grounding of an electrical circuit.

short course: To take a shortcut that bypasses a part of the official course in an off-road racing event.

Short Course Off-Road Enterprises *(SCORE):* More commonly known as *SCORE International,* an organization involved in long- distance desert competition events.

short cycling: **1.** A rapid cycling of the clutch resulting in poor cooling condition of the air conditioner that can be caused by poor refrigerant and/or air circulation or a maladjusted thermostat. **2.** An electrical condition where the device goes on and off as cycled by a circuit breaker due to a malfunction.

short deck engine: An engine block designed to accommodate a short-stroke *crankshaft.*

shorted circuit: A circuit that allows current to bypass part of the normal path.

short gear: A final drive with low gearing.

short, long-arm suspension: A conventional front-suspension system that uses a short

A B C D E F G H I J K L M N O P Q R S T U V W X Y Z

upper-control arm and a long lower-control arm.

short side radius: The small radius in a port between the bottom of the port runner of the *intake manifold* and the bowl area.

short ton: A standard English weight equal to 2,000 pounds.

short track: An oval race track that is less than one kilometer (5/8 mile).

shot: Pellets used with air pressure to clean parts.

shot bag: A leather bag filled with #9 birdshot used by metal workers to help shape and form metal panels.

shotgun motor: A rare Ford big block 429 cid Hemi, offered in 1969 and 1970.

shot peen: To harden the surface of a metal part by bombarding it with metal shot using high air pressure.

shot rodder: One who is a discredit to the sport of hot rodding.

show car: A custom-built car for show and not for driving, though driveable.

show 'n' shine: To display a hot rod or custom car in a car show.

showroom stock: A factory-stock vehicle with minor modifications and safety equipment under *SCCA* rules.

show through: The appearance of a sanding pattern after the vehicle has been painted.

shrink: To reduce an area of a piece of metal by heating and hammering it.

shrink fit: A tight or snug fit accomplished by shrinking a part.

shrink-fit tubing: An insulated tubing that shrinks to about half its original diameter when heated.

shrinking hammer: A special hammer that is used to shrink spots that have been stretched by excessive hammering.

shrink wrap: A *shrink-fit tubing* used to protect wires, wire splices, and terminals of an automotive electrical system.

shroud: **1.** A metal or plastic duct that directs *ambient air* to the radiator cooling fan. **2.** A hood-like device placed around an engine fan to improve air flow.

shrouding: An obstruction, such as carbon buildup, around a valve in the combustion chamber that interferes with the proper air flow.

shunt: **1.** A British term for vehicle accident. **2.** To bump or shove another car in an oval track or road race. **3.** A parallel electrical connection or circuit.

shunt winding: Bypass winding found on some alternators.

shut down: **1.** To defeat a competitor. **2.** To stop an engine.

shut down valve: A valve which is used to shut down an engine by interrupting the fuel supply; a safety requirement for many competitive vehicles.

shut off: To slow a vehicle by releasing the accelerator or throttle.

shut off area: An area beyond the measured distance of competition where the vehicle may be safely brought to a stop.

shut the gate: **1.** To pass a competitor in closed-course racing then immediately cut in front. **2.** To prevent an opponent from passing on the inside of a turn by blocking the *apex*.

SI: **1.** An abbreviation for *spark ignition*. **2.** An abbreviation for *Système Internationale des Unités*.

siamesed: **1.**Two exhaust pipes joined together. **2.** Cylinders in an engine block that are cast so close together that there is no room for a coolant passage between them. **3.** Two adjacent valves in a cylinder head that is served by a single port.

siamese ports: Intake or exhaust ports inside the cylinder head where two cylinders are feeding through the one port.

side bolt mains: Side-mounted main bearing bolts that increase the rigidity of the lower end.

side clearance: The clearance between the cheeks of the crankshaft journal and the connecting rod.

side-dash components: The installation of heating and air-conditioning components that have the evaporator mounted on the curb side of the firewall in the engine compartment and the heater core in the duct in the passenger compartment.

side-draft carburetor: A carburetor having one or more horizontal barrels.

side gears: Bevel gears that transfer power from the differential pinion gears to the splined axle shafts, providing differential action during turns.

side guard door beam: The structural member of a vehicle door that prevents it from being pushed inward if struck.

side lead: The effects of centrifugal force as a vehicle rounds a turn.

side marker light: Lamps installed in all vehicles sold in the United States since 1969 that permit the vehicle to be seen when entering a roadway from the side and to provide a means for other drivers to determine vehicle length (clearance).

side molding: The trim on the sides of a vehicle that offer protection or to improve the appearance.

side-mount battery: A battery having terminals on the side.

side oiler: A Ford big block *V–8* having a main oil gallery relocated on the low left side of the block.

side shift: A steering-column mounted transmission gear-shift lever.

side step: The act of slipping one's foot on the clutch pedal suddenly while revving the engine in drag racing.

side valve: An engine having intake and exhaust valves in the block beside the cylinders.

sidewall: The side of a tire between the bead and the tread.

sidewall information: Certain information required by the Department of Transportation to be imprinted on each tire, such as size, load rating/inflation pressure maximums, generic name of each cord material in the sidewalls and tread areas, actual number of plies in the sidewall and tread area, the words tubeless or tube type, as applicable, the word radial, if applicable, and the manufacturing code to determine who made the tire, where it was made, and when it was made.

sidewinder: A vehicle having an engine that is mounted transversely.

SIGMA: An abbreviation for *Society of Independent Gasoline Marketers of America.*

sight glass: A glass window in the liquid line or top of the receiver-drier used to observe the liquid refrigerant flow in an air-conditioning system.

Significant New Alternatives Policy *(SNAP)*: A rule established in 1994, under Section 612 of the 1990 *Clean Air Act (CAA)* to initiate a program in which the *Environmental Protection Agency (EPA)* is to evaluate applications for use of substitute chemicals and technology designated to replace ozone depleters in specific uses, including flammability, chemical toxicity, global-warming potential, exposure of workers, consumers, the general public, and aquatic life.

silencer: A device, such as a muffler, designed to reduce noise.

silent block: A rubber mount designed to reduce noise due to vibration.

silhouette car: A modern, front-drive 4- or 6-cylinder car that has been converted to a rear-drive *V–8* car.

silica gel: A drying agent used in many automotive applications, such as an air-conditioner *desiccant* because of its ability to absorb large quantities of water.

silicon: A chemical element (Si).

silicone: A group of organic compounds based on the non-metallic element, silicon (Si).

silicone controlled rectifier *(SCR)*: A semiconductor diode device used in controlling large amounts of DC current or voltage.

silicone fluid: Fluid made of chemicals with a silicone carbon bond; adaptable for many industrial uses.

silicone jackets: High-temperature coverings for *spark plugs.*

silicon-killed steel: A steel alloy that has been *killed* with silicon in the molten stage to refine its grain structure.

silver solder: A filler alloy that contains up to 45% silver (Sn), that melts at 1,120°F (604°C) and flows at 1,145°F (618°C).

silver tape: A term used for *duct tape.*

simple planetary gear set: Gear set with a sun gear, planetary pinions, planetary carrier, and ring gear.

Simpson gear set: A gear set having two simple planetary gears mounted on a common sun gear.

single-board engine controller *(SBEC)*: A single microprocessor used to control engine functions.

single chamber capacity: The measure of the maximum volume or displacement of the rotor chamber of a Wankel engine.

single leaf spring: A spring having one leaf that may by tapered thinner and wider toward the ends, allowing a variable flexing rate.

single overhead camshaft *(SOHC)*: An engine having one *overhead camshaft.*

single overhead camshaft engine: An engine having a single camshaft mounted over each cylinder head.

single pass: A term applicable to a refrigerant recovery unit that removes refrigerant from the air conditioner and passes it through only one time on its way to the recovery cylinder.

single-pivot control arm: A term used for *control arm.*

single-plane crankshaft: A *crankshaft* having *throws* on the opposite side of the same plane, at 180 degrees.

single-plane manifold: An *intake manifold* having a single *plenum* between the *carburetor* and *intake ports.*

single rear wheels *(SRW)*: Single wheels on either side of the drive axle of a rear-drive vehicle.

single reduction axle: An axle assembly that has but one gear reduction through its differental carrier assembly.

single-wire circuit: A circuit using a single wire that relies on the metal structures of the vehicle, such as the frame or body, as a ground.

single wrap: A transmission brake with a circular steel strap that is lined internally with friction material.

sintered: A somewhat porous metallic bearing formed by pressing particles of powdered metal to a temperature at which they adhere to each other.

sintered brake lining: A term used for *metallic brake lining.*

six holer: A six-cylinder engine.

six pack: A carburetor setup having a total of six barrels.

six-stroke cycle: The stroke cycle of an experimental engine by Toyota where combustion occurs every third revolution of the crankshaft.

6×4: A six-wheel truck, two in front and four in the rear, that is driven by the rear four.

6×6: A six-wheel truck driven by all six of its wheels, two in front and four in the rear.

sixty foot time: The time it takes to cover the first sixty feet (18.3 meters) from the starting line in a drag race.

SKD: An abbreviation for *semiknocked down.*

skid control: A device that prevents wheel lockup during braking, to prevent skidding.

skid control system: A system designed to respond to a locking wheel by relieving hydraulic pressure to the locking brake.

skid pad: A flat area of pavement with a painted circle 300 feet (91 meters) in diameter used to determine a vehicle's lateral grip and *lateral acceleration.*

skid plate: A shield under the power train of an off-road vehicle to protect the engine and transmission.

skin: **1.** The surface hardening of *RTV* or some body fillers. **2.** The outer sheet metal of a vehicle.

skin effect: **1.** The tendency of current to crowd toward the outer surface of a conductor; increases with conductor diameter and frequency. **2.** A thin, unburned layer of *air/fuel mixture* next to the combustion camber surface.

skinnies: Narrow front wheels and tires.

skinning: The removal of insulation from electrical conductors before making splices or connections.

skins: Tires.

skirt: A term used for *piston skirt.*

SK steel: A shortened term for *silicon-killed steel.*

skull: The unmelted residue from a filler metal when welding.

skunk works: A small, secretive group within the research and development department of a large organization that focuses on more advanced research and development.

slalom: A contest of speed and maneuverability through a tight course marked off with pylons.

slam: To modify a vehicle's suspension to make it lower.

slam the door: **1.** Beating another vehicle in competition. **2.** Preventing a competitive vehicle from passing.

slant six: A six-cylinder engine design offered on some Chrysler car lines in the 1960s and 1970s.

slave valve: An air-pressure operated device that helps to protect gears and components in the transmission's auxiliary section by permitting range shifts to occur only when the transmission's main gearbox is in neutral.

sled runner: The taper of the skirt of a piston to compensate for heat expansion.

sleeper: A vehicle that looks ordinary but is actually a high performer.

sleeve: A term used for *cylinder sleeve.*

sleeve bearing: A plain bearing.

slick: A smooth, treadless racing tire.

slider clutch: A special clutch used in drag race cars that slips to allow the engine to rev up before engaging.

slide travel: The distance that a *sliding fifth wheel* is designed to move.

sliding caliper: A disc-brake caliper that has a piston(s) on one side of the disc only that moves sideways on machined "ways" or keys to press the pad on the other side against the disc.

sliding fifth wheel: A specialized *fifth wheel* design that incorporates provisions to readily relocate the kingpin center forward and rearward, which affects the weight distribution on the tractor axles and/or overall length of the tractor and trailer.

sling: A loop of rope, cable, or chain used in hoisting heavy material.

slinger: A metal disc attached to the *crankshaft* to keep engine oil away from the front seal.

slip angle: The difference in the path the wheels follow during a turn compared to the actual direction they are pointing, caused by centrifugal force at higher speeds.

slip-in: A part that fits without modification or adjustment.

A B C D E F G H I J K L M N O P Q R S T U V W X Y Z

slip joint: A term used for *slip yoke.*

slipout: A condition that occurs with a tractor-trailer when pulling with full power or decelerating with the load; tapered or worn clutch teeth tend to "walk" as they rotate, causing the sliding gear and clutch to slip out of engagement.

slipper skirt: A term used for *slipper skirt piston.*

slipper skirt piston: A *piston* that has a cutaway skirt so that the *piston* can come closer to the counterweights reducing the overall size of the engine. Also known as *slipper skirt.*

slippery: A streamlined, aerodynamically efficient vehicle.

slip rings: The electrical contact area for the brushes in an alternator.

slip stream: A partial vacuum that is created behind a vehicle traveling at a high speed.

slip yoke: A component having internal splines that slide on the transmission output-shaft external splines, allowing the drive line to adjust for variations in length as the rear axle assembly moves.

sloper: A fastback body type.

slow reverse: A simple planetary gear combination with sun gear as input, planetary carrier held, and ring gear as output.

sludge: A buildup of combustion by-products that can clog oil lines and interfere with proper lubrication.

slug: A piston.

slugging: **1.** The return of liquid refrigerant or oil to the compressor. **2.** The act of adding a separate piece(s) of material in a joint before or during welding that results in a welded joint not complying with design, drawing, or specification requirements.

slush pump: An *automatic transmission.*

small block *(SB):* A *V–8* engine of 400 cid (6.5 liters) or less.

smog: Air pollution, especially the photochemical variety formed when sunlight causes chemical reactions in air pollutants, resulting in the formation of ozone and other compounds.

smog check: To measure the emissions level of a vehicle's exhaust gases.

smog motor: A vehicle engine with exhaust emissions controls.

smog pump: An *air-injection system pump.*

smog test: A term used for *smog check.*

smoke: **1.** To defeat a racing competitor. **2.** The color of the vapor coming out of the vehicle's exhaust system.

smoke in exhaust: A visible blue or black substance often present in the vehicle exhaust.

smoke it over: To discuss, analyze, and give careful consideration to a concept, idea, or problem.

smoke off: To leave the starting line in a drag race with the rear wheels smoking.

smoker: A vehicle that has an engine that smokes excessively.

smoking rules: Rules concerning smoking, such as, NO SMOKING or SMOKE ONLY IN DESIGNATED SMOKING AREAS. Fumes in the shop may be ignited if these rules are not followed.

smoothed: A body panel with trim removed and the holes filled.

SNAP: An acronym for *Significant New Alternatives Policy.*

snap ring: A circular retaining clip used inside or outside a shaft or part to secure a shaft, such as a floating wrist pin.

Snell Foundation: An organization that sets safety standards for racing helmets that are adhered to by most race-sanctioning bodies.

sniffer: An exhaust-gas analyzer used in making a *smog check.*

snipe: A pipe that is placed on a wrench to increase leverage; generally an unsafe practice.

snorkel tube: A long, narrow tube attached to the air cleaner, used to direct air into the air filter.

snotty: A slippery surface.

snow-and-mud tire: A tire, identified with an "MS" suffix, having treads designed to provide traction when driving in mud or snow; available in various ply and belt designs.

Society of Automobile Engineers *(SAE)*: A professional organization established in 1905 and now known as the *Society of Automotive Engineers.*

Society of Automotive Engineers *(SAE)*: A professional organization of the automotive industry founded in 1905, the *SAE* is dedicated to providing technical information and standards to the automotive industry.

Society of Collision Repair Specialists *(SCRS)*: A trade association of those involved in the collision-repair industry.

Society of Independent Gasoline Marketers of America *(SIGMA)*: A trade association of wholesale and retail private-brand gasoline marketers.

sodium bicarbonate: Baking soda (NaHCO) used to neutralize battery acid.

sodium-cooled valve: A partially hollow valve containing metallic sodium that melts at a low temperature (208°F) 97.8°C and when in its liquid state at operating temperatures splashes around inside the valve, transferring heat away from the valve head.

sodium hydroxide: A caustic soda (NaOH) that makes a good parts cleaner when mixed with water.

sodium silicate: A white, grey, or colorless compound sometimes used to seal small cracks or leaks in the cooling system.

soft plug: A term used for *core plug.*

soft solder 95/5: A lead-free metallic alloy of 95% tin (Sn) and 5% antimony (Sb); used to repair or join ferrous metal parts for temperatures below 350°F (176°C).

soft top: **1.** A convertible top. **2.** A vehicle having a convertible top.

SOHC: An abbreviation for *single overhead camshaft.*

solar cell: A *semiconductor* that converts sunlight to electrical energy.

solar power: Electricity that is generated by solar cells.

solder: A filler metal used in joining two or more parts that has a liquidus state not exceeding 840°F (450°C).

soldering: A welding process that produces consolidation of materials by heating them to the proper temperature and using a filler metal having a liquidus not exceeding 840°F (450°C) and below the solidus of the base metals.

solenoid: An electro-mechanical device used to effect a push-pull mechanical operation using electrical current.

solenoid relay: A relay that connects a solenoid to an electrical circuit, such as a starter-motor solenoid relay.

solenoid switch: An electrical switch that is opened and closed mechanically by the movement of a solenoid core.

solid axle: A term used for *beam axle* or *rigid axle.*

solid-axle suspension: Suspension system in which the wheels are mounted at each end of a solid, or undivided, axle or axle housing.

solid pushrod: A *pushrod* made from solid stock.

solids: **1.** A type of paint pigment. **2.** A solid or mechanical valve lifter.

solid state device: A *semiconductor.*

solid state ignition: An ignition system using diodes and transistors to control spark timing.

solid state regulator: An alternator regulator having no moving parts.

solidus: The highest temperature at which a metal or an alloy is completely solid.

solid valve lifter: A term used for *mechanical valve lifter.*

solid wire: A single stranded conductor, usually insulated.

solo: **1.** By oneself. **2.** A run made by a single car during a drag-race elimination.

soluble: A substance that will dissolve in a *solvent.*

solvent: **1.** A liquid substance, such as water, in which other substances can dissolve. **2.** A petrochemical liquid that will dissolve oil and grease. **3.** A paint cleaner and thinner.

sonic testing: A procedure for testing the integrity of engine blocks, using sound waves.

SOP: **1.** An abbreviation for *standard operating procedure.* **2.** An abbreviation for *seat of the pants.*

SOP rally: A competitive event in which instruments cannot be used by the contestant to check time or distance.

sorbead: A *desiccant.*

sorted out: Corrected, such as a problem that has been corrected.

soup: **1.** A special racing-fuel mixture. **2.** To increase the output of an engine.

Southern California Timing Association *(SCTA):* A sanctioning body concerned with the annual Bonneville speed trials.

south pole: The pole or end at which magnetic lines of force enter a magnet.

space frame: A light-weight race car frame constructed of small-diameter metal tubing that is welded together in such a manner as to provide high rigidity.

spacer: A device, such as a *shim* or *washer,* that is used to increase the space between two mating surfaces or parts.

space-saver spare tire: **1.** A deflated, *compact spare tire* that must be inflated to 35 *psi* (241 *kPa*) with a vehicle-battery powered air compressor or a can of compressed air. **2.** An inflated spare tire which is smaller and narrower than those on the vehicle that is to be used in an emergency only.

space-saver tire: A term used for *space-saver spare tire.*

spalling: A condition where chips, flakes, or scales of metal break off a part due to fatigue rather than by wear.

span: The width of an *air foil.*

spanner: The British term for *wrench.*

spare: **1.** Not in regular use or immediately needed. **2.** Extra or reserve. **3.** A term used for *spare tire.*

spare tire: A full-size replacement tire or a compact space-saver tire, generally for emergency service, but available for use when needed.

spark advance: The moving ahead of the ignition spark in relation to the *piston* position.

spark advance curve: The rate at which *ignition timing* advances as plotted on a graph; the line rises from some initial amount of advance and levels off at the maximum advance.

spark decel valve: A vacuum valve, located in the line between the *distributor* and *carburetor,* to advance the *spark* during deceleration, to reduce emissions.

spark delay valve *(SDV):* A vacuum valve acting like a restrictor, used in the vacuum line between the *distributor* and *carburetor,* to delay vacuum-timing advance under certain driving conditions to reduce NO_X *(oxides of nitrogen)* emissions.

spark duration: The time a spark is established across the gap of a *spark plug.*

spark ignition *(SI):* An engine-operating system where the *air/fuel mixture* is ignited by an electrical spark.

spark knock: A term used for *detonation* or *ping*.

spark line: The line on an *oscilloscope* that indicates the *voltage* required to fire the *spark plug* and the number of degrees the *distributor* turns while the spark exists.

spark plug: An ignition component threaded into the cylinder head that contains two electrodes extending into the cylinder that form a gap across which high-voltage electricity arcs to ignite the compressed *air-fuel mixture*.

spark plug fouling: An accumulation of deposits on the lower, exposed end of the *spark plug* that act as an electrical conductor, thereby creating a path for electricity to leak to ground rather than jump across the electrode gap.

spark plug heat range: The temperature limits: hot, normal, or cold, within which a *spark plug* is designed to operate. It is determined by varying the length of the exposed lower section of the plug ceramic insulator.

spark plug, resistor type: A *spark plug* having an electrode resistance of 5,000 to 10,000 ohms to increase electrode life and suppress radio interference.

spark plug well: The recess in a *cylinder head* for a *spark plug*.

spark plug wire: A special high-voltage wire from the *distributor cap* to the *spark plug*.

spark test: A quick check of the *ignition system* made by carefully placing a metal end of one of the spark plug wires close to the *engine* while cranking the *engine* to see if there is a spark gap and to determine its intensity.

spatter: The metal particles expelled during *fusion welding* that do not form a part of the *weld*.

spatter paint: The technique of applying two separate colors of paint simultaneously to provide a speckled finish.

spec car: **1.** A term used for *specification car*. **2.** One car in a group of identical *specification cars* that form their own racing category with a focus on driving skills.

special: A high-performance, individually built car, such as a *prototype*.

Specialty Equipment Market Association *(SEMA)*: A trade association for the automotive aftermarket industry.

specialty repair shop: A term used for *specialty service shop*.

specialty service shop: A repair shop that specializes in certain vehicle components, such as engine rebuilding, brake repair, radiator repair, and so on. Also known as *specialty shop*.

specialty shop: A term that may be used for *specialty service shop*.

specialty tire: Any of several types of tires such as, all-season tire, puncture-sealing tire, snow-and-mud tire, or studded tire.

specification car: One car in a group of identical cars that form their own racing category with a focus on driving skills. Also known as *spec car*.

specifications: Technical data, numbers, clearances, and measurements used to diagnose and adjust automobile components supplied by the manufacturer.

specs: A term used for *specifications*.

speed: A rate-of-motion measured in miles per hour or kilometers per hour.

speed adjustment: A term used for *idle-speed adjustment*.

speedbowl: A short, oval dirt track.

speed flare up: An operating condition where the engine speeds up without an increase in vehicle speed.

speedometer: An instrument, usually dash mounted, used to measure the speed of a vehicle.

speed rating: A tire rating that indicates the maximum safe vehicle speed that a tire will withstand.

A B C D E F G H I J K L M N O P Q R S T U V W X Y Z

speed ratio: A comparison of the difference in speed between two moving parts such as *impeller* speed and *turbine* speed.

speed sensor: An electrical device that can sense the rotational speed of a shaft or member and transmit this information to another device, such as a readout.

speed shift: **1.** To upshift a manual transmission without releasing the *accelerator.* **2.** The art of upshifting a manual transmission without using the *clutch.*

speedster: An open, two-passenger roadster.

speedway: A large oval-track racing facility.

Speedway: The *Indianapolis Motor Speedway.*

SPFI: An abbreviation for *sequential-port fuel injection.*

spherical joint: A term used for *ball joint.*

spider: A set of gears in the *differential* that allow the rear wheels to rotate at different speeds as the vehicle is cornering.

spin: To skid out of control by 180 degrees or more.

spindle: **1.** A shaft or stub axle upon which the wheel hub and bearing rides. **2.** A shaft which carries either cutting tools or work that is being machined.

spinner: A term used for *knock-off hub.*

spinning balancer: A balancer that rotates the tire and wheel to determine the amount and location of any unbalanced condition.

spin on filter: An *oil filter* having a threaded attachment for the ease and speed of replacement.

spiral bevel gears: A differential ring gear with helical gear teeth.

spiral-grooved shock absorber: An arrangement that is used to reduce the mixture of air with the fluid as it passes through the valves by breaking up the air bubbles and, at the same time, reducing lag.

spiral spring: Springs formed from flat strip of wire wound in the form of a spiral, loaded by *torque* about an axis, normal to the plane of the spiral.

splash-feed oil system: A type of engine lubrication system in which oil is splashed onto the engine parts to be lubricated.

splash lubrication: A non-pressurized system of lubrication, same as *splash-feed oil system.*

splash shield: A stamped sheet-metal deflector plate located behind the disc-brake rotor providing component protection from dirt and water and improving the flow of air over the rotor.

splined discs: Multiple-disc clutch components splined internally to the side gears or over a special hub on single-pack types.

splined yoke: A consideration to allow the driveshaft length to increase and decrease to accommodate the movement of the rear axle.

splines: The internal or external longitudinal grooves in a gear or shaft that mesh when assembled, causing the gear to turn with the shaft, but allow lateral movement.

split brake system: A service-brake system having two or more separate fluid, electrical, mechanical, or other circuits. If one circuit fails, full or partial brake actuating capability is retained.

split crankcase: An *oil pan* split horizontally, in the same plane as the crankshaft, into two or more pieces to simplify service.

split guide ring: Part of a torque converter assembly designed to reduce fluid turbulence and improve efficiency.

split pin: A round, split-spring steel tubular *roll pin* used for locking purposes, such as a gear to a shaft.

split sump: A term used for *split crankcase.*

split torque converter: A design having a simple planetary gear set to divide engine torque between mechanical and hydraulic operation.

split valve guide: A two-piece valve guide used in some older flathead engines with mushroom valves.

spoiler: **1.** An aerodynamic device that "spoils" the airflow over the vehicle. **2.** A device attached below the front bumper to reduce drag by deflecting air away from the vehicle. **3.** A device mounted on the rear deck to provide a downward force.

spoke: The wire-like bracing between the hub and rim of a wire wheel.

spokes: Wire wheels.

spongy brake pedal: A condition where the brake pedal is not solid when depressed, but bounces softly. This is caused by air in the hydraulic lines, distortion or stretching of the connecting parts, or swelling of the hydraulic hoses.

spongy pedal: A term used for *spongy brake pedal.*

spontaneous combustion: The process by which a material ignites and burns by itself.

spook: To *psych,* distract, or unnerve a competitor at the start of a race.

spool: **1.** A final drive without any *differential* action. **2.** A term used for *locked rear end.*

spool valve: A rod used to control oil flow in an *automatic transmission.*

sports car: **1.** A two-seated vehicle with a manual transmission built for performance rather than passenger comfort. **2.** A vehicle built to incorporate the appearance, performance, and handling of a race car, but retain the qualities and requirements for regular road use.

Sports Car Club of America *(SCCA):* A major road-racing sanctioning body.

Sports Sedan *(SS):* A *pony car* under *USAC* stock car rules.

sports vehicle: A multipurpose vehicle for road and off-road service, generally appealing to the outdoor enthusiast, available in two-wheel drive, four-wheel drive, or all-wheel drive having a pickup-truck body design with standard cab or extended cab area. Van designs are considered sport vehicles, but for family use.

sport utility vehicle *(SUV):* A very popular truck-like vehicle with a box-like open or enclosed body, generally having a short wheel base, a manual four- or five-speed transmission, and four-wheel drive.

sporty car: A compact two-door, four-passenger sports-styled coupe or convertible having a long hood and a short deck.

spot on: **1.** A critical measurement. **2.** Absolutely correct or accurate.

spot putty: A body filler for slight dents and scratches.

spot repair: A small repair such as a scratch or dent.

spot weld: A weld made between or upon overlapping members in which bonding may start and occur on the mating surfaces or may proceed from the outer surface of one member.

sprag: A one-way clutch used in an *automatic transmission.*

sprag clutch: **1.** A device containing numerous oblong parts, called *sprags,* that operate by tilting between an inner and outer *race* to lock up when the outer *race* is turned in one direction and to slip when it is turned in the opposite direction. **2.** A one-way clutch having cam-profiled locking elements that engage cylindrical outer and inner races.

spray: The atomization of a liquid into a fine mist.

spread bore carburetor: A four-barrel carburetor having small primary and large secondary barrels.

spread tandem suspension: A two-axle assembly in which the axles are spaced to allow maximum axle loads under existing regulations, usually more than 55 inches (140 cm).

spring: **1.** Small steel coils that are used to close the *intake* and *exhaust valves* when the cam lobes release pressure on the valve stem. **2.** A steel or composite elastic leaf- or coil-like device that compresses as it absorbs energy and returns to its original position when it releases that energy.

spring back: The tendency of a material to return to its original shape or near its original shape after being bent.

spring bind: A term used for *coil bind.*

spring break chamber: **1.** A device used with brake-foundation assemblies as a fail safe unit that automatically applies the truck's service brakes in the event of an air loss. **2.** A device used as a *parking brake* and is used with cam and wedge-type foundation brakes.

spring break valve: **1.** A device that limits the hold-off pressure to the spring-brake chambers via a relay valve or quick release valve. **2.** A device that provides a modulated spring-brake application to the front axle proportional to the service braking pressure whenever a loss of pressure occurs.

spring chair: A term used for *axle seat.*

spring energized: A *one-way roller clutch* that has the rollers held in contact with the *cam* and *race* by individual springs, or by means of a spring-actuated cage.

spring eye: The main spring-leaf end that is formed in an O-shape for placement of the rubber spring-mounting bushings.

spring hanger: The vehicle-frame bracket for the eye at the front of the rear leaf spring.

spring leaf: A long, flat section of spring steel making up all or part of a leaf spring.

spring leaf insert: Small, replaceable composition pads placed between the spring leaves near the ends to aid in slippage between the leaves while they are flexing.

spring load: A measure, in pounds or kilograms, of how much weight a spring can support, generally at the *installed height.*

spring oscillation: Continued compression and rebound motion of a spring, after the wheel has encountered a bump or hole on the road surface, which diminishes gradually, depending on the condition of the shock.

spring pin: A small piece of metal rolled in a pin-like manner that is used to dowel small parts that may occasionally have to be disassembled and reassembled.

spring rate: The relationship of spring deflection to load applied, such as the amount of weight, in pounds per inch or newtons per millimeter, required to deflect the rear spring.

spring retainer: The element which locates and provides reaction for the spring in a *one-way roller clutch.*

spring sag: The loss of *spring load* due to overloading and/or metal fatigue.

spring seat: The recess in a chassis where a coil spring is mounted.

spring shackle: A small, swing-arm attachment at the rear of the leaf spring to allow the spring to flex.

spring torque windup: A term used for a *Hotchkiss drive.*

spring walk: The tendency of a valve spring to bounce around in its seat at high-engine speed.

spring windup: The slight S-shape assumed by the leaf spring during extreme acceleration and braking that may be controlled by traction bars on high-performance vehicles.

springy thingy: A dragster that has a light, flexible structure to allow for maximum weight transfer.

sprint car: A single-seat, front-engine car designed for short oval tracks.

sprint race: A short race of a few laps on an oval track.

sprocket: A sheet-like disk with teeth around its outer perimeter that mesh with a belt or chain.

sprocket pitch: The dimension between the centers where the rollers would be bedded against the bottoms of adjacent tooth spaces.

sprocket-pitch diameter: The pitch diameter of a sprocket used in a synchronous belt drive that coincides with the belt pitch line and is always greater than the sprocket outside diameter.

sprung weight: The mass of the vehicle that is supported by the springs, including the body, *engine,* and *transmission.*

spun bearing: Any bearing on the *crankshaft* that has seized on the journal and turned in the *housing bore.*

spur gear: A transmission or differential gear having teeth cut straight across its face, parallel to the rotational axis.

spyder: **1.** A light, two-person, horse-drawn carriage. **2.** A light, two-person sports roadster.

square engine: An engine in which the bore and stroke dimensions are the same.

squash area: A term used for *quench area.*

squat: The tendency of the rear end of a vehicle to press down on its springs during hard acceleration.

squeak: A high-pitched noise of short duration.

squeal: A continuous high-pitched noise.

squeegee: **1.** A flexible rubber block used to apply glazing putty and light coats of body filler. **2.** A metal-backed rubber blade having a handle used to clean windshields.

squirrel: A driver that cannot handle a vehicle very well.

squirrelly: Poor handling.

squirt hole: **1.** A hole in the side of a connecting rod in an *OHV* engine which squirts oil toward the *camshaft.* **2.** A hole in the pin end of a connecting rod to squirt oil to the underside of a *piston* for cooling.

squirt racing: A term used for *drag racing.*

squish: The action where some compressed *air/fuel mixture* is pushed out of a decreasing space between the *piston* and *cylinder head* of the *combustion chambers* in some engines.

SR: An abbreviation for *street roadster.*

SREA: An abbreviation for *Street Rod Equipment Association,* now known as *Street Rod Market Association.*

SRMA: An abbreviation for *Street Rod Market Association.*

SRW: An abbreviation for *single rear wheel.*

SS: **1.** An abbreviation for *stainless steel.* **2.** An abbreviation for *Super Stock, Sports Sedan,* and *Showroom Stock.*

SSA: An abbreviation for the *Suspension Specialists Association.*

SSDA: An abbreviation for the *Service Station Dealers of America.*

stabilize bar link: A device that connects the *lower control arm* to the *stabilizer bar.*

stabilizer: A device that uses the torsional resistance of a steel bar to reduce the roll of a vehicle and prevent too great a difference in the spring action at the two front wheels.

stabilizer bar: A long, spring-steel bar attached to the cross member and interconnects the lower control arm that twists like a *torsion bar* during turns to transmit cornering forces from one side of the vehicle to the other to help equalize wheel loads and prevent excessive leaning.

stabilizing ball joint: A term for *non-load-carrying ball joint.*

stacked pattern: A term used for *raster pattern.*

stacks: A term used for *velocity stacks.*

stage: To place a competition vehicle in a proper starting position.

staged turbos: Turbochargers in series, one feeding the other.

staggered timing camshaft: A *camshaft* ground so as to provide a longer duration for the cylinders farthest away from the main intake tract to equalize the amount of fuel mixture reaching all of the *cylinders.*

staging area: The area between the pits and starting line in drag racing where the cars are lined up.

A B C D E F G H I J K L M N O P Q R S T U V W X Y Z

staging lane: Lanes within the staging area where cars are grouped according to class.

staging lights: The light beams used in drag racing to guide the front wheels of a car into proper position.

stainless steel *(SS)*: A steel alloy that is highly resistant to rust and corrosion.

staking: Securing a part, such as a *bearing,* in place by exerting pressure around the opening by punching or peening.

stall: **1.** A work area. **2.** To come to a stop because the engine failed. **3.** The disruption of air flow around an *air foil.* **4.** A condition where the engine is operating and the *transmission* is in gear, but the drive wheels are not turning because the *turbine* of the *torque converter* is not moving.

stall test: A starter-motor test of the current draw made when the motor is stalled.

stall-torque test: A starting system test that determines pound-feet or newton-meters of torque developed by the armature when restrained from rotating by a special measuring instrument attached to the *drive pinion gear.*

standard bolt-and-nut torque specifications: A chart showing the standard torque recommendations for standard English and metric sizes of bolts and nuts.

standard English measurements: The English system of measure that includes feet, inches, miles, gallons, pounds, ounces, and so on.

standard operating procedures *(SOP)*: **1.** The way things are generally done. **2.** The routine of things.

standard temperature and pressure *(STP)*: A set of engineering reference conditions for dealing with gases with temperature equal to 273.15°K and pressure equal to 101,325 pascal.

standard ton: A term often used for *short ton.*

standard transmission: A transmission that must be manually shifted into the various gears.

standing kilometer: The international standards for acceleration records for a kilometer taken from a dead stop over a measured distance.

standing mile: The international standards for acceleration records for a mile taken from a dead stop over a measured mile.

standing quarter mile: **1.** The length of a typical drag-race course. **2.** The acceleration from a dead stop to the end of a measured quarter mile.

stand off: The fine mist of *air/fuel mixture* that may be forced back into the *carburetor* during timing overlap when the intake and exhaust valves are open.

stand on it: To hold the *accelerator* to the floor.

Stand On It: A popular novel based on actual events, credited to a fictitious driver, Stroker Ace, and believed to be hilarious by most race enthusiasts.

standpipe: A type of check valve that prevents a reverse flow of the lubricant that becomes liquid due to heat generated during operation. It prevents the loss of lubrication at start up when the lubricant has cooled and is solidus.

standpipe assembly: A term used for *vapor-liquid separator.*

star connection: The interconnection of 3 electrical-equipment windings in star or *wye* fashion, as in an *alternator.*

star fastener: A term used for *Torx fastener.*

starter: A small electric motor used to turn the engine for starting.

starter circuit: The electrical system that carries the low and high current power to the starter components required to start the vehicle.

starter drive: The part of the *starter motor* that engages the armature-mounted pinion gear to the engine-flywheel ring gear.

starter motor: A small electric motor used to turn the engine for starting.

starter motor drive: A term used for *starter drive.*

starter relay: A magnetic switch, generally operated by the *ignition switch,* that uses low current to close a circuit to control the flow of very high current to the starter.

starter ring gear: A relatively large diameter starter-driven gear attached over the *torque converter* or flywheel flex-drive plate.

starter safety switch: An electrical switch that prevents the starting of a vehicle with an automatic transmission while in gear.

starter solenoid: A magnetic switch, energized by the *ignition switch,* that shifts the starter-drive pinion gear into the *flywheel ring gear,* then closes the electric circuit to the starter.

starting motor: A term used for *starter motor.*

starting motor drive: A term used for *starter drive.*

starving: A condition where: **1.** The *evaporator* does not get enough refrigerant to properly function. **2.** The engine does not get enough fuel and/or air to function.

star wheel adjusting screw: A toothed wheel that is manually or automatically rotated to expand the *brake shoes* outward as the lining wears.

state of the art: **1.** A phrase that implies being up-to-date in technology. **2.** Pertaining to the latest technology.

static: **1.** The state in which a quantity does not change appreciably within an arbitrarily long period of time. **2.** A body at rest.

statically balanced: A tire or other rotating device that is balanced at rest so that it will not turn when free to do so because its weight mass is distributed evenly around the circumference of the tire or the axis of rotation.

static balance: A process of checking and/or correcting weight distribution of interrelated parts while they are not in motion.

static friction: The friction between two bodies at rest.

static pressure: The pressure of a vapor or liquid in a container or system under ambient temperature conditions with no load imposed.

static pressure seal: No movement between the sealing surfaces, such as the oil pan to engine.

static system pressure: The pressure of a vapor or liquid in a system under ambient temperature conditions with no load imposed.

static tension: The tension in the belt when the drive is at rest, a factor that determines drive efficiency, service life, and other operating characteristics.

static timing: To set the ignition timing on an engine that is not running.

stationary fifth wheel: A non-movable fifth wheel whose location on the tractor frame is fixed.

station wagon: An enclosed two- to nine-passenger vehicle which is characterized by its roof, extending for the length of the vehicle, allowing a spacious interior cargo area in the rear and having a rear door or hatch to provide access.

stator: **1.** A reaction device or member unit inside the *torque converter,* mounted on a one-way roller or sprag clutch, having vanes to redirect the fluid flow that discharges from the turbine center back toward the *impeller* at an angle most advantageous to *torque* multiplication. **2.** A stationary field having three windings to provide overlapping, three-phase electrical pulses to smooth the current flow.

stator assembly: The reaction member or torque multiplier that is supported in a free-wheel roller race splined to the valve and front support assembly.

stator support: A part of the transmission-pump housing in which the stator assembly in a *torque converter* locks.

steady rest: A journal support on a *camshaft* or *crankshaft* grinder to prevent vibration.

steam engine: An external-combustion engine operated by steam generated from a boiler.

steam holes: Passages designed to permit the flow of steam in an engine cooling system in hot spot areas where steam is expected to collect.

steam rollers: Huge drag-racing slicks.

steelies: Wheels made of a ferrous material.

steel shim gasket: The same as a *corrugated metal gasket.*

steering-and-ignition lock: A device that locks the *ignition* open and, at the same time, locks the *steering wheel* in position so it cannot be turned.

steering arm: An arm that is attached to the *steering knuckle* that turns the knuckle and wheel for steering.

steering axis: The vertical line through the centerline of the upper and lower pivot- or ball-joints on a steered wheel.

steering axis inclination: The angle of a line through the center of the upper strut mount and lower ball joint in relation to the true vertical centerline of the tire, viewed from the front of the vehicle.

steering brake: A braking system that provides separate control for the left and right rear wheels as an aid in steering around curves and/or correcting a pull to either side in some off-road vehicles.

steering column: Tubing through which the steering shaft mounts and rotates, providing a surface for the mounting of the hazard switch, turn signal switch, *ignition switch,* and transmission selector or shifter.

steering column shift: An arrangement where the transmission shifter is mounted on the *steering column.*

steering drift: The tendency of the steering to gradually drift to either side when the vehicle is driven straight ahead on a smooth, level road surface, generally caused by improper *caster* or *camber* or an under-inflated tire. Also known as *steering pull.*

steering gear: The assembly that converts motion from the steering column to the *pitman arm.*

steering geometry: The relationship of the steering linkage and the wheels to the road affected by *caster, camber, scrub radius, steering offset, toe in,* and *toe out.*

steering kickback: The shock felt in the *steering wheel* as the front wheels encounter obstructions in the road.

steering knuckle: The part around which each front wheel pivots as it is steered.

steering linkage: The assembly of *tie rods, idler arms,* and *links* that transfer steering motion from the steering gear box to the steering spindles with the rods, pivoting joints, and supporting parts that transfer steering motion from the *pitman arm* to both *knuckle arms.*

steering lock: A locking device on the *steering column* that prevents steering wheel rotation and/or selector lever motion unless unlocked using an ignition key.

steering offset: A term used for *scrub radius.*

steering pull: The tendency of the steering to gradually pull to the right or left when the vehicle is driven straight ahead on a smooth, level road surface, generally caused by improper caster or camber or an underinflated tire. Also known as *steering drift.*

steering ratio: **1.** The ratio of the worm to the sector. **2.** The ratio of the rack to the pinion.

steering sector: A term used for *sector gear.*

steering shaft: A shaft that extends from the *steering wheel* toward the *gearbox* through the *steering column.*

steering spindle: A term used for *spindle.*

steering system: The mechanism that permits the driver to change vehicle direction by turning a wheel inside the vehicle.

steering terminology: Terms that apply to steering, such as *bump steer, memory steer, steering pull, steering wander,* and *torque steer.*

steering wander: The tendency of the steering to pull to the right or left when the vehicle is driven straight ahead on a smooth road surface that may be caused by improper *caster* adjustment.

steering wheel: The wheel, located at the top of the *steering shaft,* which the driver uses to steer the vehicle.

steering wheel centering: The procedure of turning both *tie-rod couplings* equally in the proper direction to correctly position the steering-wheel spokes, and placing the steering gears on their high (center) position.

Stellite: A tradename for a very hard alloy made from cobalt (Co), chromium (Cr), and tungsten (W) used for valve-seat inserts.

stemming: A condition where the valve radius section has corroded to the extent that it has a smaller diameter than the stem.

stem-type service valve: A service valve requiring a special wrench be affixed to a stem for opening and closing.

step: **1.** The raised portion of a chassis providing added clearance over the axle. **2.** A term used for *stepped flywheel.* **3.** A raised portion on one part so another part can be joined to it.

stepless transmission: A transmission without gears that goes from *low gear* to *overdrive* without meshing gears.

stepped flywheel: A *flywheel* having a ledge to which a pressure plate is attached.

stepped resistor: **1.** A resistor having two or more fixed-resistance values. **2.** A resistor assembly having a switch that is wired in series to increase/decrease the circuit resistance thereby controlling an electrical motor speed.

step ratio: A *transmission* with steps or *gear ratios,* such as four steps for a four-speed transmission.

Step side: The tradename of a popular pickup truck by Chevrolet.

stethoscope: A medical-type listening device used to detect and isolate noises within an engine while it is running.

stick: **1.** A *camshaft.* **2.** A term used for *stick shift.*

stick shift: A *manual transmission.*

stiffness: A measure of the dynamic elongation of a belt under tension.

stinger: A slightly conical pipe used as an exhaust resonator to which individual headers feed, such as at the top of a high-performance Volkswagen Beetle engine.

Stirling engine: A type of internal-combustion engine where the *piston* is moved by changes in pressure of the alternately heated and cooled working gas.

stock: A factory-manufactured vehicle having standard design, parts, and color.

stock car: **1.** A regular production vehicle. **2.** A production vehicle modified to *NASCAR* standards for racing.

stocker: **1.** A *stock car.* **2.** A *stock car* that has been modified for racing.

stoichiometric: A term often used for *stoichiometric ratio.*

stoichiometric ratio: The ideal air to fuel ratio of 14.7:1, in terms of mass, to achieve the most complete combustion possible in an *internal-combustion engine.*

stoichiometry: The state of having a stoichiometric *air/fuel mixture.*

stoked: Excited or enthused.

stone: Poor performing engine or vehicle.

stoned: One who is intoxicated and should not be driving.

stoplights: Lights at the rear of a vehicle that are illuminated when the driver applies the brakes to slow or stop the vehicle. Also known as brake lights.

stoplight switch: An electrical device used to illuminate the stoplights when the brakes are applied.

stopping distance: The distance required to stop a vehicle based on speed, reaction time, and road conditions.

stop solenoid: A term used for *idle-stop solenoid.*

A B C D E F G H I J K L M N O P Q R S T U V W X Y Z

storage battery: A device that stores electrical energy in chemical form and produces that electrical energy when required.

storm: To perform very well.

stovebolt: Originally, a 1929 Chevrolet; now any Chevrolet vehicle or engine.

STP: **1.** A brand name for a popular oil or fuel additive. **2.** An abbreviation for *standard temperature and pressure.*

straight: **1.** Moving or continuing in one direction without turning. **2.** Not curved or bent. **3.** A straight stretch in a closed race course.

straightaway: A straight stretch in a closed race course such as the front and back straightaway at *Indy*.

straight cut gear: A term used for *spur gear.*

straightedge: A metal bar used to check the engine block deck and cylinder head for warpage.

straight eight: An inline, eight-cylinder engine.

straight flexible hose: A term used for *flexible hose.*

straight four: An inline, four-cylinder engine.

straight in: Hitting the wall nose first in closed-course racing.

straight in damage: **1.** Damage caused by one vehicle hitting another vehicle directly or straight on. **2.** Damage caused by hitting the wall nose first in closed-course racing.

straightness: A condition where all elements of a surface or an axis are in a straight line.

straight six: An in-line, six-cylinder engine.

stranded wire: A conductor made up of several small wires twisted together.

stratified: To layer or to have in layers.

stratified charge: A type of combustion having a small amount of rich *air/fuel mixture* near the *spark plug* with a leaner mixture throughout the remainder of the *combustion chamber.*

stratified-charge engine: An engine in which each cylinder has two combustion chambers connected by a small passage; the smaller prechamber contains the *spark plug* and receives a rich mixture while the main chamber receives a lean mixture which is ignited by a flame front from the prechamber.

stratosphere: An upper portion of the atmosphere that extends 10 to 30 miles (16 to 48 km) above the surface of the Earth.

stratospheric ozone layer: A layer extending from 6 to 15 miles (9.7 to 24.1 km) above Earth's surface, protecting the Earth from ultraviolet (UV) rays from the sun.

streamliner: **1.** A specific class vehicle with a fully enclosed body, including the wheels, for dry-lakes racing. **2.** A racing car with an aerodynamically enclosed body.

streamlining: The shaping of a vehicle body so as to minimize air resistance.

street legal: A vehicle which is driven legally on a public street or highway.

street machine: A custom-built vehicle or hot rod built for street use.

street racing: Racing done on public streets. It is usually illegal.

street roadster *(SR)*: **1.** A hot rod with roadster body work built for street use. **2.** A specific category of drag racer having pre-1937 roadster body work.

street rod: A hot rod built for street use based on a pre-1949 car or light truck.

Street Rod Equipment Association *(SREA)*: An association now known as *Street Rod Market Association.*

Street Rod Market Association *(SRMA)*: A street rod division of *SEMA*.

streetside: The left-hand side of a vehicle.

stress relieve: An engine block that has been relieved of the stress caused by casting and machining.

stress riser: An area of a part that is most likely to crack due to mishandling and/or misuse.

stretchout: A luxury sedan that has been lengthened and made into a limousine.

stringer bead: A type of *weld bead* made without appreciable weaving motion.

striping: Thin paint or decal stripes added to the paint work.

stripped: A vehicle, sometimes stolen, from which parts have been removed for resale.

stripper: **1.** The lowest price vehicle in the line. **2.** A chemical used to remove paint.

strobe: Short term for *stroboscope.*

stroboscope: An instrument used in the study of the rapid revolutions or vibrations of a body by rendering it visible at frequent intervals with a flash of light.

stroke: **1.** The distance traveled by the piston from *top dead center (TDC)* to *bottom dead center (BDC).* **2.** To increase the stroke in an engine. **3.** To drive carefully or treat gently.

stroker: **1.** An engine that has been stroked. **2.** A *camshaft* used to *stroke* an engine.

stroker kit: A special *crankshaft* and connecting rod kit used to increase the displacement of an engine by lengthening the *stroke* of the *pistons.*

strut: Components connected from the top of the steering knuckle to the upper strut mount that maintain the knuckle position and act as *shock absorbers* to control spring action in a vehicle's suspension system. This is used on most front wheel drive cars and some rear wheel drive cars.

strut actuator: An electronically controlled actuator that controls strut firmness in a *computer-controlled suspension system.*

strut adjusting knob: In an adjustable strut, an eight-position adjusting knob which can vary the strut orifice opening, generally accessible without having to raise the vehicle.

strut cartridge: A self-contained unit with a pressure tube and a piston rod assembly, factory sealed and calibrated.

strut rod: A heavy steel rod in the suspension system located ahead or behind a lower control arm. A strut attached between car framework and lower control arm outer end on many vehicles to determine fore-and-aft position of the outer arm.

strut suspension, MacPherson: See *MacPherson strut.*

stub axle: The short shaft upon which the wheel hub and bearings ride.

stub frame: A term used for *sub frame.*

stud: A round, bolt-like metal fastener with threads on both ends.

studded tires: Tire that provide improved traction on ice, but are prohibited by law in many states because their use resulted in road-surface damage.

stud-mounted rocker arms: Individually mounted rocker arms with the use of a stud and ball.

stuffer: A supercharger.

stumble: **1.** The tendency of an engine to falter and catch resulting in a hesitation. **2.** A momentary abrupt deceleration during acceleration.

subassembly: A self-contained group of parts or electrical components that are a part of the whole overall assembly.

subcooler: A section of liquid line used to ensure that only liquid refrigerant is delivered to the metering device; may be a part of the condenser or may be placed in the drip pan of the evaporator.

subcooling: The cooling of a liquid below its condensing temperature.

Sub-EGR Control Valve: A vacuum valve used on Chrysler/Mitsubishi 2.6 liter engines that is operated mechanically by means of the throttle linkage, so it varies the signal to the exhaust-gas recirculation valve according to the position of the *accelerator pedal.*

sub frame: A partial front or rear chassis frame often used in unibody design to support engine or suspension assemblies.

A B C D E F G H I J K L M N O P Q R S T U V W X Y Z

submerged arc welding *(SAW)*: A process where a weld is made by feeding a bare wire into a blanket of grandular fusible flux.

substrate: The supporting structure in a catalytic converter where the catalyst is applied.

Suburban: A large truck-based station wagon marketed by Chevrolet and GMC.

suction accumulator: A term used for *accumulator.*

suction accumulator/drier: A term used for *accumulator.*

suction line: **1.** The line connecting the evaporator outlet to the compressor inlet. **2.** A low-pressure line in a system usually used as a return line.

suction-line regulator: A device used to regulate the pressure in the suction line.

suction manifold: The point where vapor enters a compressor.

suction pressure: Compressor inlet pressure.

suction service valve: A term used for *low-side service valve.*

suction side: That portion of a system that is under low pressure.

suction throttling valve: A back-pressure regulated device that prevents freeze-up of the evaporator core.

suicide doors: Vehicle doors that are hinged at the rear and open from the front.

suicide front axle: A special spring and front axle design for early Fords.

suicide front end: A term used for *suicide front axle.*

sulfated battery: A battery condition where sulfate has built up on the plates to the extent that it can no longer be charged.

sulfation: The lead sulfate that builds up on a battery as a result of battery action that produces electric current.

sulfuric acid: An *electrolyte* used in batteries.

sulfur oxides: Small amounts of acids (SO_X) that forms as a result of a reaction between the hot exhaust gas and the catalytic converter catalyst.

sump: **1.** The bottom part of some compressors that contains oil for lubrication of the moving parts of the compressor. **2.** The reservoir for oil at the bottom of an engine.

sun gear: Central gear the planet gears mesh with and revolve around.

sunload: Heat intensity and/or light intensity produced by the sun.

sunload sensor: A sensor placed on the *dashboard* to determine the amount of sun coming into the vehicle; a device that senses heat and/or light intensity.

supe: A term for *supercharger.*

Super Car: **1.** A term used for the early *muscle car.* **2.** An ultra-powerful, expensive, limited-edition vehicle such as a Ferrari.

supercharger: A compressor which pumps air into the engine's induction system at a pressure much higher than atmospheric pressure.

super-duty parts: High-performance heavy-duty parts.

superheat: The heat intensity added to a gas after the complete evaporation of a liquid.

superheated vapor: Vapor at a temperature higher than its boiling point for a given pressure.

superheat switch: An electrical switch activated by an abnormal temperature-pressure condition, such as a superheated vapor, used for system protection.

superspeedway: A high-banked, paved track at least 1 mile (1.6 kilometers) long.

Super Stock *(SS)*: An American factory-production vehicle that meets *NHRA* drag rules and performance ratings.

super-stock eliminator: Any vehicle running under the *NHRA* index/handicap system.

super tape: *Duct tape.*

supplemental brake system: An additional brake system used to assist the service-brake system in retarding a vehicle.

supporting career: A career path that is related to auto mechanics, such as careers in claims adjusting, vocational teaching, auto-body repair, frame and alignment repair, and specialty repair shops.

Sure-Grip: A limited-slip differential by Chrysler.

surface finish: The roughness or smoothness of a surface.

surface ignition: The ignition of the *air/fuel mixture* in a *combustion chamber* by heated metal or carbon deposits.

surface loading: The transfer of pressure as one part acts on another.

surfacing: The application by welding, brazing, or thermal spraying of a layer of material onto a surface to obtain desired properties or dimensions.

surge: A condition where the engine speed increases and decreases slightly with no throttle action.

surge bleeding: A brake-bleeding technique designed to dislodge air from wheel cylinders by creating turbulence in the wheel cylinder.

surge tank: The reservoir of a cooling system where *coolant* condenses before being returned to the *radiator.*

suspendability: The ability of a fluid to suspend heavier dirt particles rather than letting them fall to the bottom.

suspension: The system that supports the weight of the vehicle and provides for a comfortable and safe ride for the driver and passengers.

suspension arm: An arm pivoted on the frame at one end and on the steering-knuckle support on the other end.

Suspension Specialists Association *(SSA):* A trade association of heavy-duty service facilities and suppliers.

suspension system: Components that support the total vehicle, including *front* and *rear suspensions, springs, shock absorbers, torsion bars, axles, MacPherson strut system,* and connecting linkages.

SUV: An abbreviation for *sports utility vehicle.*

SUVA: A trade name for duPont's new generation of ozone-friendly refrigerants. *R-134a* is a *SUVA* refrigerant recommended for automotive use.

S/V ratio: The ratio of the surface area of a *combustion chamber* to its volume with the *piston* at *TDC.*

swaged end: A tube or pipe having an inside diameter that has been stretched to accept the outside diameter of another tube or pipe that is the same size.

swaging: To reduce or taper.

swap: **1.** To replace one component with another. **2.** To trade components or vehicles.

swap ends: To spin a vehicle a full 180 degrees.

swash plate: **1.** An angular plate attached to the bottom of the four pistons on a Stirling engine. As the pistons move downward, the swash plate is turned. **2.** A mechanical system that is used for pumping, having an angled plate attached to a center shaft, and pistons that are attached to the plate along the axis of the shaft. As the shaft rotates, the pistons move in and out of a cylinder, producing suction and pressure.

swash-plate compressor: A compressor in which the pistons are driven by an offset (swash) plate affixed to the main shaft, such as the six-cylinder air-conditioning compressor.

sway bar: A bar on the suspension system that connects the two sides together. It is designed so that during cornering, forces on one wheel are shared by the other.

sway-bar link: A connector from the lower control arm to the sway bar.

sweeping: A term used for *purge.*

swing attachment: A *leaf-spring shackle.*

swing axle: A drive system used with independent rear suspension systems.

swing pedals: Pedals that are suspended from beneath the dash, such as *clutch, brake,* and *accelerator.*

swirl: A cylinder-head design that causes the *air/fuel mixture* to enter the *combustion chamber* at a high rate of speed, increasing its *atomization.*

switch: An electrical device that controls the on and off of a sub-system or system.

swivel foot: A valve-adjusting screw having a ball that swivels when it contacts the valve stem.

symmetrical: Corresponding in size, form, and relative position on opposite sides of a line, plane, point, or axis.

symmetrical camshaft lobes: Camshaft lobes having identical opening and closing ramps.

symmetrical rear-leaf spring: A term used for *rear-leaf spring.*

Synchro: The trade name for an all-wheel drive system by Volkswagen.

synchromesh: A type of manual transmission where the *synchronizer* is used to bring a selected gear up or down to the speed of the main shaft.

synchromesh transmission: A transmission having a *synchronizer.*

synchronize: To cause two or more events to operate at the same time and/or the same speed.

synchronizer: A device used in a manual transmission to bring a selected gear up or down to the speed of the main shaft.

synchronous belt: A belt having cogs or teeth that mesh in mating cogs or teeth of a pulley.

synfuel: A term for *synthetic fuel.*

synthetic fuel: A fuel made by liquefying coal or by extracting oil from shale or tar sands.

synthetic oil: **1.** A type of engine lubricant consisting of highly polymerized chemicals. **2.** A non-mineral based lubricant for use in automotive air conditioners.

system-dependent recovery system: Refrigerant recovery system that relies on system components, such as the compressor, to remove the *refrigerant* from the system.

Système Internationale des Unités *(SI):* The international metric system of measurement.

system overcharge: A term used for an overcharged system.

system pressure: The average pressure in a system, such as the fuel-injection system.

system protection valves: A device that is used to protect the brake system against accidental loss of air pressure, buildup of extreme pressure, or backflow and reverse air flow in a truck braking system.

T

T: **1.** Model T Ford. **2.** Any hot rod based on Model T stock, repro chassis, and/or body.

t: Tan.

TAC: An abbreviation for *thermostatically controlled air cleaner.*

tach: An abbreviation for *tachometer.*

tach-dwell meter: An instrument used to measure engine speed and distributor dwell.

tachometer: An instrument used to measure engine speed.

tack cloth: A varnish-coated cheese cloth having a tacky surface used to clean metal prior to painting.

tack rag: A term used for *tack cloth.*

tack weld: A weld made to hold the parts of a weldment in proper alignment until the final welds are made.

tag axle: The rearmost axle of a tandem-axle tractor used to increase the load-carrying capacity of the vehicle.

tailgate: **1.** The rear opening of a station wagon or a van. **2.** To follow another vehicle closely.

tail lights: Lights used on the rear of a vehicle.

tailpipe: **1.** The pipe from the *muffler* or *catalytic converter* to the rear of the vehicle. **2.** The outlet pipe from the *evaporator* to the *compressor.*

tall block: An engine assembly consisting of a *short block* and *cylinder heads.*

tall-deck engine: An engine-block design height that permits a longer crank throw, or stroke.

tall gear: A final drive assembly with high gearing.

tandem: One directly in front of the other and working together, such as the rear wheels of a tractor.

tandem-axle drive: A type of drive that combines two single axles through the use of an interaxle differential or power divider and a short shaft that connects the two.

tandem-axle suspension: A suspension consisting of two axles with a means for equalizing weight between them.

tandem drive: A two-axle drive combination.

tandem master cylinder: A *master cylinder* that consists of a single bore with two pistons and separate fluid compression chambers used with "split" braking systems to ensure that there will be some braking power in one braking circuit, even if there is a severe fluid loss in the other.

tang: A projecting piece of metal placed on the end of the *torque converter* on *automatic transmissions,* used to rotate the oil pump.

tank: A container for vapor or fluid such as an *expansion tank* or a *header tank.*

tank unit: That part of a fuel-indicating system that is mounted in the *fuel tank.*

tank-vapor space: The area above the fuel level provided as a breathing space for the *liquid-vapor separator* and to compensate for expansion of fuel by heat.

taper: A cylindrical part, solid or hollow, that is larger at one end than at the other.

taper-bored piston pins: Piston pins having tapered bores to reduce weight and add strength.

tapped resistor: A fixed resistor having two or more taps to provide different resistance values, as for fan speed control.

A B C D E F G H I J K L M N O P Q R S **T** U V W X Y Z

tappet: A mechanical, solid-valve lifter located between the *camshaft* and *valve stem,* usually on L-head engines.

TARA: An abbreviation for *Truck-Frame and Axle Repair Association.*

Targa: A two-passenger convertible with a roll bar behind the seat, introduced by Porsche in 1965.

tarp: An abbreviation for *tarpaulin.*

tar paper: A vinyl roof covering.

tarpaulin: **1.** A cover used to protect the passenger compartment of an open car when the seat is not occupied. **2.** A cover used to protect the cargo bed of a pickup truck.

taxable horsepower: The power of a vehicle engine calculated by a formula providing a uniform comparison that is used for licensing vehicles in some areas.

taxi: Short for *taxicab.*

taxicab: **1.** A vehicle for hire. **2.** A racing vehicle that looks like an ordinary passenger car.

TBI: An abbreviation for *throttle-body injection.*

T-bone: For one vehicle to strike another broad side.

T-bucket: A Model T Ford *roadster.*

TCS: An abbreviation for *transmission controlled spark.*

TDC: An abbreviation for *top dead center.*

TE: An abbreviation for *top eliminator.*

tech: Short for *technical inspection* or *technician.*

technical inspection: The examination of vehicles before an event to ensure that they conform to the rules.

technical service bulletin *(TSB):* The periodic information provided by the vehicle manufacturers regarding any problems encountered on current-year model vehicles.

technician: One involved in the design, operation, or repair of a vehicle.

tee pee exhaust: An exhaust system for the Volkswagen Beetle engine.

teething troubles: Having difficulties with a new part or with a new vehicle.

Teflon pin button: A Teflon piston-pin retainer that fits between the pin and cylinder wall.

TEL: An abbreviation for *tetraethyl lead.*

telescoping shock absorber: A shock having certain working parts that move in and out during compression and rebound.

television-radio-suppression cables: A term used for *spark plug wires* or *high tension cables.*

temper: **1.** The measure of a metal's hardness. **2.** To heat treat metal to reduce brittleness.

temperature: A measure of heat intensity.

temperature gauge: A device that provides a visual indication of *temperature.*

temperature glide: The range of an evaporating or condensing temperature for a given pressure.

temperature indicator: A gauge or lamp to warn of an overcooling or overheating condition.

temperature rating: The temperature-resistance ratings that indicate a tire's ability to withstand heat.

temperature-regulated valve: A term used for *hot-gas bypass valve.*

temperature sending unit: A device in contact with the engine coolant that changes resistance, depending on its temperature, and sends a voltage signal to the gauge or lamp.

temperature sensing bulb: A term used for *remote bulb.*

temperature stick: A stick-like crayon designed to melt at a specific temperature when rubbed on a heated surface.

temporary strainer: A device installed during the run-in period of a new or rebuilt system, intended to remove coarse debris particles present in the system. After a suitable period it is removed, cleaned, and stored for future use.

temporary-use-only spare tire: A term used for *compact spare tire.*

temp stick: Short for *temperature stick.*

tensile failure: Failure in a synchronous belt drive caused by mishandling, misalignment, overloading, shock loads, or by large debris entering into the drive.

tensile member: The muscle of the belt whose purpose is to withstand the tension that is imposed to transmit the desired power.

tensile strength: The maximum tension a metal can withstand without breaking.

tension-loaded ball joint: A suspension component mounted in the lower control arm with the ball-joint stud facing upward into the knuckle so the vehicle weight tends to pull the ball out of the joint.

ten tenths: Great performance by an individual.

tenths: A measure of personal performance as rated on a scale of one-tenth to *ten-tenths.*

terminal: **1.** A battery post or connector. **2.** An electrical connector.

terminal speed: The maximum speed, in drag racing, that a vehicle achieves when passing through the traps at the end of the strip.

ternary: Having three parts, elements, or divisions.

test lamp: A term used for *test light.*

test light: A small, non-powered or self-powered light used to determine electrical continuity in a circuit.

test wheel: A controlled procedure by the *National Highway Traffic Safety Administration (NHTSA)* to perform laboratory tests for temperature-resistance testing of automotive tires as determined by how long the tire lasts on the test wheel. The ratings are A, B, or C, with an A rating having the best temperature resistance.

tetraethyl lead *(TEL)*: A chemical, $Pb(C_2H_5)$, once added to gasoline to increase its octane and aid in lubrication of the valves. Lead damages *catalytic converters* and *oxygen sensors* and therefore cannot be used in vehicles designed to operate on unleaded fuel. Leaded gasoline is no longer sold in the United States.

tetrafluoroethane: The chemical name for the "ozone-friendly" refrigerant commonly known as *HFC-134a* or, more simply, *R-134a.* Its chemical symbol is CH_2FCF_3.

T-head engine: A flathead engine having *intake valves* on one side and *exhaust valves* on the other side of the cylinders.

Thermactor: An air pump or air-aspirator, air-injection system by Ford.

Thermactor pump: A device that injects *ambient air* into the *exhaust system.*

thermal: Pertaining to heat.

thermal cutting: A cutting process that removes metal by localized melting, burning, or vaporizing of the work piece.

thermal delay fuse: A device once used in conjunction with the compressor protection switch that heats and blows a fuse to stop a compressor action during abnormal operation.

thermal efficiency: The difference between potential and actual energy developed in a fuel measured in Btu's per pound or gallon (kilogram or liter).

thermal fuse: A temperature-sensitive fuse link designed so that it melts at a certain temperature and opens a circuit.

thermal limiter: An electrical or mechanical device, similar to a fuse, that opens at 300°F (149°C) to control the intensity or quantity of heat.

thermal reactor: An obsolete early attempt of an emissions-control device comprising a

A B C D E F G H I J K L M N O P Q R S T U V W X Y Z

large, heavy, exhaust manifold in which hydrocarbons and carbon monoxide that escape from the cylinders are oxidized.

thermal stability: The ability of a device to reach and maintain a specific temperature for a long period of time.

thermistor: A resistor that changes resistance values depending on its temperature.

thermit reaction: The chemical reaction between metal oxide and aluminum that produces superheated molten metal and a slag containing aluminum oxide.

thermostat: **1.** A driver-operated device used to cycle the clutch to control the rate of refrigerant flow as a means of temperature control. **2.** A temperature-sensitive component located in a housing at the coolant outlet of the engine that restricts coolant flow to the radiator to maintain the desired engine-operating temperature. Often referred to as *Ranco control.*

thermostatic air cleaner: An engine air-cleaner assembly that controls the temperature of the intake air by blending relatively cool underhood or outside air with relatively hot air picked up from a shroud over the *exhaust manifold.*

thermostatically controlled air cleaner *(TCAC):* A device used to regulate the temperature of the air entering the engine air cleaner as an aid in reducing emissions.

thermostatic clutch control: A method of temperature control using a *thermostat* to cycle a clutch on and off.

thermostatic de-icing switch: A *thermostat* that prevents the *evaporator* from icing up due to low temperature conditions.

thermostatic expansion valve: The component of a refrigeration system that regulates the rate of flow of refrigerant into the *evaporator* as governed by the action of the remote bulb-sensing tailpipe temperatures.

thermostatic gauge: A type of gauge used to indicate fuel level, *oil pressure,* and *engine-coolant temperature.*

thermostatic pressure valve: A valve that opens and closes in relation to temperature changes.

thermostatic switch: A term used for *thermostat.*

thermostatic vacuum switch *(TVS):* A valve that controls the passage of vacuum according to temperature, blocking vacuum until a certain coolant temperature is reached, at which point it opens.

thermostatic vacuum valve *(TVV):* A valve that is operational with a vacuum signal that is proportional to an applied temperature.

thermosyphon: The flow of coolant in a heat exchanger, such as a *radiator,* that is possible by the difference in densities of heated and cooled fluid.

thin wall guide: An insert that may be used to repair a damaged valve guide.

thinner: A solvent used to reduce the *viscosity* of a liquid.

third gauge: A low-side gauge used to check pressure drop across a suction-pressure regulator at the compressor inlet on some car lines.

third law of motion: For every action there is an equal and opposite reaction.

third member: A differential.

thou: A term used for *thousandths,* as in a measurement or *thousands,* as in revolutions per minute.

thousand: A number equal to ten times one hundred.

thousandths: One part of a *thousand.*

thread class: A standard designation to indicate the closeness of fit between two threaded parts, such as a bolt and nut.

threaded fastener: Any of several types of components that hold parts together in assembly by interference of screw threads with mating threads in another component, such

such as capscrews, self-threading screws, stud-and-nut, and bolt-and-nut assemblies.

threaded insert: A threaded coil that is used to repair a damaged internal thread.

thread series: A standard designation of the number of threads and their pitch per unit, inch or millimeter.

three-mode cycle: A quick test procedure to study the causes of high emissions performed on a dynamometer under load or in a service bay without a load.

three on the tree: A three-speed manual transmission having a column-mounted shifter.

three phase: A type of electrical power such as that produced by an *alternator.*

three piece valve: An intake or exhaust valve made of three different types of material in the head, stem, and tip.

three-speed differential: A type of two-speed axle arrangement with the capability of operating both axles at a different speed at the same time. The third speed is actually an intermediate speed between the high and low range.

three-speed transmission: A transmission providing three forward-gear ratios, one reverse gear ratio and neutral.

three-way catalyst *(TWC):* A *catalytic converter* that oxidizes hydrocarbons and carbon monoxide and reduces *oxides of nitrogen* (NO_X) emissions. It has a chamber upstream handling reduction, and one downstream handling oxidation.

three-window coupe: A coupe without rear-quarter windows.

threshold limit value: The percentage, in parts per million (ppm), of refrigeration in atmospheric air above which a human can become drowsy or have loss of concentration.

throat: **1.** The lower part of a connecting rod. **2.** A carburetor barrel.

throttle: **1.** A hand control to adjust engine speed. **2.** An accelerator pedal.

throttle body: The housing of a *throttle-body injection system* that contains the valves.

throttle-body injection: A term used for *throttle-body injection system.*

throttle-body injection system *(TBI):* A fuel-injection system that resembles a *carburetor* and has fuel injectors located in a common throttle body. It provides many of the advantages of *fuel injection,* such as easier starting and lower emissions, without the cost and complexity of a multiport-injection system.

throttle cut-out relay: A term used for *wide-open-throttle cut-out relay.*

throttle plate: A term used for *throttle valve.*

throttle position sensor: A variable three-wire resistor-type electrical sensor which sends a signal to the electronic control unit relative to the throttle position.

throttle return check: A *dashpot.*

throttle solenoid positioner: An electric device that holds the throttle plate in the hot-idle position and closes it when the *ignition switch* is turned off.

throttle valve: A flap valve that controls the amount of air admitted into the induction system.

throttling valve: A term used for *throttle valve, suction throttling valve,* or *evaporator pressure regulator.*

throw: **1.** A connecting *rod journal* on a *crankshaft.* **2.** The number of output circuits on a switch.

throw a rod: **1.** A loose connecting-rod bearing. **2.** A broken connecting rod that has been forced through the *block* or *oil pan.*

throwout bearing: The *clutch-release bearing.*

thrust bearing: A bearing or a part of the main bearing that limits end-to-end movement of the *crankshaft.*

A B C D E F G H I J K L M N O P Q R S T U V W X Y Z

thrust load: Load placed on a part that is parallel with the center of the axis.

thrust plate: A retainer that positions the *camshaft* in an *OHV* engine and limits its end-to-end movement.

thrust surface: The area of a crank or block that absorbs end-to-end thrust pressure.

thrust washer: A washer that is capable of supporting a *thrust load.*

thumbnail grooves: Small grooves in a *thrust bearing* that provide a path for lubrication to the *thrust surfaces.*

tie rod: The linkage between the *idler arm* or *pitman arm* and the *steering arm.*

tie-rod coupling: A threaded sleeve between the *tie rod* and the *tie-rod end* providing lengthwise adjustment to set front-wheel toe in.

tie-rod end: A pivoting ball-and-socket joint located near the outer end of the *tie rod* used to connect the *tie rods* to the center link and to the *steering arms.*

TIG: An abbreviation for tungsten inert gas.

TIG welding: A welding process using tungsten (W) and an inert gas such as Argon (Ar).

Tijuana chrome: Silver, chrome, or aluminum spray paint.

tilt steering wheel: A type of steering wheel that can be tilted to various angles to provide for driver comfort and convenience.

time guide: A reference, providing average time required to perform specific repairs or service to a motor vehicle, that may be used as a labor charge based on an hourly rate.

time, speed, and distance *(TSD):* A type of *rally* where the object is to cover the greatest distance in the shortest period of time without exceeding the speed limit.

time trials: Timed runs for individual race vehicles, usually to determine starting positions for a race.

timing: **1.** The combustion spark delivery in relation to the piston position. **2.** The procedure of marking the appropriate teeth of a gear set prior to installation and placement in proper mesh.

timing belt: The belt through which the *crankshaft* drives the *camshaft* or *camshafts* in an *overhead valve* or overhead cam engine.

timing chain: The chain through which the *crankshaft* drives the *camshaft* or *camshafts* in an *overhead valve* or overhead cam engine.

timing diagram: A graphical method used to identify the time in which all of the events of the four-stroke engine operate.

timing gear: The gear or gears through which the *crankshaft* drives the *camshaft* or *camshafts* in an *overhead valve* or overhead cam engine.

timing light: A stroboscopic tool used to check the precise timing of a conventional ignition, adjusted by loosening the holddown bolt and turning the *distributor.*

timing valve: A device in the fuel-injector pump which times the delivery of fuel to the injectors.

tin: **1.** An element (Sn). **2.** A term often used for metal.

tin indian: A Pontiac.

tin knocker: One who repairs auto bodies.

tinny: A station wagon having an all metal body.

tin top: A vehicle with a fully enclosed body.

tin work: Auto bodywork.

tin worker: One that works on vehicle bodies.

tip insert: A term used for *leaf-tip insert.*

tip the can: To increase the ratio of nitro in a racing-fuel mixture.

tire: **1.** An air-filled or solid covering for a wheel, normally of rubber. **2.** A device made of rubber, fabric and other materials that, when filled with fluid or gas under pressure and mounted on a wheel, cushions and sustains the imposed load. Tires contribute to the ride and steering quality of a vehicle and play a significant role in vehicle safety. Tires must be designed to carry the weight of the vehicle, transfer braking and driving torque to the road, and withstand side thrust over varying speeds and conditions.

tire aspect ratio: A term used for *aspect ratio.* Also known as *series number.*

tire balancing: The procedure of using special equipment for identifying the lighter portions of a tire and adding weights until opposite tire sections weigh the same.

tire bead: A term used for *bead.*

tire belt: A term used for *belt.*

tire black: A black liquid dressing used to refresh the appearance of tires.

tire carcass: The plies that make up the underbody of the tire.

tire casing: Layers of cord called *plies* on which the rubber tread is applied.

tire chain: Specialty chains which may be placed over the tires to improve traction when driving on ice or snow; typically used in emergency situations such as driving on snow-covered or ice-covered mountain roads.

tire coding: Information required by federal legislation to be placed on all tires, such as: manufacturer and tire name, size designation, maximum load-carrying characteristics, limit, and range, a ten-digit *Department of Transportation* serial number indicating where and when it was made, and the letter A, B, or C, indicating conformity to a uniform tire quality grading system.

tire conicity: A condition where the plies and/or belts are not level across the tire tread and are somewhat cone shaped. This causes a pull to one side as the car is driven straight ahead if the tires are on the front of the vehicle.

tire construction: A typical, modern, tire-construction design has two wire beads, bead filler, liner, steel reinforcement in the sidewall, sidewall with hard side compound, rayon carcass plies, steel belts, jointless belt cover, hard under-tread compound, and hard high-grip tread compound.

tire contact area: The footprint, patch, or patch area of the tire that is in contact with the road surface when the tire is supporting the vehicle weight.

tire deflection: The difference between the free diameter and the rolling diameter of the tire.

tire design: The three basic tire construction designs are based on the arrangement of the body plies: bias, belted radial, and belted bias. A bias-ply tire is constructed with the plies arranged on a bias crossing each other. A belted radial-ply tire is constructed with plies that run at right angles to the circumference of the tire. A belted bias-ply tire is constructed with reinforcing belts beneath the tread section.

tire free diameter: The free diameter of a tire is the distance of a horizontal line through the center of the spindle and wheel to the outer edges of the tread.

tire functions: **1.** To cushion the vehicle ride to provide a comfortable ride for the occupant; reduces jolts to the vehicle caused by road irregularities. **2.** To support the vehicle weight. **3.** To develop traction to drive and steer the vehicle under a wide variety of road conditions. **4.** To contribute to the directional stability of the vehicle, and absorb all the stresses of accelerating, braking, and centrifugal force in turns.

Tire Industry Safety Council *(TISC)*: A public education organization.

tire inflation: Air pressure inside the tire, measured in pounds per square inch *(psi)* or kilopascals *(kPa),* checked when tire is cool or has been driven less than one mile.

tire life expectancy: The expected service life of a tire based on its tread wear rating and uniform tire quality grading.

tire manufacturer's code: A term for *manufacturer's code.*

tire mixing: Having more than one size tire on a vehicle. Tires should be replaced only with those recommended by the vehicle manufacturer. Using incorrect or improper tires can cause the vehicle to handle improperly and may result in an accident.

tire motion forces: Forces that affect a tire in motion, including centrifugal, acceleration, and deceleration forces.

tire out of roundness: The *lateral runout* or *radial runout* of a tire.

tire performance criteria: Information molded on the tire sidewall indicating that the tire meets the manufacturer's performance standards for traction, endurance, dimensions, noise, handling, and rolling resistance.

tire placard: A label that is permanently attached to the rear face of the driver's door of most vehicles providing information such as maximum vehicle load, tire size, including spare, and cold inflation pressure, including spare.

tire profile: The profile of a tire is based on the width and height of a cross section of the tire determined by an aspect ratio equal to the section height divided by the section width.

tire pucker: A tire tread that has pulled up on the edges due to mounting on a too-narrow wheel.

tire purposes: A term used for *tire functions.*

tire rim: A term used for *wheel rim.*

tire rolling diameter: The distance of a perpendicular straight line through the center of the spindle to the outer edges of the tread when the tire is supporting the vehicle weight.

tire rotation: Swapping tires around to various positions, as prescribed by the tire manufacturer, on the vehicle to equalize tire wear.

tire runout: The *lateral runout* and *radial runout* of a tire.

tire's area of contact: A term used for *footprint* or *tire contact area.*

tire scrub: The sliding of a tire at an angle to the direction that it is pointed.

tire shimmy: An out-of-balance condition called *wheel shimmy.*

tire sidewall: That area of the tire between the "bead" and the "tread pattern" made from an abrasive-resistant blend of rubber, which absorbs shocks and impacts from road irregularities and prevents damage to the plies.

tire-speed rating: A tire rating that specifies the speed at which the tire can be safely used and does not suggest that a vehicle can be driven safely at the designated speed, because of various road and weather conditions that may be encountered. Also the condition of the vehicle may affect high-speed operation.

tire stagger: Providing different size tires on each side of the axle of an oval- track race car to compensate for all left turns.

tire tramp: A term used for *wheel tramp.*

tire tread: A term used for *tread.*

tire trueing: Procedure of "rounding" a tire by removing rubber so the tread face forms a true circle. This is done by removing high-tread areas while revolving the tire.

tire valve stem: A term used for *valve stem.*

tire waddle: The side-to-side movement at the front and/or rear of the vehicle caused by the steel belt not being straight within the tire. This is most noticeable at low speeds of 5 to 30 mph (8 to 48.3 km/hr).

tire wear pattern: A term used for *wear pattern.*

TISC: An abbreviation for *Tire Industry Safety Council.*

TMC: An abbreviation for *Tune-Up Manufacturers Council.*

toe: The end of the brake shoe against the anchor.

toe in: The amount by which the front of a front wheel points inward.

toe out: The amount by which the front of a front wheel points outward.

toe out on turns: The difference between the angles each of the front wheels makes during a turn.

tolerance: The allowable variation from a standard specification.

toluene: A colorless, flammable liquid hydrocarbon, $C_6H_5CH_3$, used as a cleaning solvent.

ton: **1.** A speed of 100 miles-per-hour. **2.** A unit of weight regarded as 2,000 pounds (907.2 kilograms).

tongue weight: The load applied to the hitch by the trailer tongue, equal to about 10–15 percent of the trailer gross weight.

tonneau cover: An up term used for *tarpaulin.*

ton of refrigeration: The effect of melting one ton of ice in 24 hours: equal to 12,000 *Btu* per hour.

tooling: A set of required standard or special tools needed to produce a particular part, including jigs, fixtures, gauges, and cutting tools, but excluding machine tools.

tooth: A projection on a gear rim or *synchronous belt* that meshes or engages with another component.

tooth cracks: In a synchronous belt drive, tooth cracks and eventual tooth separation can be caused by under tensioning, over tensioning, or using a backside idler with too small a diameter or as a result of an under-designed belt drive.

tooth form: The shape of the working surface of a sprocket tooth from the bottom of the seating curve up through the working faces to the tip of the tooth.

tooth separation: A *tooth crack.*

topcoat: Usually the final paint film applied to a surface.

top dead center *(TDC)*: The *piston* position at the top of its stroke.

top eliminator *(TE)*: The overall winner in a series of drag races.

top end: **1.** High engine rpm; a point that horsepower is the greatest. **2.** The far end of a quarter-mile drag strip.

top-end power: The engine output at high speed.

top fuel: The hottest category in drag racing having cars capable of a quarter mile in less than five seconds, and a top speed of over 300 mph (482.7 km/h).

top inch: The first downward inch of a piston stroke where most wear occurs.

top loader: A Ford four-speed transmission of the mid 1960s to early 1970s.

top-mount battery: A battery having terminals located at the top of the case.

top ring: The top ring of a piston.

top time *(TT)*: The speed at which a vehicle passes through the traps at the end of a quarter mile.

top U-bolt plate: A plate located on the top of the spring and held in place when the U-bolts are tightened, clamping the spring and axle together.

torching: **1.** The cutting or burning of a valve face caused by excessive detonation. **2.** To flame-cut metal.

torque: The measure of a force-producing tension and rotation around an axis.

torque and twist: The forces developed in the trailer and/or tractor frame that are transmitted through the *fifth wheel* when a rigid trailer, such as a tanker, is required to negotiate bumps, such as street curbs.

torque converter: Unit that transfers power from the engine to the transmission input shaft by directing and redirecting fluid flow.

Torqueflite: A three-speed, plus reverse, automatic transmission used by Chrysler.

A B C D E F G H I J K L M N O P Q R S T U V W X Y Z

torque-limiting clutch brake: A system designed to slip when loads of 20–25 pounds–feet (27–33.8 N•m) are reached, to protect the brake from overloading and the high heat damage.

torque multiplication: Torque increase as a result of converter action that allows the *turbine* to revolve slower than the *impeller* during *acceleration* and heavy-load conditions at a ratio as much as 2:1.

torque plate: A thick, metal plate bolted to the deck of a block during cylinder boring and honing to reproduce the stress caused when the heads are installed and the bolts are properly torqued.

torque rod shims: Thin metal shims used to rotate the axis pinion to change the operating angle of the *U-joint.*

torque sequence: The order in which a series of bolts or nuts should be tightened.

torque spec: The force required to tighten a nut or bolt.

torque steer: An outside influence, such as uneven front tire-tread wear, causing the *steering wheel* to turn right or left during hard acceleration, relating to the dual torque and steering requirements that are applied to the drive axle of front-wheel-drive vehicles.

torque test: A starting motor test to determine *torque* produced and current required while the specified voltage is applied.

torque-to-yield bolt: A bolt that has been tightened to a specified yield or stretch point.

torque tube: A drive-shaft system that is enclosed in a tube extending from the transmission tailshaft to the rear-axle housing.

torque wrench: A wrench that indicates the amount of torque being exerted when used, to enable threaded parts to be tightened a specified amount.

torsional balancer: A term used for *vibration damper.*

torsional rigidity: The resistance of a structure to twisting and flexing.

torsional rings: Rings that have a slight twist when placed within the cylinder wall, made by adding a chamfer or counterbore on the ring.

torsional stabilizer bar: A *stabilizer bar* made of spring steel reduces the body roll, thereby reducing the effect of *centrifugal force* when a vehicle rounds a corner.

torsional vibration: A rotary vibration that causes a twist-untwist action on a rotating shaft.

torsion bar: A long, spring-steel bar replacing the coil springs, connected from the chassis to the lower control arm, to provide the necessary springing effect on the front suspensions by twisting rather than compressing or bending.

torsion-bar spring: A term used for *torsion bar.*

torsion-bar steering gear: A rotary valve power-steering gear.

torsion-bar suspension: A suspension featuring a *torsion bar* rather than a *coil* or *leaf spring.*

Torx fastener: A six-point fastener that is easy to grip and drive without slippage. Sometimes called a "star" fastener, this relatively new fastener is used on most late model cars in luggage racks, headlights, tail-light assemblies, mirror mountings, and exterior trim.

total heat load: The amount of heat to be removed or added, based on all conditions.

totalled: The condition of a vehicle that has been damaged beyond repair.

total pedal travel: The complete distance a clutch pedal must move to disengage the *flywheel.*

total volume: Volume of space in the *cylinder* and *combustion chamber* above the *piston* at *bottom dead center (BDC).*

town car: A four-door, chauffeur-driven sedan with an open cockpit and an enclosed passenger compartment.

tow truck: A truck set up to pull a trailer or other vehicle.

tow vehicle: A vehicle set up to pull a trailer or other vehicle.

toxicity: A measure of how poisonous a substance is.

toy: Any vehicle driven for fun.

Toy: An abbreviation for Toyota.

TPC designation: A term used for *tire performance criteria.*

TPI: An abbreviation for *tuned port injection.*

trace: **1.** A colored dye introduced into a system to detect leaks. **2.** To follow logically from beginning to end. **3.** A bar hinged to two parts to transfer motion between them.

track: **1.** A term used for *tread.* **2.** The roadway or pathway a vehicle travels during a competitive event. **3.** To follow. **4.** A belt or tread that guides lines of motion.

track bar: A steel bar attached to the *axle housing* on one side and the frame on the other side of the vehicle to maintain sideways alignment between the axle housing and frame on some coil-spring equipped models.

tracking: Rear wheels following directly behind the front wheels.

traction: A body's friction on a surface such as a vehicle tire on a roadway.

traction bar: A device used on a rear-wheel-drive vehicle to prevent axle twist and spring wind up during hard acceleration.

Traction-Lok: A *limited-slip differential* by Ford.

traction rating: A tire rating that indicates the traction capabilities of the tire to the consumer.

tractive effort: The pushing force exerted by the vehicle's driving wheels against the road's surface.

tractor: **1.** A term used for *truck tractor.* **2.** A motor vehicle having a *fifth wheel* used for pulling a semitrailer. **3.** A truck or short wheel-base vehicle used for pulling a trailer or semitrailer. **4.** A self-propelled vehicle having tracks or wheels.

tractor breakaway valve: A device that automatically seals off the tractor air supply from the trailer air supply when the tractor system pressure drops to 30 or 40 *psi* (206.8 to 275.8 *kPa*).

tractor protection valve: A term used for *tractor breakaway valve.*

tractor/trailer lift suspension: A single-axle air-ride suspension with lift capabilities commonly used with steerable axles for pusher and tag applications.

trail braking: A driving technique where the brakes are applied after entering a turn to help rotate the vehicle.

trailer **1.** A platform or container on wheels pulled by a truck or tractor. **2.** A non-powered vehicle used to haul cargo. **3.** To *put on the trailer.*

trailer charge: To fill the trailer air tanks with air by means of a dash control valve, tractor protection valve, and a trailer relay emergency valve.

trailer hand-control valve: A device located on the dash or steering column used to apply only the trailer brakes; primarily used in jack-knife situations.

trailer slider: A movable suspension frame that is capable of changing the trailer wheelbase by "sliding" and locking into different positions.

trailer tongue: An extension at the front of a trailer that attaches to a hitch ball on a tow vehicle.

trailing arm: A suspension arm that attaches to the chassis ahead of the wheel, generally for the rear wheels on a front-wheel drive vehicle.

trailing shoe: A shoe, toward the rear of a non-servo brake assembly, which pivots around a fulcrum in a direction opposite to normal drum rotation.

tramp: A form of wheel hop where a pair of wheels on an axle hop in opposite phase due to *axle windup.*

tranny: A term for *transmission.*

trans: A term for *transmission.*

Trans-Am: An annual road-racing series, the oldest in the United States, for pony cars and small sedans.

transaxle: A combination transmission-and-axle assembly, common in front-engine, front-drive vehicles.

transaxle pinion gear shaft: The shaft on which a *pinion* is found in a *transaxle.*

transducer: A device that converts an input of one form into an output of another form, such as a motor that converts an electric signal to a rotary motion.

transfer case: A small transmission used on four-wheel-drive vehicles to transfer engine torque to the front and/or rear axles.

transfer gears: Gears used to transfer engine torque from one level to another in a *transaxle.*

transistor: A semiconductor device for controlling the flow of current between two terminals, the emitter and the collector, by means of variations in the current flow between a third terminal, the base, and one of the other two.

transistorized ignition: An *ignition system* having conventional breaker points with transistor regulation of voltage.

transitional spring coil: Coils that become inactive when compressed to their maximum load-carrying capacity.

transmission: A gearing device of a vehicle that provides variable ratios between the engine output and the differential input.

transmission controlled spark *(TCS)*: An emissions-control system to prevent distributor vacuum advance at normal operating temperature until the *transmission* has shifted into high gear by using a transmission-mounted electric switch controlling a solenoid-actuated vacuum valve.

transmission oil cooler: Heat exchanger located in the radiator outlet end section through which transmission fluid flows for cooling purposes on most automatic transmission cars.

transmission-regulated spark system *(TRS)*: A system that allows distributor vacuum advance only when the *transmission* is in high gear.

transplant: **1.** An automobile factory operated by an automaker in a country other than its own, such as United States automaker's plants in Canada. **2.** An engine swap.

transverse crack: A crack with its major axis oriented approximately perpendicular to the weld axis.

transverse vibration: A vibration caused by an unbalanced drive line.

traps: **1.** The measured section at the end of a drag strip. **2.** The fastest part of an oval track or road racing course.

trash box: **1.** An engine built from junk parts. **2.** A car built from junk parts.

travel: **1.** A term used for *rebound travel.* **2.** To move or go from one place to another.

travel-sensitive strut: A strut with the capability to adjust its firmness in relation to the amount of piston travel within.

travel trailer: A trailer with living accommodations.

traverse engine: An engine mounted sideways in a vehicle, such as in most front-drive vehicles.

traverse leaf spring: A leaf spring mounted sideways, such as in the Model T, or Corvette.

tread: That portion of a tire that comes into contact with the surface of the road and is designed to allow the air flow to cool the tire and to channel water during wet weather.

tread compound: A blend of synthetic rubber used in the *tire tread* material.

tread distortion: *Tread* shape change on certain tires as rotating contact is made with the road surface; less severe in belted tires.

tread groove: The recessed portion of the *tire tread* between the ribs.

treadle: A dual, heavy-truck brake valve that: **1.** Releases air from the service reservoir to the service lines and *brake chambers.* **2.** Opens ports to service lines to vent air in the primary and secondary systems.

tread rib: The raised portion of the *tire tread* between the grooves.

tread rubber: A blend of up to 30 different synthetic, and eight natural rubbers used in a *tire tread.*

tread-wear indicator: Projections of rubber, 1/16 inch (1.59 mm) high at the bottom of the *tread grooves,* spaced around the tire to identify a tire worn to the recommended safe limit.

tread-wear rating: A tire rating indicating the wear capabilities of the *tread* that allow consumers to compare tire life expectancy.

tree: A term used for *Christmas tree.*

tree diagnostics chart: A flow chart that suggests logical procedures to follow when troubleshooting a problem.

triangulation: The arrangement of frame members for greater durability and rigidity.

triaxle suspension: A suspension consisting of three axles with a means of equalizing weight between them.

Tri-Chevy: 1955, 1956, and 1957 Chevrolets.

trickle charger: A battery charger with a very low output.

tri-coat: A painting system having three layers of *topcoat* paint.

tri-oval track: A racing course with three high-banked turns such as at the *Daytona International Speedway.*

triple evacuation: The process of evacuating a system that involves three pump downs and two system purges with an *inert gas,* such as dry nitrogen (N).

tripod: The central part of certain *CV joints* having three arms or *trunnions* with needle bearings and rollers running in grooves or races in the assembly.

tri-power: A triple two-barrel *carburetor* setup.

trophy: An award for achievement, such as winning a race.

trophy dash: A short match, at the start of the event, among the fastest qualifying cars in an oval track race.

trophy run: A match for an individual class trophy in drag racing.

trouble code: A number generated by a *computer* to indicate a failure in a sensor, *circuit,* or the *computer* itself. The number may be communicated to the technician by the flashing of a dash light when the diagnostic mode is entered. Also known as *failure code.*

trouble diagnosis: The work required to find the cause of a problem.

troubleshooting: The location of a problem or malfunction by a process of elimination.

TRS: An abbreviation for *transmission-regulated spark system.*

truck: A large vehicle generally used for hauling cargo or for pulling a trailer.

Truck-Frame and Axle Repair Association *(TARA):* A trade association of those engaged in repair of truck and heavy equipment frames.

truck tractor: A large truck with a *fifth wheel* used to pull a *semitrailer.*

trueing: **1.** Straightening bent shafting or parts. **2.** Dressing a grinding wheel. **3.** Setting up a *flywheel* or pulley so that it runs evenly without sideways motion at its rim. **4.** Making an edge straight or bringing two parts square with each other.

trunk: A storage compartment at the rear of most passenger cars.

trunk unit: An early automotive air-conditioning evaporator that mounted in the trunk compartment and was channeled through the package tray.

trunnion: The ends of a universal joint cross where the bearings ride.

try-Y headers: An *exhaust manifold* setup that pairs *cylinders* according to their firing order to provide better exhaust scavenging.

TSB: An abbreviation for *technical service bulletin.*

TSD: An abbreviation for *time, speed, and distance.*

TT: An abbreviation for *top time.*

T-top: A partially removable car roof with panels above both front doors that may be lifted out.

T-tub: A term used for *T-bucket.*

tub: **1.** A term used for *phaeton.* **2.** A *phaeton* body. **3.** The external structure of an auto body.

tubbed: A standard passenger-car body with cut-out rear wheel wells to accommodate oversize rear wheels.

tube-and-fin condenser: A type of *heat exchanger* that consists of tubes to which *fins* are attached to facilitate the transfer of heat from the fluid passing through it to the air passing over it.

tube-and-fin evaporator: A type of *heat exchanger* that consists of tubes to which *fins* are attached to facilitate the transfer of heat from the air passing over it to the fluid passing through it.

tube-and-fin radiator: A type of *heat exchanger* that consists of tubes to which *fins* are attached to facilitate the transfer of heat from the fluid passing through it to the air passing over it.

tubeless tire: A tire having an integral inside rubber liner to retain air; mounted directly on the rim.

tubular core: A radiator or heater core made up of tubular cells.

tubular pushrod: A term used for *hollow pushrod.*

tucked and rolled: A type of upholstery pleating.

Tudor: An early Ford designation for a four-door sedan.

Tuftride (210): The tradename of a liquid chemical used to surface-harden metal parts, such as *crankshafts.*

tules: The boondocks or the back country.

tulip valve: An *intake* or *exhaust valve* with a tulip-shape on the backside of the valve head.

tuned exhaust: An exhaust system having equal length passages.

tuned header: A term used for *tuned exhaust.*

tuned intake: A term used for *ram tuning.*

tuned port injection *(TPI):* A multiport fuel-injection system used on 5.0 and 5.7 Liter General Motors V-8 engines featuring tuned intake runners from a common *plenum.*

tunes: Slang for a stereo sound system.

tune up: A routine maintenance procedure that includes replacing the *rotor* and *spark plugs;* in older cars, replacing the *rotor, points, capacitor, spark plugs* and adjustment of the timing.

Tune-Up Manufacturers Council *(TMC):* A trade association for the manufacturers of ignition parts.

tune-up specifications: Specifications used during a *tune-up* procedure on a vehicle.

tungsten electrode: A non-filler metal electrode used in *arc welding, arc cutting,* and plasma spraying, made principally of tungsten (W).

tungsten inert gas welding: A welding process using tungsten (W) with an *inert gas,* such as argon (Ar), as a shield.

tunnel ram manifold: An *intake manifold* having a large *plenum* and long runners to the *intake ports,* improving output at high *rpm.*

turbine: The driven member inside a *torque converter* consisting of many vanes that receive fluid flow from the *impeller.*

turbine engine: An *engine* in which gas pressure is created by *combustion* to spin a *turbine.*

turbo: An abbreviation for *turbocharger.*

turbo car: A car with a *turbine engine.*

turbocharger: A turbine-type supercharger driven by exhaust gases.

Turbo Hydra-Matic: An *automatic transmission* by General Motors.

Turbo Hydro: A term used for *Turbo Hydra-Matic.*

turbo intercooler: A term used for *intercooler.*

turbo lag: The short delay in *engine* response when a driver punches the *throttle* in a turbocharged vehicle.

turbosupercharger A term used for *turbocharger.*

turbulence: A violent disturbance such as the rapid swirling motion of the *air/fuel mixture* entering a *combustion chamber.*

turkey pan: *Flanges* that are installed as deflectors in an *engine* to reduce oil splash in the lifter areas.

turn: **1.** To change directions. **2.** A specific speed, as in "to turn 90 mph."

turning circle: A term used for *turning diameter.*

turning diameter: The diameter of a circle a vehicle would make if the steering were locked.

turning torque: Amount of *torque* required to keep a shaft or gear rotating.

turn ratio: The ratio between the number of turns between windings in a transformer, primary to secondary.

turn signal: Lights located at the four corners of a vehicle to signal a turn.

TVS: An abbreviation for *thermostatic vacuum switch.*

TVV: An abbreviation for *thermostatic vacuum valve.*

TWC: An abbreviation for *three-way catalyst.*

twenty five ampere rate: A battery rating to indicate the time a fully charged battery at 80°F (26.7°C) can deliver 25 amperes before the cells drop to 1.75 volts.

twenty-hour rating: The amount of current a battery can deliver for 20 hours at 80°F (26.7°C) without the cell voltage dropping below 1.75 volts.

twice pipes: A dual-exhaust system on a *hot rod.*

twin: **1.** A two-cylinder engine. **2.** An identical pair.

twin cam: Dual overhead camshafts.

Twin-Grip: A *limited-slip differential* by AMC.

twin I-beam: A front suspension using two I-beams, each attached to the *chassis* at the end opposite the wheel, and a coil spring at the wheel end.

twin-plug head: A cylinder head having provisions for two *spark plugs* per *cylinder.*

twin-plug ignition system: An *ignition system* having two distributors, two coils, and two plugs per cylinder.

twin torsion bar: A suspension system having two *torsion bars,* one placed above the other.

two-bolt main: An *engine block* in which the mains are held in place with two bolts each.

two cycle: A term used for *two-stroke cycle.*

two in the glue: A two-speed *automatic transmission.*

two-piece drive shaft: A type of *drive shaft* having two sections requiring a center support bearing mounted on the vehicle frame and universal joints at both ends and in the center.

two-piece piston: A *piston* having a removable *skirt.*

two-piece valve: A *valve* having a head and stem made of two different materials.

two-plane manifold: An *intake manifold* with two *plenums.*

two plus two: **1.** A term used for *club coupe.* **2.** A four or five passenger two-door auto body with limited rear-seat space.

two speed axle assembly: A heavy-duty, vehicle-axle arrangement having two different output ratios from the differential that are controlled from the cab or the truck.

two-stroke cycle: An *engine* in which the four events, *intake, compression, combustion,* and exhaust, take place in two strokes of the piston.

two-way catalyst: A *catalytic converter* that oxidizes hydrocarbons and carbon monoxide, but has little effect on *oxides of nitrogen.*

2WD: An abbreviation for two-wheel drive.

U-bolt: U-shaped bolts to fit around the rear-axle housing for clamping the leaf spring to the housing.

U-joint: A short term for *universal joint.*

ultraviolet: An invisible spectrum with wavelengths shorter than visible light but longer than X-rays.

ultraviolet light: A light source used in non-destructive testing for cracks in metal parts.

umbrella seal: An umbrella-shaped valve stem seal used to direct oil away from the stem and guide.

unburned hydrocarbons: HC is a major pollutant in exhaust gases as a result of the partially unburned fuel after *combustion.*

undercarriage: The *chassis* of a vehicle.

undercharge: An air-conditioning system that is short of *refrigerant,* resulting in improper cooling.

under-dash unit: A term used for *hang on unit* such as for a CB radio or other aftermarket accessory.

underfill: A depression on the weld face or root surface extending below the adjacent surface of the base metal.

under-hood lubrication services: The work performed under the hood during a chassis lubrication such as lubricating the distributor, manifold heat-control valve, upper suspension, hood hinges and latch, checking the fluid level in the engine crankcase, brake master cylinder, radiator, battery, and power steering reservoir, plus inspecting belts and safety-related items.

underinflation: A condition with respect to tire inflation that decreases the rolling diameter and increases the contact area, which results in excessive sidewall flexing and *tread* wear causing excessive heat buildup, ultimately leading to severe tire damage.

undersize *(US):* A part of item that is smaller than that specified.

underslung suspension: A suspension system in which the spring is positioned under the axle.

understaging: A vehicle placed at the beginning of a drag race, behind the normal starting point.

understeer: A condition where the front of the vehicle tends to break out and slide toward the outside of a turn.

undertread compound: A blend of synthetic rubbers used in the material under the tire tread.

under-vehicle lubrication services: Work performed under the vehicle during a chassis lubrication such as changing oil and filter, checking or changing transmission fluid, and gear oil or differential gear oil, lubricating the suspension and steering system, universal joints, clutch and/or transmission linkage, front-wheel bearings, and inspection of safety-related items.

UNEP: An abbreviation for *United Nations Environment Program.*

unequal A-arms: A suspension system where one arm is larger than the other.

unequal length control-arm suspension: Front suspension-control arms of unequal length to compensate for jounce and rebound.

unfair advantage: Not really "unfair," the term given for that extra margin realized by a racing team with careful planning and attention to detail.

unglued: An *engine* or *transmission* that has blown and is destroyed.

unibody: A short term for *unit body.*

unicast: A component or part cast as a single unit.

unicast rotor: Type of rotor used on many late-model cars having the rotor and hub cast as one integral component, often from gray iron rather than cast iron.

uniform pitch spring: A valve spring having uniformly spaced coils.

uniform tire quality grading: Information required by the Department of Transportation to be molded into the sidewall of each tire to provide information relating to tread wear, traction, and temperature ratings.

unit: **1.** One. **2.** An assembly or device that can perform its function without outside interference or assistance.

unit body: A form of vehicle construction that includes the chassis, frame, and body in one unit.

unit cable: A cable having pairs of cables placed into groups (units) of a given quantity, then these groups form the core.

unit distributor: A breakerless ignition distributor used by General Motors containing the ignition coil, magnetic pickup coil, and timer core.

United Nations Environment Program *(UNEP)*: A program or agreement mandating the scheduled phaseout of *CFC refrigerants.*

United States Auto Club *(USAC)*: A sanctioning body for oval-track racing, such as the Indianapolis 500.

United States Customary System (USC measurements, *USCS*): The type of measuring system used in the United States, such as gallon, inch, mile, and pounds.

unitized construction: A type of vehicle construction whereby the frame and body parts are welded together to form a single assembly.

unitized distributor: An electronic ignition-system magnetic-pulse distributor that combines the distributor, electronic module, pickup coil, cap, and coil into one compact assembly.

universal flexible hose: A flexible radiator hose.

universal joint: A jointed connection in the drive line that permits the driving angle to change. Also known as *U-joint.*

unleaded fuel: Fuel that does not contain *tetraethyl lead.*

unleaded gasoline: Gasoline that contains less than 0.0018 ounces (0.05 grams) of lead per gallon (3.785 liters).

unloader: A device linked to the throttle valve to open the choke valve when in the wide-open throttle position.

unloading solenoid: An electrically controlled valve for operating the throttling valve or bypass valve in some applications.

unobtanium: A mythical alloy that is lighter, stronger, more flexible and, at the same time, more rigid than anything available.

unsprung weight: The vehicle weight not supported by the springs, including the steering knuckle, brake assembly, tire and wheel.

updraft: A device or system having an upward airflow.

updraft carburetor: A *carburetor* having an updraft airflow.

upper A-arm: A term used for *upper-control arm.*

upper control arm: A front-suspension component bolted to the frame between the pivoting attachment point on the vehicle-frame crossmember and the upper ball joint, which is fastened to its outer end having bushings that are attached to the control arm.

upper radiator hose: The radiator hose from the engine-coolant outlet to the radiator inlet.

upshift: To shift from a lower gear to a higher gear ratio.

upstream blower: A blower arranged in the duct system so as to push air through the heater and/or air-conditioner core(s).

US: An abbreviation for *undersize.*

U.S.: An abbreviation for United States.

USA: An abbreviation for United States of America.

USAC: An abbreviation for *United States Auto Club.*

USC measurements: **1.** A term used for *United States Customary System.* **2.** The standard English system of measurement.

USCS: An abbreviation for *United States Customary System.*

used-car price guide: Either of several periodically published sources, generally available at libraries, for the used-car buyer or seller to determine the retail and wholesale value of a vehicle in the regional marketplace.

used-car value sources: A *used-car price guide.*

ute: An abbreviation for *utility vehicle.*

utility body: A replacement for the cargo bed of a pickup truck that includes compartmental storage for parts and tools.

utility vehicle *(ute):* A passenger-car-based pick up truck.

utilization equipment: Equipment that uses electric energy for mechanical, chemical, heating, lighting, or other useful purposes, such as an electric motor.

U-tube: A term used for *manomometer.*

UV: An abbreviation for ultraviolet.

UV light: That portion of the light spectrum beyond violet.

v: Violet.

vacuum: An enclosed space from which all air has been removed, having an absolute atmospheric pressure of near zero.

vacuum advance: The principle-advancing ignition timing by using the vacuum generated by the engine, accomplished through the use of a mechanism attached to the distributor which moves the breaker point or pickup coil plate when it receives vacuum.

vacuum-advance control: An emissions-control system component that allows vacuum advance only during certain modes of engine operation.

vacuum-advance solenoid: An electrically operated two-position valve that provides or prevents a vacuum signal to the distributor vacuum-advance unit.

vacuum amplifier: A term used for *venturi vacuum amplifier.*

vacuum booster: A power-brake actuating mechanism that uses vacuum on one side of a diaphragm as a power source to amplify braking force.

vacuum brake booster: A device that uses vacuum to amplify the braking force.

vacuum canister: A tank-like enclosure to serve the requirements of vacuum-actuated accessories.

vacuum chamber: A term used for *vacuum canister* or *vacuum reserve tank.*

vacuum-check relay: A mechanical, air-operated device that closes off a vacuum line to a pot whenever the manifold vacuum pressure falls below the applied vacuum pressure.

vacuum-check valve: **1.** An air-operated mechanical device that closes a vacuum line to the vacuum reserve tank whenever the manifold vacuum pressure falls below the reserve vacuum pressure. **2.** In a power-brake system, a one-way valve on the power cylinder vacuum-supply hose connection to maintain system vacuum under certain conditions.

vacuum-controlled temperature sensing valve: A valve that connects the manifold vacuum to the distributor advance under hot-idle conditions.

vacuum delay valve: An orifice-controlled valve which delays a vacuum signal to a diaphragm, such as in the distributor vacuum-advance unit.

vacuum diverter valve: A term used for *vacuum pot* or *vacuum motor.*

vacuum gauge: An analog or digital instrument used for measuring vacuum.

vacuum hose: A rubber-like, plastic or nylon hose used to transmit vacuum.

vacuum hydraulic power unit: A unit consisting of a vacuum brake cylinder or chamber, hydraulic cylinder(s), and control valve in which driver effort is combined with force from the cylinder piston or chamber diaphragm to displace fluid under pressure for actuation of the brake(s).

vacuum line: A rubber-like, plastic or nylon tube used to transmit a vacuum reading from one place to another.

vacuum modulator: A component of the air-temperature sensor that regulates the amount of vacuum signal that is applied to the *servo* motor.

vacuum motor: **1.** A device having a chamber containing a vacuum diaphragm used to create movement in a component, such as a heated air-intake blend door, when engine vacuum is routed to the chamber. **2.** A device designed to provide mechanical control by the use of a vacuum.

vacuum pot: A term used for *vacuum motor.*

vacuum power unit: A device for operating the doors and valves of an air-conditioner control or duct system using vacuum as a source of power.

vacuum programmer: A device with a bleed valve that changes vacuum pressure by bleeding more or less air, thereby controlling the vacuum signal.

vacuum pump: An electro-mechanical device used to rid a system of excess moisture and air.

vacuum reserve tank: A container used to store and provide reserve engine vacuum.

vacuum secondaries: The secondary bores of a four-barrel carburetor that are opened by a vacuum.

vacuum selector valve: A vacuum valve that controls the vacuum motors, which, in turn, operate the airflow-control doors.

vacuum-suspended power brake: A type of power brake in which both sides of the piston are subject to a *vacuum.*

vacuum switch: An electrical switch controlled by a *vacuum.*

vacuum tank: A term used for *vacuum canister, vacuum chamber,* or *vacuum reserve tank.*

vacuum tap: A point at which a vacuum can be released from an enclosure.

vacuum test: A test to determine the presence of a vacuum.

validity list: A list of valid bulletins supplied by the manufacturer.

value guides: A *used-car price guide.*

valve: A circular-stemmed device used to control the flow of *air/fuel mixture* in and the flow of burned gasses out of the engine.

valve angle: The angle at which the valve face is machined.

valve arrangement: The way valves are placed in an engine, such as valves-in-head or valves-in-block.

valve assembly: **1.** A tire valve stem that fits through the wheel to allow the tire to be filled with air. A cap seals the valve and prevents foreign debris from entering and damaging the pin. **2.** A term used for *valve body assembly.*

valve body assembly: An assembly containing the parts that act as the "brain" of an automatic transmission, receiving messages on gear selected, vehicle speed, throttle opening and engine load, then operating automatically by sending fluid to apply the appropriate bands and/or clutches.

valve clearance: A term used for *valve lash.*

valve face: The tapered section of the valve head making contact with the valve seat.

valve float: **1.** A valve that does not close completely. **2.** A valve that does not close at the proper time.

valve guide: The bore in the cylinder head through which the valve stem passes.

valve head: A term designating the enlarged end of the valve.

valve-in-head engine: An overhead-valve engine with the valves in the head.

valve job: The reconditioning of a cylinder head, including the valves.

valve keeper: A small part that fits into the retainer groove located near the tip of the valve stem to secure the valve and valve spring.

valve key: A term used for *valve keeper.*

valve lash: The specified clearance between a valve stem end and a rocker arm on an OHV engine or valve stem end and the camshaft on an OHC engine; necessary to allow for heat expansion.

valve lift: Distance that the valve moves from the closed to the open position.

valve lifter: A cylindrically shaped hydraulic or mechanical device in the valve train that rides on the camshaft lobe to lift the valve off its seat.

A B C D E F G H I J K L M N O P Q R S T U V W X Y Z

valve-lifter foot: The end of a *valve lifter* that rides on the *camshaft.*

valve overlap: The time that the closing of the *exhaust valve* overlaps the opening of the *intake valve* at the end of the exhaust stroke and at the beginning of the intake stroke, when the intake and exhaust valves are partially open at the same time depending on the spacing of the lobe centers on the *camshaft* and the cam's duration.

valve plate: A plate containing suction and/or discharge valves located under the compressor heads.

valve pocket: An area cut into the top of a piston to provide valve clearance.

valve position sensor: A sensor that relays information, in electronic form, to a processor relative to the position of a valve.

valve rack: A holder that keeps valves in proper order when removed, to ensure proper reassembly.

valve relief: A term used for *valve pocket.*

valve retainer: A term used for *valve keeper* or *valve key.*

valve rotator: A device that rotates the valve while the engine is running.

valve seat: The ring of hard metal to which the valve seals.

valve-seat insert: A replacement valve seat.

valve-seat recession: The tendency for some valves to contact the seat in such a manner as to wear away the seat.

valves-in-receiver: An assembly containing the *expansion valve, suction throttling valve, desiccant,* and *receiver.*

valve spool: A spool-shaped valve such as that in a power-steering unit.

valve spring: A small coil spring that closes the valve and keeps the lifter in contact with the *camshaft.*

valve-spring retainer: A device on the valve stem that holds the spring in place.

valve-spring retainer lock: A device on the valve stem that locks the *valve-spring retainer* in place.

valve-spring seat: The area in the cylinder where the fixed end of the valve spring is attached.

valve stem: **1.** A device mounted in the rim or inner tube to provide a method of increasing or decreasing air pressure inside a tire. **2.** The long, slim, round part of a valve.

valve-stem groove: That part of the valve stem used to position the keepers and locate the valve retainer.

valve-stud boss: The boss in the cylinder head that supports and holds the rocker-arm stud.

valve tappet: A term used for *valve lifter.*

valve temperature: The operational temperature of a valve that may reach 1,400°F (760°C) at the head and be as low as 100°F (37.8°C) at the stem.

valve timing: The actual opening and closing of the valves in relationship to the number of degrees of crankshaft rotation.

valvetrain: The many parts making up the valve assembly and its operating mechanism.

van: A box-shaped, light truck with a forward cab.

van conversion: A van that has been customized as a camping vehicle or for some other specific use.

vane: A flat, extended surface such as that on an impeller of a pump.

vane pump: A pump having small vanes in an elliptically shaped housing.

vanes: The scientifically designed and positioned fins that direct or redirect fluid flow inside the *torque converter.*

vapor: A gaseous form of matter.

vapor/fuel separator: A term used for *vapor/liquid separator.*

vaporization: The process of turning a liquid, such as gasoline, into a vapor which is often accomplished after the atomized fuel leaves the carburetor.

vapor lines: Lines used to carry gas or vapor.

vapor/liquid separator: Part of the evaporative-emissions control system that prevents liquid (gasoline) from flowing through the vent line to the charcoal canister.

vapor lock: A condition in carbureted engines where excessive heat has caused the fuel in the fuel line or fuel pump to boil, blocking the flow of fuel to the carburetor and preventing the engine from starting.

vapor recovery system: **1.** That part of the evaporative-emissions control system that prevents gasoline vapors from escaping to the atmosphere by trapping them in the charcoal canister to be drawn into the engine and burned when the engine is started. **2.** The provisions of a fuel nozzle that prevent gasoline vapors from entering the atmosphere when a vehicle is being fueled.

vapor return line: A hose used to carry fuel vapor from the fuel pump or the liquid vapor separator back to the fuel tank.

vapor saver system: A term used for *vapor recovery system.*

variable-aperture carburetor: A term used for *variable venturi carburetor.*

variable-displacement engine: An engine designed so that displacement can be altered by activating and deactivating cylinders.

variable-pitch spring: A valve spring with unevenly spaced coils.

variable-pitch stator: A torque-converter design for transmissions used in off-highway, special-equipment applications.

variable-rate coil spring: A spring having an average spring rate based on load at a predetermined deflection in a variety of wire sizes and shapes.

variable-rate spring: A spring that increasingly resists deflection as the load is increased, such as a coil spring having its coils closer together at one end, or a single-leaf spring that is thinner near the ends.

variable-ratio power steering: A power-steering system in which the response of the vehicle wheels varies according to how much the steering wheel is turned.

variable-ratio steering: A power-steering system that varies the steering ratio at different vehicle speeds.

variable resistor: A resistor that provides for an infinite number of resistance values within a specific range, such as a rheostat or potentiometer.

variable-speed fan: A fan having a fluid-filled or thermostatic-actuated clutch that senses the temperature of the air coming through the radiator and adjusts its speed accordingly.

variable venturi carburetor: A carburetor having an arperture that increases or decreases in size as controlled by intake manifold vacuum.

varnish: Baked on oil and grease.

V-block: **1.** A term sometimes used to identify a block for a V-type engine, such as a *V–6.* **2.** An accurately machined metal block having a V-groove used for checking the out of roundness of a shaft or other device.

VE: An abbreviation for *volumetric efficiency.*

Vee Dub: A Volkswagen.

vehicle: **1.** A means of conveyance or transport, sometimes motorized, that moves on wheels, runners, or tracks. **2.** A generic term for a car or truck. **3.** The combination of binder and solvents or diluents, which are used to put the binder in a liquid, usable form.

vehicle body clearance: The distance from the inside of the inner tire to the spring or other nearest frame or body structure.

vehicle classification: Any of several standard methods of classifying vehicles such as engine type, body/frame construction, fuel consumption structure, or type of drive. The classifications most common to consumers are the use of body shape, seat arrangement, and number of doors.

vehicle construction types: Passenger cars of today use one of two types of construction: conventional body-over-frame and unitized or unibody.

vehicle emission-control information: The emission-control information shown directly on a label on each vehicle.

vehicle engine tests: A test involving engine rigs or vehicles being placed into cold rooms at –22°F to –40°F (–30°C to –40°C) that are started after a cold-soak period to test all engine components in a cold-start situation.

vehicle identification number *(VIN)*: The vehicle's identification number, located on the left front of the dashboard, which represents various data such as the model of the vehicle, year, body, style, engine type, and serial number.

vehicle off road *(VOR)*: A vehicle that is out of service for lack of parts to repair it.

vehicle on-board radar *(VORAD)*: A vehicle warning system that provides drivers additional reaction time to respond to a potential danger.

vehicle retarder: An optional type of brake system used to supplement or assist the regular service brakes on a heavy duty truck.

vehicle traction: A term used for *traction.*

vehicle vapor recovery *(VVR)*: A term used for *vapor recovery system.*

V–8 engine: An engine having two banks of four cylinders each set at an angle to form a V.

velocity: The rate of motion in a particular direction.

velocity stacks: The short tubes attached to *carburetors* that are tuned to a resonate frequency that forces more air into the *carburetors.* Also referred to as *stacks.*

Velvetouch: A brand of sintered-metal brake linings.

V-engine: An engine having two banks or rows of cylinders set at an angle to form a V.

vent: A controlled opening through which air or other vapors can escape.

ventilated rotor: A disc-brake rotor whose friction surfaces are separated by cooling fins having open, ventilated passages between each face for heat dissipation.

ventilate the block: To throw a rod through the block.

ventilation: To circulate fresh air through a space or system.

venturi: **1.** A narrow area in a pipe through which a liquid or a gas is permitted to flow. **2.** The constriction of the air passageway between the choke and throttle plates in a *carburetor,* so as air flows past this constriction, its velocity increases and a partial vacuum is produced, thereby promoting a flow of gas from the main gas nozzle.

venturi principle: The condition of a gas or liquid flowing through a pipe and the pipe diameter is narrow in one place, the flow speeds up in the narrow area, causing a lower pressure in the narrow area, which may be defined as a venturi.

venturi vacuum amplifier: An exhaust-gas recirculation system having a device that uses the weak venturi vacuum signal to regulate the application of strong manifold vacuum to the *EGR valve,* usually including a reservoir that supplies sufficient vacuum when the engine is producing too little for proper operation.

'vert: A term used for *convertible.*

vertical load: The weight on a tire that serves as the input for a tire's performance.

vertical-load capacity: The maximum recommended vertical downward force that can be safely applied to a coupling device.

vertical position: The position of welding in which the weld axis is approximately vertical.

'vette: A Corvette.

vibration: A rapid back and forth motion.

vibration damper: A device designed to dampen or reduce crankshaft torsional (twisting) vibrations that might otherwise cause crankshaft cracks or breakage.

vibrator coil: An induction coil constructed so that the magnetism of the core operates the make and break or vibrator of the primary circuit.

vicky: Short for *Victoria.*

Victoria: **1.** A four-wheel carriage with a two-seat passenger compartment and a separate, high seat in the front for the driver, first used to transport England's Queen Victoria. **2.** A close-coupled, two-door sedan first offered by Ford in the early 1930s.

victory lane: The area at a race track where the winner of an event is honored.

VIN: An abbreviation for *vehicle identification number.*

vintage car: A historic car; one built before 1925.

viscose friction: Friction between moving parts, such as in a compressor, resulting from using an oil that is too heavy or thick.

viscosity: The resistance of a fluid to flowing.

viscosity grade: The numerical rating of a fluid's resistance to flow.

viscosity index: A number to indicate the change in viscosity of an oil when heated.

viscosity rating: A numerical indicator of the viscosity of an engine oil established by the *American Petroleum Institute.*

viscous: **1.** Thick. **2.** Tendency to resist flowing.

viscous coupling: A coupling device having input and output shafts fitted with thin, alternating discs in a closed chamber filled with fluid.

viscous damper: A type of crankshaft *harmonic balancer* that contains a liquid to control engine vibration.

viscous-drive fan clutch: A belt-driven, water-pump mounted, liquid-filled or thermostatically controlled cooling fan clutch that drives the fan fast at high temperatures and slow at low temperatures.

viscous fan: A cooling fan attached to a viscous coupling.

viscous friction: The friction between layers of a liquid.

VOC: An abbreviation for *volatile organic compound.*

voice synthesizer: A computer-controlled phoneme generator capable of reproducing the phonemes used for basic speech, putting them into the right combination to create words and sentences.

voice warning system: A warning system that uses a *voice synthesizer* to alert the driver of monitored conditions.

voiture omnibus: A French term for carriage for all, loosely translated to *bus.*

volatile: Evaporating rapidly.

volatile memory: Memory that does not retain data when power is interrupted to the computer.

volatile organic compound *(VOC):* A term that relates to chemicals that are negligibly photochemically reactive.

volatility: The measure of how rapidly a liquid evaporates.

Volksrod: A Volkswagen hot rod.

volt: A unit of measure of electrical force required to produce a current of one ampere through a resistance of one ohm.

voltage: A specific quantity of electrical force.

voltage drop: A reduction of voltage in an electrical circuit.

voltage-generating sensor: A device having a junction of dissimilar metals that provide a small voltage output.

voltage potential: The electrical pressure at a particular point.

voltage regulator: An electrical device to control the output of a generator or alternator.

voltmeter: An instrument used to read the pressure behind the flow of electrons.

volumetric efficiency *(VE)*: A percentage ratio that varies with the engine's *rpm* measuring the difference between the *air-fuel mixture* actually entering a cylinder and the amount that can enter under ideal conditions.

VOR: An abbreviation for *vehicle off road.*

VORAD: An abbreviation for *vehicle on-board radar.*

vortex oil flow: **1.** The circular oil-flow path as it leaves the impeller, travels to the turbine, then through the stator and back into the impeller. **2.** A mass of whirling fluid within a *torque converter* when impeller speed is high and turbine speed is low.

V-pulley: Used in automotive applications to drive the accessories, such as a water pump, generator, alternator, power steering, and air-conditioner compressor.

V-type engine: An engine having two banks or rows of cylinders set at an angle to form a V.

vulcanizing: A process of treating raw rubber with heat and pressure.

VVR: An abbreviation for *vehicle vapor recovery.*

w: White.

waddle: A term used for *lateral runout* or *tire waddle.*

wagon: A *station wagon.*

wail: To perform at peak efficiency.

wander: The abnormal tracking or steering of a vehicle.

Wankel engine: An engine concept developed in Germany in the 1950s having a three-sided rotor in a slightly hourglass-shaped oval chamber.

warm up: To allow an engine to reach its normal operating temperature.

warm-up regulator: A device in a *fuel-injection system* to adjust the *air/fuel mixture* while the engine is warming up.

warning blinker: A term used for *hazard system.*

warning flasher: A device found in the turn signal and hazard flasher circuit that causes the warning lamps to flash on and off.

warning light: A light on the dash to warn of a problem.

warp: A slight twist or curve in a surface.

washboard: The corrugated surface of an unpaved road.

washer: A round, metal device with a hole in the middle to help secure a nut or a bolt.

wastegate: A turbocharger relief valve to prevent the buildup of too much pressure.

Waste Oil Heating Manufacturers Association *(WOHMA):* An association of waste oil heater manufacturers that promote recycling used motor oil as a heating fuel.

waste spark: A spark occurring during the exhaust stroke on a computerized ignition system.

water brake: A type of absorption unit found on some *dynamometers.*

water burnout: The application of bleach or water to the rear wheels prior to a *burnout* to clean, scuff, and heat the tire surfaces for better traction immediately before a drag race.

water column: A *manometer.*

water control valve: A mechanically operated or vacuum-operated shutoff valve that stops the flow of hot water to the heater core.

water cooled: Using water as a heat transfer medium.

water-cooled system: A term used to identify a liquid cooling system such as one that uses water and antifreeze.

water diverter: A device used to direct the flow of coolant in a head or block.

water filter: A replaceable filter used to remove impurities from an engine-cooling system.

water glass: A common term for *sodium silicate.*

water-heated choke: A bimetallic spring in the choke assembly that opens the throttle valve during warm up when the engine coolant reaches a specific temperature.

water jacket: **1.** The hollow passages inside the cylinder head and engine block through which coolant flows to carry away the heat. **2.** The open spaces within the cylinder block and cylinder head where coolant flows.

water pump: A device, usually located on the front of the engine and driven by one of the accessory drive belts, that circulates the coolant by causing it to move from the lower radiator-outlet section into the engine by centrifugal action of a finned impeller on the pump shaft.

water soluble: Any material that will dissolve in water.

water valve: An electrical-, mechanical-, or vacuum-operated device that controls the flow of coolant to the heater core.

water wash: The forcing of exhaust air and fumes from a spray booth through water so that the vented air is free of thermal-sprayed particles or fumes.

watt: A unit of measure of electrical power.

wattage: A specific quantity of electrical power.

Watt's linkage: A three-bar arrangement of a live or *de Dion rear axle* to prevent lateral movement.

wave scavenging: Internal exhaust resonating that increases the extraction of exhaust gases.

wax: **1.** A compound to shine the painted surface. **2.** To beat a competitor in a race.

wax pellet-type thermostat: A term used for *pellet thermostat.*

wear compensator: A device mounted on the clutch cover having an actuator arm that fits in a hole in the release-sleeve retainer.

wear limit: A manufacturer's specifications as to the durability of a part in terms of serviceability.

wear mated parts: A condition that exists when two parts rub against each other.

wear pattern: Visible wear.

wear sensor: A projection on an inboard brake pad that causes a squealing sound when the brake pads are worn thin.

weather modulator: A term used for *cold-weather modulator.*

weather-pack connector: A connector, having rubber seals on the terminal ends and on the covers of the connector half, used on computer circuits to protect the circuit from corrosion, which may result in a voltage drop.

weave bead: A type of weld bead made with transverse oscillation.

wedge: To shift vehicle weight in oval-track racing by raising or lowering the springs with small blocks or wedges.

wedge-actuated brakes: A braking system that uses air pressure and air brake chambers to push a wedge and roller assembly into an actuator that is located between the adjusting and anchor pistons.

wedge combustion chamber: A combustion chamber resembling a wedge.

wedge head: A cylinder head with wedge-shaped combustion chambers to provide a large quench area.

wedge-shaped combustion chamber: A type of combustion chamber that is shaped similar to a wedge or "V," designed to increase the movement of air and fuel to aid in mixing.

wedging factor: A factor that takes into account the multiplication of force between the V-belt and pulley-groove surfaces that occurs because of the wedging action of the belt in the groove.

weedburner: The exhaust headers on a drag racer that swoop close to the ground and to the rear.

weekend warrior: A part-time, weekend racer having a regular weekday job.

weenie: A tire.

weight distribution: The percentage of the vehicle gross weight on the front wheels and on the back wheels, or the percentage on each wheel.

weight mass: A concentration of weight around the tire; may or may not be equally distributed.

weight-saver spare tire: A *compact spare tire.*

weight-to-power ratio: The ratio of a vehicle's weight to its horsepower.

weight transfer: The momentary shift of a vehicle's weight forward or rearward.

weld: A localized merging of metals or nonmetals produced either by heating the materials to the welding temperature, with or without the application of pressure, or by the application of pressure alone and with or without the use of filler material.

weldability: The capacity of material to be welded under the imposed fabrication conditions into a specific, suitably designed structure and to perform satisfactorily in the intended service.

weld axis: A line through the length of the weld, perpendicular to and at the geometric center of its cross section.

weld bead: A weld resulting from a pass.

weld crack: A crack located in the weld metal or heat-affected zone.

welding: A joining process that produces merging of materials by heating them to the welding temperature, with or without the application of pressure or by the application of pressure alone, and with or without the use of filler material.

welding arc: A controlled electrical discharge between the electrode and the work piece that is formed and sustained by the establishment of a gaseous, conductive medium.

welding filler metal: The metal or alloy added in making a weld joint that alloys with the base metal to form weld metal in a fusion-welded joint.

welding rod: A form of welding filler metal, normally packaged in straight lengths, that does not conduct the welding current.

welding tip: That part of an oxyfuel gas welding torch from which the gases issue.

weldment: An assembly of two or more metal parts that have been joined by fusing metal with the application of heat.

weld metal: The portion of a fusion weld that has been completely melted during welding.

weld pool: The localized volume of molten metal in a weld prior to its solidification as weld metal.

Western Union splice: The electrical connection made by paralleling the bared ends of two conductors and then twisting these bared ends, each around the other.

wet bulb: A device, such as a thermometer, having a wet sock over its sensing element.

wet-bulb temperature: The ambient temperature measured with a wet-bulb thermometer.

wet-disk clutch: A clutch having a friction disk that operates in a bath of oil.

wet liner: A cylinder liner that is in contact with the coolant.

wet sleeve: A term used for *wet liner.*

wet tank: A supply reservoir.

whale tale: A large, horizontal spoiler at the rear of a vehicle.

wheel: A circular frame or hub of an axle to which a tire is attached.

wheel adapter: A metal plate that permits the use of a wheel having a different bolt pattern.

wheel alignment: A condition where the wheels and tires are in proper position on the vehicle.

wheel-and-axle speed sensors: Electromagnetic devices used to provide wheel speed information for an anti-lock brake system.

wheel balance: The equal distribution of weight of a wheel with a mounted tire.

wheel base: The distance from the center of the front wheels to the center of the rear wheels.

wheel cans: The wheel wells on a race car having a full width body.

wheel centerline: An imaginary line through the center of the tire, a vertical line if the tire is exactly in an upright position.

wheel cylinder: A hydraulic cylinder device at each wheel of a drum-brake system that transfers hydraulic pressure developed in the master cylinder to the brake shoes.

wheel cylinder piston: The device in a wheel cylinder that expands, causing the brake shoe to contact the drum.

wheel estate: A mobile home.

wheelie: A wheelstand, lifting the front wheels off the pavement.

wheelie bars: A pair of long bars with wheels extending from the rear of the vehicle to prevent wheelstanding.

wheel meter: A chassis *dynamometer.*

wheel nut: Threaded nuts used to retain the wheel to the studs on the hub assembly.

wheel offset: The wheel rim offset from the center of the mounting flange.

wheel rebound: Downward wheel and suspension movement.

wheel rim: Circular steel, aluminum, or magnesium components on which the tires are mounted, manufactured from stamped, or pressed steel discs that are riveted or welded together to form the circular rim.

wheels: A car.

wheel shimmy: **1.** The wobbling motion during rotation of a dynamically unbalanced tire. **2.** Rapid inward and outward oscillations of the front wheels.

wheel-slip brake-control system: A system which automatically controls rotational wheel slip during braking.

wheel-slip sensor: A device used in combination with the wheel slip brake-control system to sense the rate of angular rotation of the wheel(s) and transmit signals to the logic controller.

wheel-speed sensor: A sensor on each wheel used to monitor speed for anti-lock braking systems.

wheel spindle: The *spindle* on which a *wheel* is mounted.

wheel tramp: The wheel-lifting action or hopping motion caused by static unbalance. Motion may be up-and-down or forward-and-backward, caused by centrifugal force acting on a heavy tire section located near the tread-face center.

wheel tubs: Wheel wells installed to accommodate oversized tires.

wheel weights: Weights used during the balancing process to equalize the weight mass around and across the tire; attached to the rim with clips or special adhesive.

white flag: A signal to a closed-course race driver that she or he is about to begin the last lap.

whoop-dee-doo: A bump or dip severe enough to make a vehicle airborne in off-road racing.

wide-open throttle *(WOT):* Having the throttle wide open, at full speed.

wide-open-throttle cut-out relay: A relay that cuts power to the compressor clutch or other accessories during heavy acceleration.

wide ratios: A transmission setup with wide spreads between speeds.

wienie: A tire. Also, *weenie.*

wienie roaster: A jet-powered car for drag racing or lakes competition.

wild: Great in performance and/or appearance.

winch: A power-driven spool having a wire cable.

windage tray: A metal meshed screen in the oil pan to deflect oil away from the *crankshaft.*

winded time: A quarter-mile drag racing time recorded with the advantage of a tail wind.

windings: **1.** The turns of wire around a core, as in a relay. **2.** The three separate bundles in which wires are grouped in a stator.

windmill: A supercharger.

window net: A net in the window of the driver's door that serves as a restraining device to keep the driver's arm inside.

window regulator: In a power-window system, the device that converts rotary motion of the motor into the linear, vertical movement of the window.

wind resistance: A term used for *air resistance.*

windscreen: The British term for *windshield.*

windshield: The forward-facing window in a vehicle.

windshield wiper: Mechanical arms that sweep back and forth across the windshield to remove water, snow, or dirt.

wind up: **1.** The tendency of the rear axle to rotate with the wheels during hard acceleration. **2.** To *rev* an engine.

wing: A wing-shaped *spoiler.*

wing nut: A nut having wing-like flanges for ease of installation and removal.

wipe: To defeat a competitor in a race.

wiped out: **1.** To be defeated in a race. **2.** To crash.

wiper: A term used for *windshield wiper.*

wire cloth: A material that may be cut to size and used for filtration.

wire placement: The position and routing of wires on an automobile, determined by factory engineers to provide a safe, practical, and economical route.

wiring diagram: A map-like drawing that shows the wiring arrangement and colors of a vehicle *wiring harness.*

wiring harness: The major assembly of a vehicle's electrical wiring system.

wiring protective devices: A device used to prevent damage to the wiring system by maintaining proper wire routing and retention with the use of special clips, retainers, straps, and supplementary insulators to provide additional protection to the conductor over what the insulation itself is capable of providing.

wishbone: A term used for *A-arm* or *A-frame.*

with tracer: A term that indicates a solid or dashed line of a contrasting color on wire insulation, used for identification purposes where many wires are involved.

witness lines: Lines scribed on the surface of adjacent parts prior to disassembly to ensure proper realignment when they are reassembled.

wobble: An unsteady movement off a normal axis.

wobble-plate compressor: A term used for *swash-plate compressor.*

WOHMA: An acronym for *Waste Oil Heating Manufacturers Association.*

WOO: An acronym for *World of Outlaws.*

Woodruff key: A half-moon-shaped key used to prevent a pulley or gear from turning on a shaft. Also known as *half-moon key.*

woody: A vehicle having wood panel or simulated wood panel bodywork

work: **1.** The change of position of an object against an opposing force. **2.** The product of a force and the distance through which it acts.

work bench: Any bench on which work is done; usually having a vise and one or more drawers.

work hardening: The brittleness of a metal part due to stress from bending, hammering, or fatigue.

working chamber: An area in a *shock absorber* where the pressure and vacuum are produced.

works team: A factory-supported racing team.

World of Outlaws *(WOO)*: A spring car-racing organization.

worm: A type of gear on which the teeth resemble large threads.

worm and sector: A type of steering-gear assembly that imparts a rotary motion to a straight-line motion.

worm bearing preload: The adjustment of a work gear bearing to prevent or reduce backlash.

worm gear: The gear at the end of the steering column in a worm-and-sector steering system.

worm hole: A galvanic reaction that erodes metal for the outer walls of wet cylinder sleeves.

worm shaft: A steering gearbox component having spiral grooves that resemble a coiled worm, to transfer motion from the steering-wheel shaft to the pitman shaft.

worm tracks: Oil flow circuits cast within a valve body.

WOT: An abbreviation for *wide-open throttle.*

woven wire cloth: A woven metal material, such as phosphor bronze, stainless steel, or monel available in a wide range of mesh sizes; widely used for filtration.

wraparound headers: Exhaust headers that fit closely to the engine block.

wraparound seat: A body-conforming bucket seat.

wrench: **1.** To twist. **2.** A hand tool for twisting and/or holding bolt heads and nuts.

wrinkled paint: A type of paint finish that has a wrinkled texture.

wrinkled walls: The wrinkling of the sidewalls on drag-racing slicks due to low air pressure.

wrist pin: A pin used to attach the connecting rod to the piston.

wrist-pin bushing: A bushing that supports the *wrist pin* in a connecting rod.

wye: A mechanical or electrical assembly connected in the form of a y.

X-axis: The longitudinal axis around which a vehicle structure rolls from side to side.

X-chassis: A conventional chassis design, used until the late 1960s, which narrows in the center, giving the vehicle a rigid structure that is designed to withstand a high degree of twist having a heavy front cross member to support the upper and lower suspension control arms and coil springs.

X-drilled crank: A term used for *X-drilled crankshaft.*

X-drilled crankshaft: A crankshaft having two oil passages at approximately 90 degrees apart in the main journals.

X-frame: A term used for *X-chassis.*

X-member: An X-shaped reinforcement member in a chassis or frame.

xylene: A chemical solvent (C_8H_{11}) used to remove grease and paint.

A B C D E F G H I J K L M N O P Q R S T U V W **X** Y Z

y: Yellow.

yaw: **1.** The *Z-axis.* **2.** The rotation of a vehicle structure around a vertical axis.

Y-axis: The lateral axis around which a vehicle pitches fore and aft.

Y-block: The block of a *V-type engine* having a deep *crankcase.*

yellow bumper: **1.** A freshman driver in *NASCAR* competition. Also known as *yellow tail.* **2.** The color of the rear bumper of a first-year driver's car in *NASCAR* competition.

yellow flag: A signal to drivers that there is a hazard on the track in closed-course racing. Also known as *yellow light.*

yellow light: A signal to drivers in closed-course racing that there is hazard on the track. Also known as *yellow flag.*

yellow line: **1.** A line that separates the *apron* from the race track. **2.** The *rev* limit of a tachometer before reaching the *red line.*

yellow tail: A freshman driver in *NASCAR* competition. Also known as *yellow bumper.*

yield strength: The amount of force that can be applied to a material before it bends or breaks.

yoke-sleeve kit: A kit that may be used to rebuild the yoke.

Y-pipe: A Y-shaped pipe, such as an exhaust pipe, that merges two passages into one.

Z

Z: Cutting Z-like notches in a frame to lower the vehicle.

zap: **1.** To defeat an opponent. **2.** To over *rev* an engine to the extent that it is damaged.

Z-axis: The vertical axis around which a vehicle structure's front and rear ends swing back and forth.

zener diode: A diode often used in electronic voltage regulators.

zerk: A term used for *zerk fitting.*

zerk fitting: A nipple-like lubrication fitting through which grease is applied to a chassis or suspension joint with a grease gun.

zero emissions: A system or device that does not emit exhaust pollutants to the atmosphere.

zero-emissions vehicle *(ZEV):* An electric vehicle.

zero-gap ring: A piston ring that does not have end clearance.

zero lash: No clearance between the valve lifter and *camshaft* lobe.

zero toe: Adjusting the wheels so they point straight ahead.

ZEV: An abbreviation for *zero-emissions vehicle.*

zinc inner liner: A thin zinc-metal leaf between the longer steel leaves of a leaf-spring suspension system to control sliding friction between the leaves, and prevent corrosion on certain models.

zing: To unintentionally over *rev* an engine.

zirc fitting: A *zerk fitting* made of zirconium alloy.

zoomies: Exhaust headers that sweep back and up toward the top of the rear tires on an open-wheeled drag racer.

zyglow: A non-destructive system using a dye penetrant and an ultraviolet light to check non-magnetic parts for faults and cracks.

Appendix A Decimal and Metric Equivalents

DECIMAL AND METRIC EQUIVALENTS

Fractions	Decimal (in.)	Metric (mm)	Fractions	Decimal (in.)	Metric (mm)
1/64	0.015625	0.397	33/64	0.515625	13.097
1/32	0.03125	0.794	17/32	0.53125	13.494
3/64	0.046875	1.191	35/64	0.546875	13.891
1/16	0.0625	1.588	9/16	0.5625	14.288
5/64	0.078125	1.984	37/64	0.578125	14.684
3/32	0.09375	2.381	19/32	0.59375	15.081
7/64	0.109375	2.778	39/64	0.609375	15.478
1/8	0.125	3.175	5/8	0.625	15.875
9/64	0.140625	3.572	41/64	0.640625	16.272
5/32	0.15625	3.969	21/32	0.65625	16.669
11/64	0.171875	4.366	43/64	0.671875	17.066
3/16	0.1875	4.763	11/16	0.6875	17.463
13/64	0.203125	5.159	45/64	0.703125	17.859
7/32	0.21875	5.556	23/32	0.71875	18.256
15/64	0.234275	5.953	47/64	0.734375	18.653
1/4	0.250	6.35	3/4	0.750	19.05
17/64	0.265625	6.747	49/64	0.765625	19.447
9/32	0.28125	7.144	25/32	0.78125	19.844
19/64	0.296875	7.54	51/64	0.796875	20.241
5/16	0.3125	7.938	13/16	0.8125	20.638
21/64	0.328125	8.334	53/64	0.828125	21.034
11/32	0.34375	8.731	27/32	0.84375	21.431
23/64	0.359375	9.128	55/64	0.859375	21.828
3/8	0.375	9.525	7/8	0.875	22.225
25/64	0.390625	9.922	57/64	0.890625	22.622
13/32	0.40625	10.319	29/32	0.90625	23.019
27/64	0.421875	10.716	59/64	0.921875	23.416
7/16	0.4375	11.113	15/16	0.9375	23.813
29/64	0.453125	11.509	61/64	0.953125	24.209
15/32	0.46875	11.906	31/32	0.96875	24.606
31/64	0.484375	12.303	63/64	0.984375	25.003
1/2	0.500	12.7	1	1.00	25.4

Appendix B Abbreviations

The following abbreviations are some of the more common ones used today in the automotive industry.

SAE J1930 Revised JUN93

Existing Usage	Acceptable Usage	Acceptable Acronized Usage
3GR (Third Gear)	*Third Gear*	*3GR*
4GR (Fourth Gear)	*Fourth Gear*	*4GR*
A/C (Air Conditioning)	*Air conditioning*	*A/C*
A/C Cycling Switch	*Air Conditioning* Cycling Switch	*A/C* Cycling Switch
A/T (Automatic Transaxle)	*Automatic Transaxle*	*A/T*
A/T (Automatic Transmission)	*Automatic Transmission*	*A/T*
AC (Air Conditioning)	*Air Conditioning*	*A/C*
ACC (Air Conditioning Clutch)	*Air Conditioning* Clutch	*A/C* Clutch
Accelerator	*Accelerator Pedal*	*AP*
ACCS (Air Conditioning Cyclic Switch)	*Air Conditioning Cycling Switch*	*A/C* Cycling Switch
ACH (Air Cleaner Housing)	Air Cleaner Housing[1]	ACL Housing[1]
ACL (Air Cleaner)	*Air Cleaner*[1]	*ACL*[1]
ACL (Air Cleaner) Element	*Air Cleaner* Element[1]	*ACL* Element[1]
ACL (Air Cleaner) Housing	*Air Cleaner* Housing[1]	*ACL* Housing[1]
ACL (Air Cleaner) Housing Cover	*Air Cleaner* Housing Cover[1]	*ACL* Housing Cover[1]
ACS (Air Conditioning System)	*Air Conditioning* System	*A/C* System
ACT (Air Charge Temperature)	*Intake Air Temperature*[1]	*IAT*[1]
Adaptive Fuel Strategy	*Fuel Trim*[1]	*FT*[1]
AFC (Air Flow Control)	*Mass Air Flow*[1]	*MAF*[1]
AFC (Air Flow Control)	*Volume Air Flow*[1]	*VAF*[1]
AFC (Air Flow Sensor)	*Mass Air Flow* Sensor[1]	*MAF* Sensor[1]
AFS (Air Flow Sensor)	*Volume Air Flow* Sensor[1]	*VAF* Sensor[1]
After Cooler	*Charge Air Cooler*[1]	*CAC*[1]
AI (Air Injection)	*Secondary Air Injection*[1]	*AIR*[1]
AIP (Air Injection Pump)	*Secondary Air Injection* Pump[1]	*AIR* Pump[1]
AIR (Air Injection Reactor)	*Pulsed Secondary Air Injection*[1]	*PAIR*[1]
AIR (Air Injection Reactor)	*Secondary Air Injection*[1]	*AIR*[1]
AIRB (Secondary Air Injection Bypass)	*Secondary Air Injection* Bypass[1]	*AIR* Bypass[1]
AIRD (Secondary Air Injection Diverter)	*Secondary Air Injection* Diverter[1]	*AIR* Diverter[1]
Air Cleaner	*Air Cleaner*[1]	*ACL*[1]
Air Cleaner Element	*Air Cleaner* Element[1]	*ACL* Element[1]
Air Cleaner Housing	*Air Cleaner* Housing[1]	*ACL* Housing[1]
Air Cleaner Housing Cover	*Air Cleaner* Housing Cover[1]	*ACL* Housing Cover[1]
Air Conditioning	*Air Conditioning*	*A/C*
Air Conditioning Sensor	*Air Conditioning* Sensor	*A/C* Sensor

[1]Emission-Related Term

FIGURE 1—CROSS REFERENCE AND LOOK UP

Reprinted with permission SAE J1930 copyright 1993 Society of Automobile Engineers, Inc.

SAE J1930 Revised JUN93

Existing Usage	Acceptable Usage	Acceptable Acronized Usage
Air Control Valve	*Secondary Air Injection* Control Valve[1]	*AIR* Control Valve[1]
Air Flow Meter	*Mass Air Flow* Sensor[1]	*MAF* Sensor[1]
Air Flow Meter	*Volume Air Flow* Sensor[1]	*VAF* Sensor[1]
Air Intake System	*Intake Air* System[1]	*IA* System[1]
Air Flow Sensor	*Mass Air Flow* Sensor[1]	*MAF* Sensor[1]
Air Management 1	*Secondary Air Injection* Bypass[1]	*AIR* Bypass[1]
Air Management 2	*Secondary Air Injection* Diverter[1]	*AIR* Diverter[1]
Air Temperature Sensor	*Intake Air Temperature* Sensor[1]	*IAT* Sensor[1]
Air Valve	*Idle Air Control* Valve[1]	*IAC* Valve[1]
AIV (Air Injection Valve)	*Pulsed Secondary Air Injection*[1]	*PAIR*[1]
ALCL (Assembly Line Communication Link)	*Data Link Connector*[1]	*DLC*[1]
Alcohol Concentration Sensor	*Flexible Fuel* Sensor[1]	*FF* Sensor[1]
ALDL (Assembly Line Diagnostic Link)	*Data Link Connector*[1]	*DLC*[1]
ALT (Alternator)	*Generator*	*GEN*
Alternator	*Generator*	*GEN*
AM1 (Air Management 1)	*Secondary Air Injection* Bypass[1]	*AIR* Bypass[1]
AM2 (Air Management 2)	*Secondary Air Injection* Diverter[1]	*AIR* Diverter[1]
APS (Absolute Pressure Sensor)	*Barometric Pressure* Sensor[1]	*BARO* Sensor[1]
ATS (Air Temperature Sensor)	*Intake Air Temperature* Sensor[1]	*IAT* Sensor[1]
Automatic Transaxle	*Automatic Transaxle*[1]	*A/T*[1]
Automatic Transmission	*Automatic Transmission*[1]	*A/T*[1]
B + (Battery Positive Voltage)	*Battery Positive Voltage*	*B+*
Backpressure Transducer	*Exhaust Gas Recirculation* Backpressure Transducer[1]	*EGR* Backpressure Transducer[1]
BARO (Barometric Pressure)	*Barometric Pressure*[1]	*BARO*[1]
Barometric Pressure Sensor	*Barometric Pressure* Sensor[1]	*BARO* Sensor[1]
Battery Positive Voltage	*Battery Positive Voltage*	*B+*
Block Learn Matrix	Long Term *Fuel Trim*[1]	Long Term *FT*[1]
BLM (Block Learn Memory)	Long Term *Fuel Trim*[1]	Long Term *FT*[1]
BLM (Block Learn Multiplier)	Long Term *Fuel Trim*[1]	Long Term *FT*[1]
BLM (Block Learn Matrix)	Long Term *Fuel Trim*[1]	Long Term *FT*[1]
Block Learn Memory	Long Term *Fuel Trim*[1]	Long Term *FT*[1]
Block Learn Multiplier	Long Term *Fuel Trim*[1]	Long Term *FT*[1]
BP (Barometric Pressure) Sensor	*Barometric Pressure* Sensor[1]	*BARO* Sensor[1]
C3I (Computer Controlled Coil Ignition)	*Electronic Ignition*[1]	*EI*[1]
CAC (Charge Air Cooler)	*Charge Air Cooler*[1]	*CAC*[1]
Camshaft Position	*Camshaft Position*[1]	*CMP*[1]
Camshaft Position Sensor	*Camshaft Position* Sensor[1]	*CMP* Sensor[1]
Camshaft Sensor	*Camshaft Position* Sensor[1]	*CMP* Sensor[1]
Canister	*Canister*[1]	Canister[1]
Canister	*Evaporative Emission* Canister[1]	*EVAP* Canister[1]
Canister Purge Valve	*Evaporative Emission* Canister Purge Valve[1]	*EVAP* Canister Purge Valve[1]
Canister Purge Vacuum Switching Valve	*Evaporative Emission* Canister Purge Valve[1]	*EVAP* Canister Purge Valve[1]

[1]Emission-Related Term

FIGURE 1—CROSS REFERENCE AND LOOK UP (CONTINUED)

SAE J1930 Revised JUN93

Existing Usage	Acceptable Usage	Acceptable Acronized Usage
Canister Purge VSV (Vacuum Switching Valve)	*Evaporative Emission* Canister Purge Valve[1]	*EVAP* Canister Purge Valve[1]
CANP (Canister Purge)[1]	*Evaporative Emission* Canister Purge[1]	*EVAP* Canister Purge[1]
CARB (Carburetor)	*Carburetor*[1]	*CARB*[1]
Carburetor	*Carburetor*[1]	*CARB*[1]
CCC (Converter Clutch Control)	*Torque Converter Clutch*[1]	*TCC*[1]
CCO (Converter Clutch Override)	*Torque Converter Clutch*[1]	*TCC*[1]
CDI (Capacitive Discharge Ignition)	*Distributor Ignition*[1]	*DI*[1]
CDROM (Compact Disc Read Only Memory)	*Compact Disc Read Only Memory*[1]	*CDROM*[1]
CES (Clutch Engage Switch)	*Clutch Pedal Position* Switch	*CPP* Switch
Central Multiport Fuel Injection	Central *Multiport Fuel Injection*[1]	Central *MFI*[1]
CFI (Continuous Fuel Injection)	*Continuous Fuel Injection*[1]	*CFI*[1]
CFI (Central Fuel Injection)	*Throttle Body Fuel Injection*[1]	*TB*[1]
Charcoal Canister	*Evaporative Emission* Canister	*EVAP* Canister[1]
Charge Air Cooler	*Charge Air Cooler*[1]	*CAC*[1]
Check Engine	*Service Reminder Indicator*[1]	*SRI*[1]
Check Engine	*Malfunction Indicator Lamp*[1]	*MIL*[1]
CID (Cylinder Identification) Sensor	*Camshaft Position* Sensor	*CMP* Sensor[1]
CIS (Continuous Injection System)	*Continuous Fuel Injection*[1]	*CFI*[1]
CIS-E (Continuous Injection System-Electronic)	*Continuous Fuel Injection*[1]	*CFI*[1]
CKP (Crankshaft Position)	*Crankshaft Position*[1]	*CKP*[1]
CKP (Crankshaft Position) Sensor	*Crankshaft Position Sensor*[1]	*CKP* Sensor[1]
CL (Closed Loop)	*Closed Loop*[1]	*CL*[1]
Closed Bowl Distributor	*Distributor Ignition*[1]	*DI*[1]
Closed Throttle Position	*Closed Throttle Position*[1]	*CTP*[1]
Closed Throttle Switch	*Closed Throttle Position* Switch[1]	*CTP* Switch[1]
CLS (Closed Loop System)	*Closed Loop*[1]	*CL*[1]
Clutch Engage Switch	*Clutch Pedal Position* Switch[1]	*CPP* Switch[1]
Clutch Pedal Position Switch	*Clutch Pedal Position* Switch[1]	*CPP* Switch[1]
Clutch Start Switch	*Clutch Pedal Position* Switch[1]	*CPP* Switch[1]
Clutch Switch	*Clutch Pedal Position* Switch[1]	*CPP* Switch[1]
CMFI (Central Multiport Fuel Injection)	Central *Multiport Fuel Injection*[1]	Central *MFI*[1]
CMP (Camshaft Position)	*Camshaft Position*[1]	*CMP*[1]
CMP (Camshaft Position) Sensor	*Camshaft Position* Sensor[1]	*CMP* Sensor[1]
COC (Continuous Oxidation Catalyst)	*Oxidation Catalytic Converter*[1]	*OC*[1]
Condenser	*Distributor Ignition* Capacitor[1]	*DI* Capacitor[1]
Continuous Fuel Injection	*Continuous Fuel Injection*[1]	*CFI*[1]

[1]Emission-Related Term

FIGURE 1—CROSS REFERENCE AND LOOK UP (CONTINUED)

SAE J1930 Revised JUN93

Existing Usage	Acceptable Usage	Acceptable Acronized Usage
Continuous Injection System	*Continuous Fuel Injection* System[1]	*CFI* System[1]
Continuous Injection System-E	Electronic *Continuous Fuel Injection* System[1]	Electronic *CFI* System[1]
Continuous Trap Oxidizer	*Continuous Trap Oxidizer*[1]	*CTOX*[1]
Coolant Temperature Sensor	*Engine Coolant Temperature* Sensor[1]	*ECT* Sensor[1]
CP (Crankshaft Position)	*Crankshaft Position*[1]	*CKP*[1]
CPP (Clutch Pedal Position)	*Clutch Pedal Position*[1]	*CPP*[1]
CPP (Clutch Pedal Position) Switch	*Clutch Pedal Position* Switch	*CPP* Switch[1]
CPS (Camshaft Position Sensor)	*Camshaft Position* Sensor[1]	*CMP* Sensor[1]
CPS (Crankshaft Position Sensor)	*Crankshaft Position* Sensor[1]	*CKP* Sensor[1]
Crank Angle Sensor	*Crankshaft Position* Sensor[1]	*CKP* Sensor[1]
Crankshaft Position	*Crankshaft Position*[1]	*CKP*[1]
Crankshaft Position Sensor	*Crankshaft Position* Sensor[1]	*CKP* Sensor[1]
Crankshaft Speed	*Engine Speed*[1]	*RPM*[1]
Crankshaft Speed Sensor	*Engine Speed* Sensor[1]	*RPM* Sensor[1]
CTO (Continuous Trap Oxidizer)	*Continuous Trap Oxidizer*[1]	*CTOX*[1]
CTOX (Continuous Trap Oxidizer)	*Continuous Trap Oxidizer*[1]	*CTOX*[1]
CTP (Closed Throttle Position)	*Closed Throttle Position*[1]	*CTP*[1]
CTS (Coolant Temperature Sensor)	*Engine Coolant Temperature* Sensor[1]	*ECT* Sensor[1]
CTS (Coolant Temperature Switch)	*Engine Coolant Temperature* Switch[1]	*ECT* Switch[1]
Cylinder ID (Identification) Sensor	*Camshaft Position* Sensor[1]	*CMP* Sensor[1]
D-Jetronic	*Multiport Fuel Injection*[1]	*MFI*[1]
Data Link Connector	*Data Link Connector*[1]	*DLC*[1]
Detonation Sensor	*Knock Sensor*[1]	*KS*[1]
DFI (Direct Fuel Injection)	*Direct Fuel Injection*[1]	*DFI*[1]
DFI (Digital Fuel Injection)	*Multiport Fuel Injection*[1]	*MFI*[1]
DI (Direct Injection)	*Direct Fuel Injection*[1]	*DFI*[1]
DI (Distributor Ignition)	*Distributor Ignition*[1]	*DI*[1]
DI (Distributor Ignition) Capacitor	*Distributor Ignition* Capacitor[1]	*DI* Capacitor[1]
Diagnostic Test Mode	*Diagnostic Test Mode*[1]	*DTM*[1]
Diagnostic Trouble Code	*Diagnostic Trouble Code*[1]	*DTC*[1]
DID (Direct Injection-Diesel)	*Direct Fuel Injection*[1]	*DFI*[1]
Differential Pressure Feedback EGR (Exhaust Gas Recirculation) System	Differential Pressure Feedback *Exhaust Gas Recirculation* System[1]	Differential Pressure Feedback *EGR* System[1]
Digital EGR (Exhaust Gas Recirculation)	*Exhaust Gas Recirculation*[1]	*EGR*[1]
Direct Fuel Injection	*Direct Fuel Injection*[1]	*DFI*[1]

[1]Emission-Related Term

FIGURE 1—CROSS REFERENCE AND LOOK UP (CONTINUED)

SAE J1930 Revised JUN93

Existing Usage	Acceptable Usage	Acceptable Acronized Usage
Direct Ignition System	*Electronic Ignition* System[1]	*EI* System[1]
DIS (Distributorless Ignition System)	*Electronic Ignition* System[1]	*EI* System[1]
DIS (Distributorless Ignition System) Module	*Ignition Control Module*[1]	*ICM*[1]
Distance Sensor	*Vehicle Speed Sensor*[1]	*VSS*[1]
Distributor Ignition	*Distributor Ignition*[1]	*DI*[1]
Distributorless Ignition	*Electronic Ignition*[1]	*EI*[1]
DLC (Data Link Connector)	*Data Link Connector*[1]	*DLC*[1]
DLI (Distributorless Ignition)	*Electronic Ignition*[1]	*EI*[1]
DS (Detonation Sensor)	*Knock Sensor*[1]	*KS*[1]
DTC (Diagnostic Trouble Code)	*Diagnostic Trouble Code*[1]	*DTC*[1]
DTM (Diagnostic Test Mode)	*Diagnostic Test Mode*[1]	*DTM*[1]
Dual Bed	*Three Way + Oxidation Catalytic Converter*[1]	*TWC + OC*[1]
Duty Solenoid for Purge Valve	*Evaporative Emission* Canister Purge Valve[1]	*EVAP* Canister Purge Valve[1]
E2PROM (Electrically Erasable Programmable Read Only Memory)	*Electrically Erasable Programmable Read Only Memory*[1]	*EEPROM*[1]
Early Fuel Evaporation	*Early Fuel Evaporation*[1]	*EFE*[1]
EATX (Electronic Automatic Transmission/Transaxle)	*Automatic Transmission* *Automatic Transaxle*	*A/T* *A/T*
EC (Engine Control)	*Engine Control*[1]	*EC*[1]
ECA (Electronic Control Assembly)	*Powertrain Control Module*[1]	*PCM*[1]
ECL (Engine Coolant Level)	*Engine Coolant Level*	*ECL*
ECM (Engine Control Module)	*Engine Control Module*[1]	*ECM*[1]
ECT (Engine Coolant Temperature)	*Engine Coolant Temperature*[1]	*ECT*[1]
ECT (Engine Coolant Temperature) Sender	*Engine Coolant Temperature* Sensor[1]	*ECT* Sensor[1]
ECT (Engine Coolant Temperature) Sensor	*Engine Coolant Temperature* Sensor[1]	*ECT* Sensor[1]
ECT (Engine Coolant Temperature) Switch	*Engine Coolant Temperature* Switch[1]	*ECT* Switch[1]
ECU4 (Electronic Control Unit 4)	*Powertrain Control Module*[1]	*PCM*
EDF (Electro-Drive Fan) Control	*Fan Control*	*FC*
EDIS (Electronic Distributor Ignition System)	*Distributor Ignition* System[1]	*DI* System[1]
EDIS (Electronic Distributorless Ignition System)	*Electronic Ignition* System[1]	*EI* System[1]
EDIS (Electronic Distributor Ignition System) Module	Distributor *Ignition Control Module*[1]	Distributor *ICM*
EEC (Electronic Engine Control)	*Engine Control*[1]	*EC*[1]

[1]Emission-Related Term

FIGURE 1—CROSS REFERENCE AND LOOK UP (CONTINUED)

SAE J1930 Revised JUN93

Existing Usage	Acceptable Usage	Acceptable Acronized Usage
EEC (Electronic Engine Control) Processor	*Powertrain Control Module*[1]	*PCM*[1]
EECS (Evaporative Emission Control System)	*Evaporative Emission* System[1]	*EVAP* System[1]
EEPROM (Electrically Erasable Programmable Read Only Memory)	*Electrically Erasable Programmable Read Only Memory*[1]	*EEPROM*[1]
EFE (Early Fuel Evaporation)	*Early Fuel Evaporation*[1]	*EFE*[1]
EFI (Electronic Fuel Injection)	*Multiport Fuel Injection*[1]	MFI[1]
EFI (Electronic Fuel Injection)	*Throttle Body Fuel Injection*[1]	*TBI*[1]
EGO (Exhaust Gas Oxygen) Sensor	*Oxygen Sensor*[1]	*O2S*[1]
EGOS (Exhaust Gas Oxygen Sensor)	*Oxygen Sensor*[1]	*O2S*[1]
EGR (Exhaust Gas Recirculation)	*Exhaust Gas Recirculation*[1]	*EGR*[1]
EGR (Exhaust Gas Recirculation) Diagnostic Valve	*Exhaust Gas Recirculation* Diagnostic Valve[1]	*EGR* Diagnostic Valve[1]
EGR (Exhaust Gas Recirculation) System	*Exhaust Gas Recirculation* System[1]	*EGR* System[1]
EGR (Exhaust Gas Recirculation) Thermal Vacuum Valve	*Exhaust Gas Recirculation Thermal Vacuum Valve*[1]	*EGR TVV*[1]
EGR (Exhaust Gas Recirculation) Valve	*Exhaust Gas Recirculation* Valve[1]	*EGR* Valve[1]
EGR TVV (Exhaust Gas Recirculation Thermal Vacuum Valve)	*Exhaust Gas Recirculation Thermal Vacuum Valve*[1]	*EGR TVV*[1]
EGRT (Exhaust Gas Recirculation Temperature)	*Exhaust Gas Recirculation Temperature*[1]	*EGRT*
EGRT (Exhaust Gas Recirculation Temperature) Sensor	*Exhaust Gas Recirculation Temperature* Sensor[1]	*EGRT* Sensor[1]
EGRV (Exhaust Gas Recirculation Valve)	*Exhaust Gas Recirculation* Valve[1]	*EGR* Valve[1]
EGRVC (Exhaust Gas Recirculation Valve Control)	*Exhaust Gas Recirculation* Valve Control[1]	*EGR* Valve Control[1]
EGS (Exhaust Gas Sensor)	*Oxygen Sensor*[1]	*O2S*[1]
EI (Electronic Ignition) (With Distributor)	*Distributor Ignition*[1]	*DI*[1]
EI (Electronic Ignition) (Without Distributor)	*Electronic Ignition*[1]	*EI*[1]
Electrically Erasable Programmable Read Only Memory	*Electrically Erasable Programmable Read Only Memory*[1]	*EEPROM*[1]
Electronic Engine Control	Electronic *Engine Control*[1]	Electronic *EC*[1]
Electronic Ignition	*Electronic Ignition*[1]	*EI*[1]

[1]Emission-Related Term

FIGURE 1—CROSS REFERENCE AND LOOK UP (CONTINUED)

SAE J1930 Revised JUN93

Existing Usage	Acceptable Usage	Acceptable Acronized Usage
Electronic Spark Advance	*Ignition Control*[1]	*IC*[1]
Electronic Spark Timing	*Ignition Control*[1]	*IC*[1]
EM (Engine Modification)	*Engine Modification*[1]	*EM*[1]
EMR (Engine Maintenance Reminder)	*Service Reminder Indicator*[1]	*SRI*[1]
Engine Control	*Engine Control*[1]	*EC*[1]
Engine Coolant Fan Control	*Fan Control*	*FC*
Engine Coolant Level	*Engine Coolant Level*	*ECL*
Engine Coolant Level Indicator	*Engine Coolant Level* Indicator	*ECL* Indicator
Engine Coolant Temperature	*Engine Coolant Temperature*[1]	*ECT*[1]
Engine Coolant Temperature Sender	*Engine Coolant Temperature* Sensor[1]	*ECT* Sensor[1]
Engine Coolant Temperature Sensor	*Engine Coolant Temperature* Sensor[1]	*ECT* Sensor[1]
Engine Coolant Temperature Switch	*Engine Coolant Temperature* Switch[1]	*ECT* Switch[1]
Engine Modification	*Engine Modification*[1]	*EM*[1]
Engine Speed	*Engine Speed*[1]	*RPM*[1]
EOS (Exhaust Oxygen Sensor)	*Oxygen Sensor*[1]	*O2S*[1]
EPROM (Erasable Programmable Read Only Memory)	*Erasable Programmable Read Only Memory*[1]	*EPROM*[1]
Erasable Programmable Read Only Memory	*Erasable Programmable Read Only Memory*[1]	*EPROM*[1]
ESA (Electronic Spark Advance)	*Ignition Control*[1]	*IC*[1]
ESAC (Electronic Spark Advance Control)	*Distributor Ignition*[1]	*DI*[1]
EST (Electronic Spark Timing)	*Ignition Control*[1]	*IC*[1]
EVAP CANP	*Evaporative Emission* Canister Purge[1]	*EVAP* Canister Purge[1]
EVAP (Evaporative Emission)	*Evaporative Emission*[1]	*EVAP*[1]
EVAP (Evaporative Emission) Canister	*Evaporative Emission* Canister[1]	*EVAP* Canister[1]
EVAP (Evaporative Emission) Purge Valve	*Evaporative Emission* Canister Purge Valve[1]	*EVAP* Canister Purge Valve[1]
Evaporative Emission	*Evaporative Emission*[1]	*EVAP*[1]
Evaporative Emission Canister	*Evaporative Emission* Canister[1]	*EVAP* Canister[1]
EVP (Exhaust Gas Recirculation Valve Position) Sensor	*Exhaust Gas Recirculation* Valve Position Sensor[1]	*EGR* Valve Position Sensor[1]
EVR (Exhaust Gas Recirculation Vacuum Regulator) Solenoid	*Exhaust Gas Recirculation* Vacuum Regulator Solenoid[1]	*EGR* Vacuum Regulator Solenoid[1]
EVRV (Exhaust Gas Recirculation Vacuum Regulator Valve)	*Exhaust Gas Recirculation* Vacuum Regulator Valve[1]	*EGR* Vacuum Regulator Valve[1]
Exhaust Gas Recirculation	*Exhaust Gas Recirculation*[1]	*EGR*[1]

[1]Emission-Related Term

FIGURE 1—CROSS REFERENCE AND LOOK UP (CONTINUED)

SAE J1930 Revised JUN93

Existing Usage	Acceptable Usage	Acceptable Acronized Usage
Exhaust Gas Recirculation Temperature	*Exhaust Gas Recirculation Temperature*[1]	*EGRT*[1]
Exhaust Gas Recirculation Temperature Sensor	*Exhaust Gas Recirculation Temperature* Sensor[1]	*EGRT* Sensor[1]
Exhaust Gas Recirculation Valve	*Exhaust Gas Recirculation* Valve[1]	*EGR* Valve[1]
Fan Control	*Fan Control*	*FC*
Fan Control Module	*Fan Control* Module	*FC* Module
Fan Control Relay	*Fan Control* Relay	*FC* Relay
Fan Motor Control Relay	*Fan Control* Relay	*FC* Relay
Fast Idle Thermo Valve	*Idle Air Control* Thermal Valve[1]	*IAC* Thermal Valve[1]
FBC (Feed Back Carburetor)	*Carburetor*[1]	*CARB*[1]
FBC (Feed Back Control)	*Mixture Control*[1]	*MC*[1]
FC (Fan Control)	*Fan Control*	*FC*
FC (Fan Control) Relay	*Fan Control* Relay	*FC* Relay
FEEPROM (Flash Electrically Erasable Programmable Read Only Memory)	*Flash Electrically Erasable Programmable Read Only Memory*[1]	*FEEPROM*[1]
FEPROM (Flash Erasable Programmable Read Only Memory)	*Flash Erasable Programmable Read Only Memory*[1]	*FEPROM*[1]
FF (Flexible Fuel)	*Flexible Fuel*[1]	*FF*[1]
FI (Fuel Injection)	Central *Multiport Fuel Injection*[1]	Central *MFI*[1]
FI (Fuel Injection)	*Continuous Fuel Injection*[1]	*CFI*[1]
FI (Fuel Injection)	*Direct Fuel Injection*[1]	*DFI*[1]
FI (Fuel Injection)	*Indirect Fuel Injection*[1]	*IFI*[1]
FI (Fuel Injection)	*Multiport Fuel Injection*[1]	*MFI*[1]
FI (Fuel Injection)	*Sequential Multiport Fuel Injection*[1]	*SFI*[1]
FI (Fuel Injection)	*Throttle Body Fuel Injection*[1]	*TBI*[1]
Flash EEPROM (Electrically Erasable Programmable Read Only Memory)	*Flash Electrically Erasable Programmable Read Only Memory*[1]	*FEEPROM*[1]
Flash EPROM (Erasable Programmable Read Only Memory)	*Flash Erasable Programmable Read Only Memory*[1]	*FEPROM*[1]
Flexible Fuel	*Flexible Fuel*[1]	*FF*[1]
Flexible Fuel Sensor	*Flexible Fuel* Sensor[1]	*FF* Sensor[1]
Fourth Gear	*Fourth Gear*	*4GR*
FP (Fuel Pump)	*Fuel Pump*	*FP*
FP (Fuel Pump) Module	*Fuel Pump* Module	*FP* Module
FT (Fuel Trim)	*Fuel Trim*[1]	*FT*[1]
Fuel Charging Station	*Throttle Body*[1]	*TB*[1]
Fuel Concentration Sensor	*Flexible Fuel* Sensor[1]	*FF* Sensor[1]
Fuel Injection	Central *Multiport Fuel Injection*[1]	Central *MFI*[1]
Fuel Injection	*Continuous Fuel Injection*[1]	*CFI*[1]
Fuel Injection	*Direct Fuel Injection*[1]	*DFI*[1]
Fuel Injection	*Indirect Fuel Injection*[1]	*IFI*[1]
Fuel Injection	*Multiport Fuel Injection*[1]	*MFI*[1]

[1]Emission-Related Term

FIGURE 1—CROSS REFERENCE AND LOOK UP (CONTINUED)

SAE J1930 Revised JUN93

Existing Usage	Acceptable Usage	Acceptable Acronized Usage
Fuel Injection	*Sequential Multiport Fuel Injection*[1]	*SFI*[1]
Fuel Injection	*Throttle Body Fuel Injection*[1]	*TBI*[1]
Fuel Level Sensor	*Fuel Level Sensor*	*Fuel Level Sensor*
Fuel Module	*Fuel Pump* Module	*FP* Module
Fuel Pressure	*Fuel Pressure*[1]	*Fuel Pressure*[1]
Fuel Pressure Regulator	*Fuel Pressure* Regulator[1]	*Fuel Pressure* Regulator[1]
Fuel Pump	*Fuel Pump*	*FP*
Fuel Pump Relay	*Fuel Pump* Relay	*FP* Relay
Fuel Quality Sensor	*Flexible Fuel* Sensor[1]	*FF* Sensor[1]
Fuel Regulator	*Fuel Pressure* Regulator[1]	*Fuel Pressure* Regulator[1]
Fuel Sender	*Fuel Pump* Module	*FP* Module
Fuel Sensor	*Fuel Level Sensor*	*Fuel Level Sensor*
Fuel Tank Unit	*Fuel Pump* Module	*FP* Module
Fuel Trim	*Fuel Trim*[1]	*FT*[1]
Full Throttle	*Wide Open Throttle*[1]	*WOT*[1]
GCM (Governor Control Module)	*Governor Control Module*	*GCM*
GEM (Governor Electronic Module)	*Governor Control Module*	*GCM*
GEN (Generator)	*Generator*	*GEN*
Generator	Generator	GEN
Governor	*Governor*	*Governor*
Governor Control Module	*Governor Control Module*	*GCM*
Governor Electronic Module	*Governor Control Module*	*GCM*
GND (Ground)	*Ground*	*GND*
GRD (Ground)	Ground	GND
Ground	*Ground*	*GND*
Heated Oxygen Sensor	*Heated Oxygen Sensor*[1]	*HO2S*[1]
HEDF (High Electro-Drive Fan) Control	*Fan Control*	*FC*
HEGO (Heated Exhaust Gas Oxygen) Sensor	*Heated Oxygen Sensor*[1]	*HO2S*[1]
HEI (High Energy Ignition)	Distributor Ignition[1]	*DI*[1]
High Speed FC (Fan Control) Switch	High Speed *Fan Control* Switch	High Speed *FC* Switch
HO2S (Heated Oxygen Sensor)	*Heated Oxygen Sensor*[1]	*HO2S*[1]
HOS (Heated Oxygen Sensor)	*Heated Oxygen Sensor*[1]	*HO2S*[1]
Hot Wire Anemometer	*Mass Air Flow* Sensor[1]	*MAF* Sensor[1]
IA (Intake Air)	Intake Air	IA
IA (Intake Air) Duct	*Intake Air* Duct	*IA* Duct
IAC (Idle Air Control)	*Idle Air Control*[1]	*IAC*[1]
IAC (Idle Air Control) Thermal Valve	*Idle Air Control* Thermal Valve[1]	*IAC* Thermal Valve[1]
IAC (Idle Air Control) Valve	*Idle Air Control* Valve[1]	*IAC* Valve[1]
IACV (Idle Air Control Valve)	Idle Air Control Valve[1]	IAC Valve[1]
IAT (Intake Air Temperature)	*Intake Air Temperature*[1]	*IAT*[1]
IAT (Intake Air Temperature) Sensor	*Intake Air Temperature* Sensor[1]	*IAT* Sensor[1]

[1]Emission-Related Term

FIGURE 1—CROSS REFERENCE AND LOOK UP (CONTINUED)

SAE J1930 Revised JUN93

Existing Usage	Acceptable Usage	Acceptable Acronized Usage
IATS (Intake Air Temperature Sensor)	*Intake Air Temperature* Sensor[1]	*IAT* Sensor[1]
IC (Ignition Control)	*Ignition Control*[1]	*IC*[1]
ICM (Ignition Control Module)	*Ignition Control Module*[1]	*ICM*[1]
IDFI (Indirect Fuel Injection)	*Indirect Fuel Injection*[1]	*IFI*[1]
IDI (Integrated Direct Ignition)	*Electronic Ignition*[1]	*EI*[1]
IDI (Indirect Diesel Injection)	*Indirect Fuel Injection*[1]	*IFI*[1]
Idle Air Bypass Control	*Idle Air Control*[1]	*IAC*[1]
Idle Air Control	*Idle Air Control*[1]	*IAC*[1]
Idle Air Control Valve	*Idle Air Control* Valve[1]	*IAC* Valve[1]
Idle Speed Control	*Idle Air Control*[1]	*IAC*[1]
Idle Speed Control	*Idle Speed Control*[1]	*ISC*[1]
Idle Speed Control Actuator	*Idle Speed Control* Actuator[1]	*ISC* Actuator[1]
IFI (Indirect Fuel Injection)	*Indirect Fuel Injection*[1]	*IFI*[1]
IFS (Inertia Fuel Shutoff)	*Inertia Fuel Shutoff*	*IFS*
Ignition Control	*Ignition Control*[1]	*IC*[1]
Ignition Control Module	*Ignition Control Module*[1]	*ICM*[1]
In Tank Module	*Fuel* Pump Module	*FP* Module
Indirect Fuel Injection	*Indirect Fuel Injection*[1]	*IFI*[1]
Inertia Fuel Shutoff	*Inertia Fuel Shutoff*	*IFS*
Inertia Fuel-Shutoff Switch	*Inertia Fuel Shutoff* Switch	*IFS* Switch
Inertia Switch	*Inertia Fuel Shutoff* Switch	*IFS* Switch
INT (Integrator)	Short Term *Fuel Trim*[1]	Short Term *FT*[1]
Intake Air	*Intake Air*	*IA*
Intake Air Duct	*Intake Air* Duct	*IA* Duct
Intake Air Temperature	*Intake Air Temperature*[1]	*IAT*[1]
Intake Air Temperature Sensor	*Intake Air Temperature* Sensor[1]	*IAT* Sensor[1]
Intake Manifold Absolute Pressure Sensor	*Manifold Absolute Pressure* Sensor[1]	*MAP* Sensor[1]
Integrated Relay Module	*Relay Module*	*RM*
Integrator	Short Term *Fuel Trim*[1]	Short Term *FT*[1]
Inter Cooler	*Charge Air Cooler*[1]	*CAC*[1]
ISC (Idle Speed Control)	*Idle Air Control*[1]	*IAC*[1]
ISC (Idle Speed Control)	*Idle Speed Control*[1]	*ISC*[1]
ISC (Idle Speed Control) Actuator	*Idle Speed Control* Actuator[1]	*ISC* Actuator[1]
ISC BPA (Idle Speed Control By Pass Air)	*Idle Air Control*[1]	*IAC*[1]
ISC (Idle Speed Control) Solenoid Vacuum Valve	*Idle Speed Control* Solenoid Vacuum Valve[1]	*ISC* Solenoid Vacuum Valve[1]
K-Jetronic	*Continuous Fuel Injection*[1]	*CFI*[1]
KAM (Keep Alive Memory)	*Non Volatile Random Access Memory*[1]	*NVRAM*[1]
KAM (Keep Alive Memory)	Keep Alive *Random Access Memory*[1]	Keep Alive *RAM*[1]
KE-Jetronic	*Continuous Fuel Injection*[1]	*CFI*[1]
KE-Motronic	*Continuous Fuel Injection*[1]	*CFI*[1]
Knock Sensor	*Knock Sensor*[1]	*KS*[1]

[1]Emission-Related Term

FIGURE 1—CROSS REFERENCE AND LOOK UP (CONTINUED)

SAE J1930 Revised JUN93

Existing Usage	Acceptable Usage	Acceptable Acronized Usage
KS (Knock Sensor)	*Knock Sensor*[1]	*KS*[1]
L-Jetronic	*Multiport Fuel Injection*[1]	*MFI*[1]
Lambda	*Oxygen Sensor*[1]	*O2S*[1]
LH-Jetronic	*Multiport Fuel Injection*[1]	*MFI*[1]
Light Off Catalyst	*Warm Up Three Way Catalytic Converter*[1]	*WU-TWC*[1]
Light Off Catalyst	*Warm Up Oxidation Catalytic Converter*[1]	*WU-OC*[1]
Lock Up Relay	*Torque Converter Clutch* Relay[1]	*TCC* Relay[1]
Long Term FT (Fuel Trim)	Long Term *Fuel Trim*[1]	Long Term *FT*[1]
Low Speed FC (Fan Control) Switch	Low Speed *Fan Control* Switch	Low Speed *FC* Switch
LUS (Lock Up Solenoid) Valve	*Torque Converter Clutch* Solenoid Valve[1]	*TCC* Solenoid Valve[1]
M/C (Mixture Control)	*Mixture Control*[1]	*MC*[1]
MAF (Mass Air Flow)	*Mass Air Flow*[1]	*MAF*[1]
MAF (Mass Air Flow) Sensor	*Mass Air Flow Sensor*[1]	*MAF* Sensor[1]
Malfunction Indicator Lamp	*Malfunction Indicator Lamp*[1]	*MIL*[1]
Manifold Absolute Pressure	*Manifold Absolute Pressure*[1]	*MAP*[1]
Manifold Absolute Pressure Sensor	*Manifold Absolute Pressure* Sensor	*MAP* Sensor[1]
Manifold Differential Pressure	*Manifold Differential Pressure*[1]	*MDP*[1]
Manifold Surface Temperature	*Manifold Surface Temperature*[1]	*MST*[1]
Manifold Vacuum Zone	*Manifold Vacuum Zone*[1]	*MVZ*[1]
Manual Lever Position Sensor	*Transmission Range* Sensor[1]	*TR* Sensor[1]
MAP (Manifold Absolute Pressure)	*Manifold Absolute Pressure*[1]	*MAP*[1]
MAP (Manifold Absolute Pressure) Sensor	*Manifold Absolute Pressure* Sensor[1]	*MAP* Sensor[1]
MAPS (Manifold Absolute Pressure Sensor)	*Manifold Absolute Pressure* Sensor[1]	*MAP* Sensor[1]
Mass Air Flow	*Mass Air Flow*[1]	*MAF*[1]
Mass Air Flow Sensor	*Mass Air Flow* Sensor[1]	*MAF* Sensor[1]
MAT (Manifold Air Temperature)	*Intake Air Temperature*[1]	*IAT*[1]
MATS (Manifold Air Temperature Sensor)	*Intake Air Temperature* Sensor[1]	*IAT* Sensor[1]
MC (Mixture Control)	*Mixture Control*[1]	*MC*[1]
MCS (Mixture Control Solenoid)	*Mixture Control* Solenoid[1]	*MC* Solenoid[1]
MCU (Microprocessor Control Unit)	*Powertrain Control Module*[1]	*PCM*[1]
MDP (Manifold Differential Pressure)	*Manifold Differential Pressure*[1]	*MDP*[1]
MFI (Multiport Fuel Injection)	*Multiport Fuel Injection*[1]	*MFI*[1]
MIL (Malfunction Indicator Lamp)	*Malfunction Indicator Lamp*[1]	*MIL*[1]
Mixture Control	*Mixture Control*[1]	*MC*[1]
Modes	*Diagnostic Test Mode*[1]	*DTM*[1]

[1]Emission-Related Term

FIGURE 1—CROSS REFERENCE AND LOOK UP (CONTINUED)

SAE J1930 Revised JUN93

Existing Usage	Acceptable Usage	Acceptable Acronized Usage
Monotronic	*Throttle Body Fuel Injection*[1]	*TBI*[1]
Motronic	*Multiport Fuel Injection*[1]	*MFI*[1]
MPI (Multipoint Injection)	*Multiport Fuel Injection*[1]	*MFI*[1]
MPI (Multiport Injection)	*Multiport Fuel Injection*[1]	*MFI*[1]
MRPS (Manual Range Position Switch)	*Transmission Range* Switch	*TR* Switch
MST (Manifold Surface Temperature)	*Manifold Surface Temperature*[1]	*MST*[1]
Multiport Fuel Injection	*Multiport Fuel Injection*[1]	*MFI*[1]
MVZ (Manifold Vacuum Zone)	*Manifold Vacuum Zone*[1]	*MVZ*[1]
NDS (Neutral Drive Switch)	*Park/Neutral Position* Switch[1]	*PNP* Switch[1]
Neutral Safety Switch	*Park/Neutral Position* Switch[1]	*PNP* Switch[1]
NGS (Neutral Gear Switch)	*Park/Neutral Position* Switch[1]	*PNP* Switch[1]
Nonvolatile Random Access Memory	*Nonvolatile Random Access Memory*[1]	*NVRAM*[1]
NPS (Neutral Position Switch)	*Park/Neutral Position* Switch[1]	*PNP* Switch[1]
NVM (Nonvolatile Memory)	*Nonvolatile Random Access Memory*[1]	*NVRAM*[1]
NVRAM (Nonvolatile Random Access Memory)	*Nonvolatile Random Access Memory*[1]	*NVRAM*[1]
O2 (Oxygen) Sensor	*Oxygen Sensor*[1]	*O2S*[1]
O2S (Oxygen Sensor)	*Oxygen Sensor*[1]	*O2S*[1]
OBD (On Board Diagnostic)	*On Board Diagnostic*[1]	*OBD*[1]
OC (Oxidation Catalyst)	*Oxidation Catalytic Converter*[1]	*OC*[1]
Oil Pressure Sender	*Oil Pressure* Sensor	Oil Pressure Sensor
Oil Pressure Sensor	*Oil Pressure* Sensor	Oil Pressure Sensor
Oil Pressure Switch	*Oil Pressure* Switch	Oil Pressure Switch
OL (Open Loop)	*Open* Loop[1]	*OL*[1]
On Board Diagnostic	*On Board Diagnostic*[1]	*OBD*[1]
Open Loop	*Open Loop*[1]	*OL*[1]
OS (Oxygen Sensor)	*Oxygen Sensor*[1]	*O2S*[1]
Oxidation Catalytic Converter	*Oxidation Catalytic Converter*[1]	*OC*[1]
OXS (Oxygen Sensor) Indicator	*Service Reminder Indicator*[1]	*SRI*[1]
Oxygen Sensor	*Oxygen Sensor*[1]	*O2S*[1]
P/N (Park/Neutral)	*Park/Neutral Position*[1]	*PNP*[1]
P/S (Power Steering) Pressure Switch	*Power Steering Pressure* Switch	*PSP* Switch
P- (Pressure) Sensor	*Manifold Absolute Pressure* Sensor[1]	*MAP* Sensor[1]
PAIR (Pulsed Secondary Air Injection)	*Pulsed Secondary Air Injection*[1]	*PAIR*[1]
Park/Neutral Position	*Park/Neutral Position*[1]	*PNP*[1]
PCM (Powertrain Control Module)	*Powertrain Control Module*[1]	*PCM*[1]
PCV (Positive Crankcase Ventilation)	*Positive Crankcase Ventilation*[1]	*PCV*[1]
PCV (Positive Crankcase Ventilation) Valve	*Positive Crankcase Ventilation* Valve[1]	*PCV* Valve[1]
Percent Alcohol Sensor	*Flexible Fuel* Sensor[1]	*FF* Sensor[1]

[1]Emission-Related Term

FIGURE 1—CROSS REFERENCE AND LOOK UP (CONTINUED)

SAE J1930 Revised JUN93

Existing Usage	Acceptable Usage	Acceptable Acronized Usage
Periodic Trap Oxidizer	*Periodic Trap Oxidizer*[1]	*PTOX*[1]
PFE (Pressure Feedback Exhaust Gas Recirculation) Sensor	Feedback Pressure *Exhaust Gas Recirculation* Sensor[1]	Feedback Pressure *EGR* Sensor[1]
PFI (Port Fuel Injection)	*Multiport Fuel Injection*[1]	*MFI*[1]
PG (Pulse Generator)	*Vehicle Speed Sensor*[1]	*VSS*[1]
PGM-FI (Programmed Fuel Injection)	*Multiport Fuel Injection*[1]	*MFI*[1]
PIP (Position Indicator Pulse)	*Crankshaft Position*[1]	*CKP*[1]
PNP (Park/Neutral Position)	*Park/Neutral Position*[1]	*PNP*[1]
Positive Crankcase Ventilation	*Positive Crankcase Ventilation*[1]	*PCV*[1]
Positive Crankcase Ventilation Valve	*Positive Crankcase Ventilation* Valve[1]	*PCV* Valve[1]
Power Steering Pressure	*Power Steering Pressure*	*PSP*
Power Steering Pressure Switch	*Power Steering Pressure* Switch	*PSP* Switch
Powertrain Control Module	*Powertrain Control Module*[1]	*PCM*[1]
Pressure Feedback EGR (Exhaust Gas Recirculation)	Feedback Pressure *Exhaust Gas Recirculation*[1]	Feedback Pressure *EGR*[1]
Pressure Sensor	*Manifold Absolute Pressure* Sensor[1]	*MAP* Sensor[1]
Pressure Transducer EGR (Exhaust Gas Recirculation) System	Pressure Transducer *Exhaust Gas Recirculation* System[1]	Pressure Transducer *EGR* System[1]
PRNDL (Park, Reverse, Neutral, Drive, Low)	*Transmission Range*	*TR*
Programmable Read Only Memory	*Programmable Read Only Memory*[1]	*PROM*[1]
PROM (Programmable Read Only Memory)	*Programmable Read Only Memory*[1]	*PROM*[1]
PSP (Power Steering Pressure)	*Power Steering Pressure*	*PSP*
PSP (Power Steering Pressure) Switch	*Power Steering Pressure* Switch	*PSP* Switch
PSPS (Power Steering Pressure Switch)	*Power Steering Pressure* Switch	*PSP* Switch
PTOX (Periodic Trap Oxidizer)	*Periodic Trap Oxidizer*[1]	*PTOX*[1]
Pulsair	*Pulsed Secondary Air Injection*[1]	*PAIR*[1]
Pulsed Secondary Air Injection	*Pulsed Secondary Air Injection*[1]	*PAIR*[1]
Radiator Fan Control	*Fan Control*	*FC*
Radiator Fan Relay	*Fan Control* Relay	*FC* Relay
RAM (Random Access Memory)	*Random Access Memory*[1]	*RAM*[1]
Random Access Memory	*Random Access Memory*[1]	*RAM*[1]
Read Only Memory	*Read Only Memory*[1]	*ROM*[1]
Recirculated Exhaust Gas Temperature Sensor	*Exhaust Gas Recirculation Temperature* Sensor[1]	*EGRT* Sensor[1]
Reed Valve	*Pulsed Secondary Air Injection* Valve[1]	*PAIR* Valve[1]

[1]Emission-Related Term

FIGURE 1—CROSS REFERENCE AND LOOK UP (CONTINUED)

SAE J1930 Revised JUN93

Existing Usage	Acceptable Usage	Acceptable Acronized Usage
REGTS (Recirculated Exhaust Gas Temperature Sensor)	*Exhaust Gas Recirculation Temperature* Sensor[1]	*EGRT* Sensor[1]
Relay Module	*Relay Module*	*RM*
Remote Mount TFI (Thick Film Ignition)	*Distributor Ignition*[1]	*DI*[1]
Revolutions per Minute	*Engine Speed*[1]	*RPM*[1]
RM (Relay Module)	*Relay Module*	*RM*
ROM (Read Only Memory)	*Read Only Memory*[1]	*ROM*[1]
RPM (Revolutions per Minute)	*Engine Speed*[1]	*RPM*[1]
SABV (Secondary Air Bypass Valve)	*Secondary Air Injection* Bypass Valve[1]	*AIR* Bypass Valve[1]
SACV (Secondary Air Check Valve)	*Secondary Air Injection* Control Valve[1]	*AIR* Control Valve[1]
SASV (Secondary Air Switching Valve)	*Secondary Air Injection* Switching Valve[1]	*AIR* Switching Valve[1]
SBEC (Single Board Engine Control)	*Powertrain Control Module*[1]	*PCM*[1]
SBS (Supercharger Bypass Solenoid)	*Supercharger Bypass* Solenoid[1]	*SCB* Solenoid[1]
SC (Supercharger)	*Supercharger*[1]	*SC*[1]
Scan Tool	*Scan Tool*[1]	*ST*[1]
SCB (Supercharger Bypass)	*Supercharger Bypass*[1]	*SCB*[1]
Secondary Air Bypass Valve	*Secondary Air Injection* Bypass Valve[1]	*AIR* Bypass Valve[1]
Secondary Air Check Valve	*Secondary Air Injection* Check Valve[1]	*AIR* Check Valve[1]
Secondary Air Injection	*Secondary Air Injection*[1]	*AIR*[1]
Secondary Air Injection Bypass	*Secondary Air Injection* Bypass[1]	*AIR* Bypass[1]
Secondary Air Injection Diverter	*Secondary Air Injection* Diverter[1]	*AIR* Diverter[1]
Secondary Air Switching Valve	*Secondary Air Injection* Switching Valve[1]	*AIR* Switching Valve[1]
SEFI (Sequential Electronic Fuel Injection)	*Sequential Multiport Fuel Injection*[1]	*SFI*[1]
Self Test	*On Board Diagnostic*[1]	*OBD*[1]
Self Test Codes	*Diagnostic Trouble Code*[1]	*DTC*[1]
Self Test Connector	*Data Link Connector*[1]	*DLC*[1]
Sequential Multiport Fuel Injection	*Sequential Multiport Fuel Injection*[1]	*SFI*[1]
Service Engine Soon	*Service Reminder Indicator*[1]	*SRI*[1]
Service Engine Soon	*Malfunction Indicator Lamp*[1]	*MIL*[1]
Service Reminder Indicator	*Service Reminder Indicator*[1]	*SRI*[1]
SFI (Sequential Fuel Injection)	*Sequential Multiport Fuel Injection*[1]	*SFI*[1]
Short Term FT (Fuel Trim)	Short Term *Fuel Trim*[1]	Short Term *FT*[1]
SLP (Selection Lever Position)	*Transmission Range*	*TR*
SMEC (Single Module Engine Control)	*Powertrain Control Module*[1]	*PCM*[1]
Smoke Puff Limiter	*Smoke Puff Limiter*[1]	*SPL*[1]

[1]Emission-Related Term

FIGURE 1—CROSS REFERENCE AND LOOK UP (CONTINUED)

SAE J1930 Revised JUN93

Existing Usage	Acceptable Usage	Acceptable Acronized Usage
SPI (Single Point Injection)	*Throttle Body Fuel Injection*[1]	*TBI*[1]
SPL (Smoke Puff Limiter)	*Smoke Puff Limiter*[1]	*SPL*[1]
SRI (Service Reminder Indicator)	*Service Reminder Indicator*[1]	*SRI*[1]
SRT (System Readiness Test)	*System Readiness Test*[1]	*SRT*[1]
ST (Scan Tool)	*Scan Tool*[1]	*ST*[1]
Supercharger	*Supercharger*[1]	*SC*[1]
Supercharger Bypass	*Supercharger Bypass*[1]	*SCB*[1]
Sync Pickup	*Camshaft Position*[1]	*CMP*[1]
System Readiness Test	*System Readiness Test*[1]	*SRT*[1]
TAB (Thermactor Air Bypass)	*Secondary Air Injection* Bypass[1]	*AIR* Bypass[1]
TAD (Thermactor Air Diverter)	*Secondary Air Injection* Diverter[1]	*AIR* Diverter[1]
TB (Throttle Body)	*Throttle Body*[1]	*TB*[1]
TBI (Throttle Body Fuel Injection)	*Throttle Body Fuel Injection*[1]	*TBI*[1]
TBT (Throttle Body Temperature)	*Intake Air Temperature*[1]	*IAT*[1]
TC (Turbocharger)	*Turbocharger*[1]	*TC*[1]
TCC (Torque Converter Clutch)	*Torque Converter Clutch*[1]	*TCC*[1]
TCC (Torque Converter Clutch) Relay	*Torque Converter Clutch* Relay[1]	*TCC* Relay[1]
TCM (Transmission Control Module)	*Transmission Control Module*	*TCM*
TFI (Thick Film Ignition)	*Distributor Ignition*[1]	*DI*[1]
TFI (Thick Film Ignition) Module	*Ignition Control Module*[1]	*ICM*[1]
Thermac	*Secondary Air Injection*[1]	*AIR*[1]
Thermac Air Cleaner	*Air Cleaner*[1]	*ACL*[1]
Thermactor	*Secondary Air Injection*[1]	*AIR*[1]
Thermactor Air Bypass	*Secondary Air Injection* Bypass[1]	*AIR* Bypass[1]
Thermactor Air Diverter	*Secondary Air Injection* Diverter[1]	*AIR* Diverter[1]
Thermactor II	*Pulsed Secondary Air Injection*[1]	*PAIR*[1]
Thermal Vacuum Switch	*Thermal Vacuum Valve*[1]	*TVV*[1]
Thermal Vacuum Valve	*Thermal Vacuum Valve*[1]	*TVV*[1]
Third Gear	*Third Gear*	*3GR*
Three Way + Oxidation Catalytic Converter	*Three Way + Oxidation Catalytic Converter*[1]	*TWC + OC*[1]
Three Way Catalytic Converter	*Three Way Catalytic Converter*[1]	*TWC*[1]
Throttle Body	*Throttle Body*[1]	*TB*[1]
Throttle Body Fuel Injection	*Throttle Body Fuel Injection*[1]	*TBI*[1]
Throttle Opener	*Idle Speed Control*[1]	*ISC*[1]
Throttle Opener Vacuum Switching Valve	*Idle Speed Control* Solenoid Vacuum Valve[1]	*ISC* Solenoid Vacuum Valve[1]
Throttle Opener VSV (Vacuum Switching Valve)	*Idle Speed Control* Solenoid Vacuum Valve[1]	*ISC* Solenoid Vacuum Valve[1]
Throttle Position	*Throttle Position*[1]	*TP*
Throttle Position Sensor	*Throttle Position* Sensor[1]	*TP* Sensor[1]
Throttle Position Switch	*Throttle Position* Switch[1]	*TP* Switch[1]
Throttle Potentiometer	*Throttle Position* Sensor[1]	*TP* Sensor[1]

[1]Emission-Related Term

FIGURE 1—CROSS REFERENCE AND LOOK UP (CONTINUED)

SAE J1930 Revised JUN93

Existing Usage	Acceptable Usage	Acceptable Acronized Usage
TOC (Trap Oxidizer-Continuous)	*Continuous Trap Oxidizer*[1]	*CTOX*[1]
TOP (Trap Oxidizer-Periodic)	*Periodic Trap Oxidizer*[1]	*PTOX*[1]
Torque Converter Clutch	*Torque Converter Clutch*[1]	*TCC*[1]
Torque Converter Clutch Relay	*Torque Converter Clutch* Relay[1]	*TCC* Relay[1]
TP (Throttle Position)	*Throttle Position*[1]	*TP*[1]
TP (Throttle Position) Sensor	*Throttle Position* Sensor[1]	*TP* Sensor[1]
TP (Throttle Position) Switch	*Throttle Position* Switch[1]	*TP* Switch[1]
TPI (Tuned Port Injection)	*Multiport Fuel Injection*[1]	*MFI*[1]
TPS (Throttle Position Sensor)	*Throttle Position* Sensor[1]	*TP* Sensor[1]
TPS (Throttle Position Switch)	*Throttle Position* Switch[1]	*TP* Switch[1]
TR (Transmission Range)	*Transmission Range*	*TR*
Transmission Control Module	*Transmission Control Module*	*TCM*
Transmission Position Switch	*Transmission Range* Switch	*TR* Switch
Transmission Range Selection	*Transmission Range*	*TR*
TRS (Transmission Range Selection)	*Transmission Range*	*TR*
TRSS (Transmission Range Selection Switch)	*Transmission Range* Switch	*TR* Switch
Tuned Port Injection	*Multiport Fuel Injection*[1]	*MFI*[1]
Turbo (Turbocharger)	*Turbocharger*[1]	*TC*[1]
Turbocharger	*Turbocharger*[1]	*TC*[1]
TVS (Thermal Vacuum Switch)	*Thermal Vacuum Valve*[1]	*TVV*[1]
TVV (Thermal Vacuum Valve)	*Thermal Vacuum Valve*[1]	*TVV*[1]
TWC (Three Way Catalytic Converter)	*Three Way Catalytic Converter*[1]	*TWC*[1]
TWC + OC (Three Way + Oxidation Catalytic Converter)	*Three Way + Oxidation Catalytic Converter*[1]	*TWC* + OC[1]
VAC (Vacuum) Sensor	*Manifold Differential Pressure* Sensor[1]	*MDP* Sensor
Vacuum Switches	*Manifold Vacuum Zone* Switch	*MVZ* Switch[1]
VAF (Volume Air Flow)	*Volume Air Flow*[1]	*VAF*[1]
Vane Air Flow	*Volume Air Flow*[1]	*VAF*[1]
Variable Fuel Sensor	*Flexible Fuel* Sensor	*FF* Sensor[1]
VAT (Vane Air Temperature)	*Intake Air Temperature*[1]	*IAT*[1]
VCC (Viscous Converter Clutch)	*Torque Converter Clutch*[1]	*TCC*[1]
Vehicle Speed Sensor	*Vehicle Speed Sensor*[1]	*VSS*[1]
VIP (Vehicle In Process) Connector	*Data Link Connector*[1]	*DLC*[1]
Viscous Converter Clutch	*Torque Converter Clutch*[1]	*TCC*[1]
Voltage Regulator	*Voltage Regulator*	*VR*
Volume Air Flow	*Volume Air Flow*[1]	*VAF*[1]
VR (Voltage Regulator)	*Voltage Regulator*	*VR*
VSS (Vehicle Speed Sensor)	*Vehicle Speed Sensor*[1]	*VSS*[1]

[1]Emission-Related Term

FIGURE 1—CROSS REFERENCE AND LOOK UP (CONTINUED)

SAE J1930 Revised JUN93

Existing Usage	Acceptable Usage	Acceptable Acronized Usage
VSV (Vacuum Solenoid Valve) (Canister)	*Evaporative Emission* Canister Purge Valve[1]	*EVAP* Canister Purge Valve[1]
VSV (Vacuum Solenoid Valve) (EVAP)	*Evaporative Emission* Canister Purge Valve[1]	*EVAP* Canister Purge Valve[1]
VSV (Vacuum Solenoid Valve) (Throttle)	*Idle Speed Control* Solenoid Vacuum Valve[1]	*ISC* Solenoid Vacuum Valve[1]
Warm Up Oxidation Catalytic Converter	*Warm Up Oxidation Catalytic Converter*[1]	*WU-OC*[1]
Warm Up Three Way catalytic Converter	*Warm Up Three Way Catalytic Converter*[1]	*WU-OC*[1]
Wide Open Throttle	*Wide Open Throttle*[1]	*WOT*[1]
WOT (Wide Open Throttle)	*Wide Open Throttle*[1]	*WOT*[1]
WOTS (Wide Open Throttle Switch)	*Wide Open Throttle* Switch[1]	*WOT* Switch[1]
WU-OC (Warm Up Oxidation Catalytic Converter)	*Warm Up Oxidation Catalytic Converter*[1]	*WU-OC*[1]
WU-TWC (Warm Up Three Way Catalytic Converter)	*Warm Up Three Way Catalytic Converter*[1]	*WU-TWC*[1]

[1]Emission-Related Term

FIGURE 1—CROSS REFERENCE AND LOOK UP (CONTINUED)